U0931099

结构防灾、监测与控制

Structural Disaster Prevention, Monitoring and Control

李宏男　伊廷华　主编

Editors in Chief: Hongnan LI & Tinghua YI

中国建筑工业出版社

CHINA ARCHITECTURE & BUILDING PRESS

图书在版编目（CIP）数据

结构防灾、监测与控制/李宏男，伊廷华主编．—北京：中国建筑工业出版社，2008

ISBN 978-7-112-10301-0

Ⅰ．结…　Ⅱ．①李…②伊…　Ⅲ．①结构工程-防灾-文集②结构工程-监测系统-文集③结构工程-控制系统-文集　Ⅳ．TU3-53

中国版本图书馆 CIP 数据核字（2008）第 131061 号

责任编辑：赵梦梅　刘婷婷
责任设计：赵明霞
责任校对：兰曼利　陈晶晶

结构防灾、监测与控制
Structural Disaster Prevention，Monitoring and Control
李宏男　伊廷华　主编
Editors in Chief：Hong nan LI & Ting hua YI

*
中国建筑工业出版社出版、发行（北京西郊百万庄）
各地新华书店、建筑书店经销
北京红光制版公司制版
北京蓝海印刷有限公司印刷
*
开本：787×1092 毫米　1/16　印张：18　字数：576 千字
2008 年 10 月第一版　2008 年 10 月第一次印刷
定价：**58.00** 元
ISBN 978-7-112-10301-0
（17104）

前言 Preface

当前，中国正在从事世界上最大规模的土木工程建设，而这一建设高潮还会随着我国全面建设小康社会进程的推进而持续相当一段时间。持续的、大规模的土木工程建设，为结构工程技术的发展、学科的建设、人才的培养、队伍的壮健，特别是创新能力的提升，提供了历史性的机遇。中国土木工程（包括结构工程）事业取得的巨大成就引起了世人的关注。可以预期，在新的世纪里，结构工程事业必将得到更加迅猛的发展。

正是基于这种时代背景，在住房和城乡建设部的支持下，由中国建筑工业出版社、同济大学《建筑钢结构进展》编辑部、香港理工大学《结构工程进展》（Advances in Structural Engineering）编委会联合主办，大连理工大学土木水利学院承办的第二届“结构工程新进展国际论坛（The 2nd International Forum on Advances in Structural Engineering）”在大连举行。

本届论坛的主题是“结构防灾、监测与控制”。我们荣幸地邀请到了10位特邀报告人，他们的报告主题涵盖了大跨空间结构、钢结构、混凝土结构、输电塔线体系结构、结构振动控制、结构健康监测、结构抗震分析与计算等领域。

我们感谢论坛特邀报告人，他们不仅在大会上做了精彩的主题报告，而且还奉献了精心准备的章节（论文），使得本书顺利出版。

感谢参加本次论坛的所有代表，正是大家的热心参与和帮助，才使得本次论坛能够如期顺利举行。

感谢住房和城乡建设部、中国建筑工业出版社、同济大学《建筑钢结构进展》编辑部、香港理工大学《结构工程进展》编辑部、高等学校学科创新引智计划资助项目（B08014）、教育部长江学者和创新团队发展计划（IRT0518）、国家自然科学国际（地区）合作交流项目、国家自然科学重点项目（50638010）、国家自然科学青年基金（50708013）以及王宽诚教育基金会的资助和支持。

目录 Contents

第 1 章 Chapter 1

大跨空间结构理论研究若干新进展

沈世钊

哈尔滨工业大学土木工程学院　哈尔滨　150090

提　要： 本文重点介绍近年来在下列三个理论研究领域的新成果或新进展：（一）网壳结构在静力作用下的弹塑性稳定性，并对稳定性验算中的安全度问题进行了讨论；（二）网壳结构的动力稳定性及其在强震下的破坏机理；（三）大跨柔性屋盖考虑流固耦合效应的风致动力响应。

关键词： 网壳结构　弹塑性稳定　动力稳定性　失效机理　索膜结构　气弹效应

本文重点汇报近年来在大跨空间结构下列三个理论研究领域取得的一些新进展。

Progress of Theoretical research for spatial structures

S. Z. Shen

(School of Civil Engineering, Harbin Institute of Technology, Harbin 150090, China)

Abstract: Theoretical studies on static stability, dynamic stability and dynamic failure mechanism of reticulated shells, and studies on wind-induced response analysis of flexible tension structures including the consideration of the possible aeroelastic effect are introduced.

Keywords: Reticulated shells, Elasto-plastic stability, Dynamic stability, Failure mechanism, Tension structure, Aeroelastic effect

1.1　网壳结构弹塑性稳定性

1.1.1　研究概况

网壳结构在静力作用下的稳定性问题自 20 世纪 80 年代后期至 90 年代中期曾是一个理论研究热点课题。当时计算机条件已相对完备，各研究者均摆脱基于连续壳假设的解析理论，转而运用非线性有限元分析方法，多数人自编程序，对网壳结构进行弹性的或弹塑性的荷载—位移全过程跟踪，从各个方面研究其稳定性能，包括初始几何缺陷和荷载分布形式等各种实际因素对其临界荷载（或称为稳定性承载力）的影响。可以认为，这一段研究取得的成果是具有突破意义的；当时已完全有可能对实际工程中大型复杂网壳结构的稳定性能，至少是弹性稳定性能，进行较精确的定量分析。正由于有了这一研究基础，我国于 20 世纪 90 年代后期着手编制的《网壳结构技术规程》(JGJ 61—2003)[1]才有可能列入建议按弹性全过程分析进行稳定性验算的条文。文献[2]还对各种形式的网壳结构进行了大规模的参数分析，共计完成 2800 余例实际尺寸网壳的弹性全过程分析，以考察各类网壳稳定性能的变化规律，包括各种实际因素对其承载力的影响；在此基础上通过回归分析提出了各种形式网壳结构稳定性承载力的实用公式。

对于由弹性全过程分析求得的稳定性承载力（临界荷载），还应除以一个“安全系数”，以求得网壳按稳定性确定的容许承载力（或容许荷载）的标准值。这个“安全系数”应考虑下列因素：(1) 荷载等外部作用和结构抗力的不确定性可能带来的不利影响；(2) 计算中未考虑材料弹塑性可能带来的不利影响；(3) 结构工作条件中的其他不利因素。关于这一系数的取值《网壳规程》(JGJ 61—2003) 中仍沿用文献[3]提出的经验值 $K=5$，因为当时还缺少足够的统计资料作近一步论证，其中主要因素之一就是当时还没有条件进行大规模的关于网壳弹塑性稳定性能的系统分析，因而对考虑材料非线性以后网壳结构稳定性承载力可能折减多少缺乏足够的定量数据。

近几年，随着计算机条件和大型通用分析软件的进一步完备，对实际尺寸网壳结构进行系统的弹塑性全过程分析具备了较好的条件，我们就着手进行这项有意义的研究工作。我们利用 ANSYS 通用软件对各种形式的单层网壳进行了大规模的参数分析，例如，对 K8 型单层球面网壳考虑了 4 种跨度（$L=40$m，50m，60m，70m）、4 种矢跨比（$f/L=1/5$，1/6，1/7，1/8）和 4 种截面尺寸，共 64 种不同网壳，各考虑 4 种初始几何缺陷值（$r=0$，$L/1000$，$L/500$，$L/300$）和 3 种荷载分布形式（$p/g=0$，1/4，1/2），部分网壳还考虑了更多的初始缺陷值；全部网壳周边固接，但部分网壳还计算了铰接支承的情形来做对比；这样共计分析了近 1200 个 K8 型单层球面网壳的算例。再加上 K6 型、短程线型、肋环型等各种形式，共计为单层球面网壳进行了 2800 余例弹塑性全过程分析。类似地，为单层柱面网壳进行了 1400 余例，为单层双曲椭圆抛物面网壳进行了 800 余例弹塑性全过程分析。除此之外，还利用 ANSYS 软件进行了同样多的弹性全过程分析作为对比。计算工作量十分浩大，但确实使我们对网壳结构复杂的全过程受力性能、其失稳的实际过程和各种因素的相互影响，有了一些实质性的了解[4]。

限于篇幅，下面以单层球面网壳为例作一扼要介绍。

1.1.2 单层球面网壳的弹塑性稳定性能

与弹性失稳的状况类似，单层球面网壳的弹塑性失稳也均从某些节点的跳跃屈曲开始，逐渐形成一个或若干个越来越大的壳面凹陷。由于塑性变形的发展，全过程曲线在越过临界点以后呈持续弱化倾向，但会经历较长的位移发展过程而始终维持一定的承载力；当初始缺陷较大时，全过程曲线可能不出现明显的极值点，而表现为在比较稳定的承载力下位移持续发展的形式（参看图 2、图 3）。

对于无缺陷的 K8 或 K6 型理想网壳，当矢跨比较小时（$f/L=1/7$，1/8），失稳多数从主肋节点开始，一般发生在从外圈算起第三个节点，最后在该处形成凹陷（图 1a），这与弹性网壳的情形十分类似。但对于矢跨比为 1/5 和 1/6 的网壳，失稳均从由中心算起第二环上各面上节点开始，并且各节点的凹陷逐渐连成一体，最后形成十分明显的环形凹槽（图 1b）；部分 $f/L=1/7$ 的网壳和少数 1/8 的网壳也表现出这种失稳形态。究其原因，这些网壳在加荷过程中，中央第二环杆受压且受力较大，应力首先达到屈服点且全截面进入塑性，因而该环杆已无力承载进一步的荷载增量，导致杆上各节点纷纷内陷。这是弹塑性网壳由于局部区域塑性发展过于集中而提前丧失承载力的一种特殊现象。我们曾尝试把中央第二环杆的截面适当加强，再进行计算，果然不再出现这种顶部环形凹槽式的失稳形态，而恢复图 1 所示的常见失稳形态，其临界荷载也相应提高。

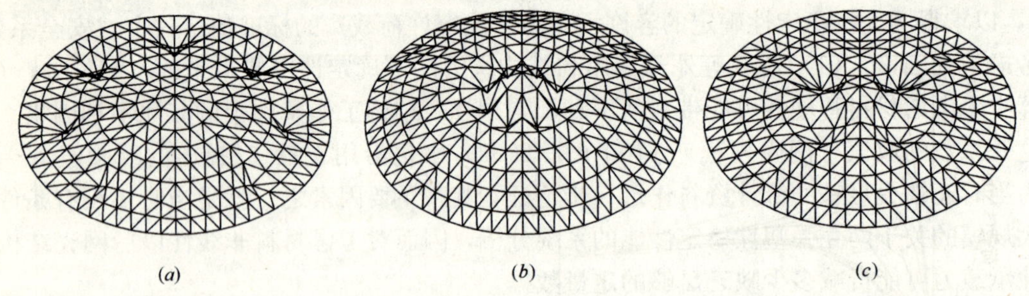

图 1　K8 型单层球面网壳弹塑性失稳后位移形态（比例放大 30 倍）

(a) $L=60$m，$f/L=1/8$，$r=0$；(b) $L=60$m，$f/L=1/5$，$r=0$；(c) $L=60$m，$f/L=1/8$，$r=6$cm

对于有缺陷的实际网壳，计算时均取结构的最低阶特征模态作为初始缺陷分布形式；分析结果表明，当最大初始缺陷取 $L/1000$ 及以上时，多数网壳的失稳形态与所采用的缺陷分布模态一致，壳面上发生多处凹陷，但最大凹陷发生在以中央第二环或第三环处的主肋节点为中心的区域（图 1c）。但也有少部分 $f/L=1/5$ 的网壳（其中央第二环杆未特意加强）仍然表现出顶部环形凹槽的失稳形态，相应的临界荷载也偏低。

单层球面网壳的弹塑性稳定性能对初始几何缺陷也比较敏感。作为例子，图 2 给出矢跨比分别为 1/5 和 1/8 的两个 K8 型网壳当具有不同大小初始缺陷时的全过程曲线及其临界荷载随初始缺陷值变化的规律；作为对比，图中也给出了弹性临界荷载的变化曲线；临界荷载曲线的纵坐标取相对值 q_{cr}/q_{cr0}（q_{cr0} 为理想弹性网壳的临界荷载）。

由图 2 可以看出，初始缺陷对单层球面网壳弹塑性稳定性能的影响规律与弹性网壳的情形类似。网壳的临界荷载随初始缺陷值的增大而降低，但当缺陷值达到 $L/300$ 时趋于稳定。《网壳规程》(JGJ 61—2003) 在建议按弹性全过程分析方法进行稳定性验算时，还建议按跨度的 1/300m 确定初始缺陷的取值；看来这一建议对弹塑性稳定性验算仍然适用。

与弹性网壳的情形类似，荷载的不对称分布对单层球面网壳弹塑性稳定性能的影响也不明显。作为例子，图 3 给出两个 K8 型网壳分别在三种不同荷载分布下的全过程曲线，且分别考察了完整网壳和具有 $L/1000$ 缺陷两种情形。在绘制这些全过程曲线时，一律以总荷载（$p+g$）作为纵坐标。可以看出，三种荷载分布下的全过程曲线几乎重合在一起；也就是说，以总荷载作为衡量指标时，不对称荷载作用对网壳的弹塑性临界荷载值几乎没有影响。

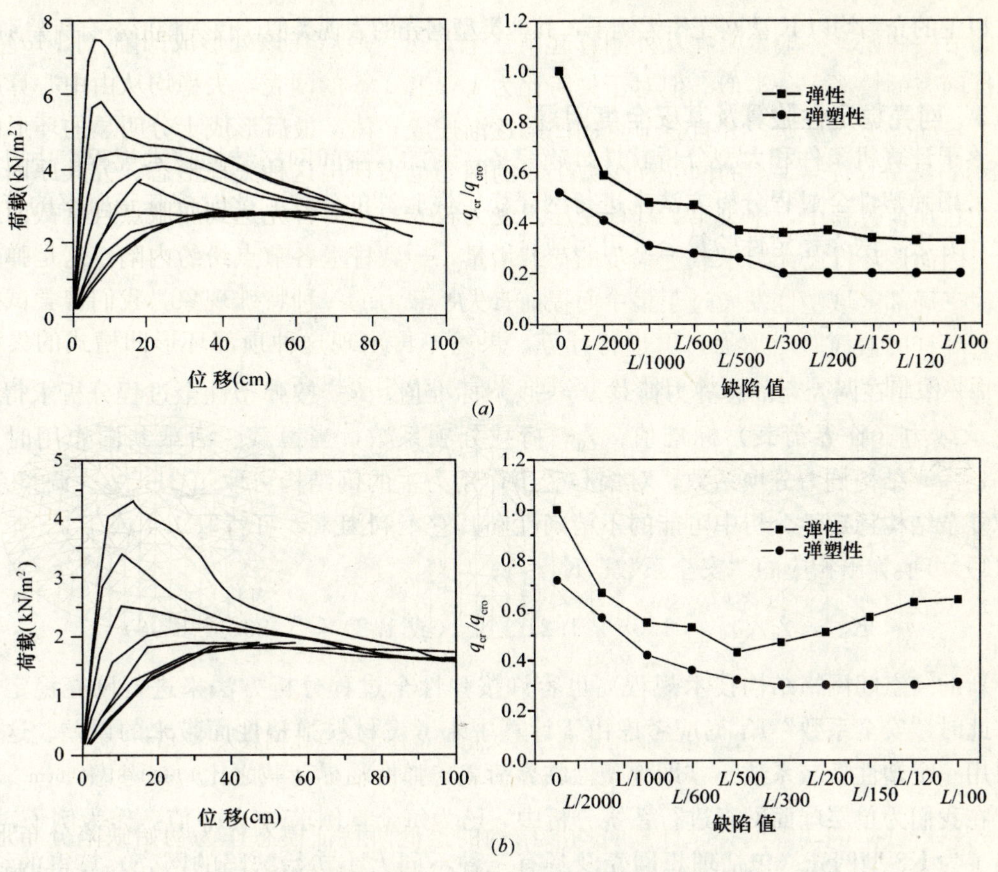

图 2　不同初始缺陷时的全过程曲线及临界荷载变化曲线

(a) K8 型网壳 $L=60\text{m}$，$f/L=1/5$，2 号截面

(b) K8 型网壳 $L=60\text{m}$，$f/L=1/8$，2 号截面

($r=0$、3、6、10、20、30、40、50、60cm)

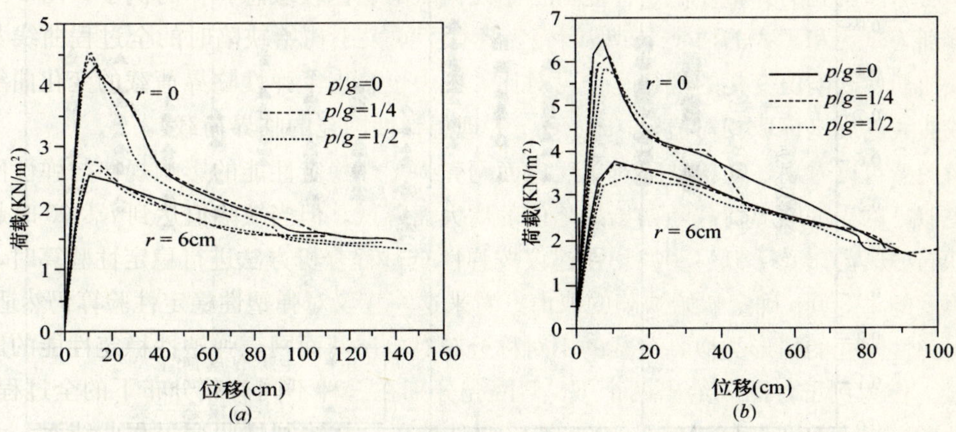

图 3　不同比例不对称荷载作用下的全过程曲线

(a) $L=60\text{m}$，$f/L=1/8$，2 号截面；(b) $L=60\text{m}$，$f/L=1/5$，2 号截面

以上的介绍均以K型网壳作为例子，其他类型网壳的表现类似，仅失稳形态有所差别。

1.1.3 网壳稳定性验算及其安全度问题

鉴于计算机条件和大型分析软件日趋完备，新的《空间网格结构技术规程》中已列入直接采用弹塑性全过程分析方法来进行网壳稳定性验算的建议。此时按基于概率的分项系数设计方法，网壳稳定性验算公式可写成：

$$\gamma_q q \leqslant \frac{R}{\gamma_0 \gamma_R} \tag{1}$$

式中 q—作用在网壳上的总静力荷载（$p+g$）标准值；R—按弹塑性全过程分析求得的稳定性承载力（临界荷载）标准值；γ_q—荷载分项系数，当恒载、活载共同作用时可取1.35；γ_R—结构抗力分项系数，对偏心受压作用为主的钢结构可取1.21；γ_0—调整系数，考虑复杂结构稳定性分析中可能的不精确性和其它不利因素，可暂取1.1或1.2。于是由式（1）可换算出相应的“安全系数”K_p：

$$K_p = \gamma_q \gamma_R \gamma_0 = 1.35 \times 1.21 \times 1.1(\text{或}\ 1.2) = 1.80(\text{或}\ 1.96) \tag{2}$$

新的《空间网格结构技术规程》也容许按弹性全过程分析方法来进行网壳稳定性验算，此时“安全系数”K_e 尚应考虑由于计算中未考虑材料弹塑性而带来的误差。这一误差可用一个塑性折减系数 c_p（即弹塑性临界荷载与弹性临界荷载之比）来考虑。

在我们为单层球面网壳进行系统分析中，已为每个算例求得了 c_p 值。作为例子，图4给出了为K8型网壳（包括理想网壳及具有三种不同大小初始缺陷的网壳）算得的结果，共计256个数据。图中还用符号分别表示不同的失稳形态。

按照规程的建议，在进行网壳的弹性全过程分析时，初始缺陷应按跨度的1/300取

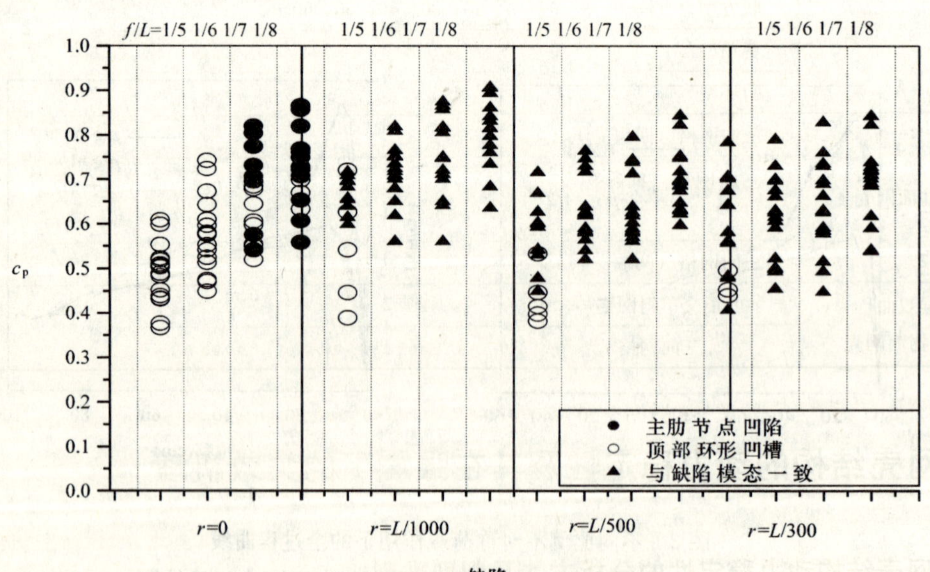

图4 K8型网壳弹塑性稳定性折减系数 c_p

值。按照这一思路，图 4 中最右一栏，即为 $r=L/300$ 的网壳算得的 c_p 数据可作为我们进行统计的依据。从图中可以看到，有少数 $f/L=1/5$ 的网壳表现出顶部环形凹槽的失稳形态，其临界荷载均偏低；如上文所述，可以通过对中央第二环杆的截面适当加强的办法来消除这种提前失稳现象，所以在统计中应该剔除这几个特殊数据。据此求得 C_p 的平均值和均方差为：$\overline{c_p}=0.641$，$\delta_{cp}=0.0992$；如果按 95％的保证率，可求得建议的折减系数为：

$$c_p=c_p-1.645\delta_{cp}=0.478 \tag{3}$$

于是当采用弹性全过程分析方法来计算网壳临界荷载时，相应的“安全系数”应取为：

$$K_e=\frac{K_p}{c_p}=\frac{1.80(或1.96)}{0.478}=3.77(或4.10) \tag{4}$$

按照上述思路和统计方法，对 8 种单层球面网壳、3 种单层柱面网壳和 2 种单层双曲椭圆抛物面网壳的“塑性折减系数”C_p 与“安全系数”K_e 的取值均进行了统计，结果如表 1 所示。可以看到，对于多数球面网壳，K_e 取 4.5 即可基本覆盖。对于结构刚度较弱的肋环型球面网壳，几何非线性起主要作用，其弹性稳定承载力较低，而材料非线性的影响反而很小，K_e 与 K_p 的差别较小。对于两端支承柱面网壳结构，由于材料非线性的影响较大，K_e 取 5.4 才能保证结构的安全性。看来，网壳结构稳定性验算方法中“安全系数”的取值应该按照具体的网壳形式来确定，本文的结果可以提供一定的参考。

单层网壳结构“安全系数”值　表 1

结构类型	结构形式	塑性折减系数	安全系数	安全系数归并
单层球面网壳	K8 型	0.478	4.10	4.5
	K6 型	0.460	4.26	
	短程线型	0.430	4.56	
	肋环双斜杆型	0.491	3.99	
	肋环单斜杆型	0.440	4.45	
	联方型	0.430	4.56	
	葵花型	0.528	3.71	
	肋环型	0.800	2.45	2.5
单层柱面网壳	三向网格（四边支承）	0.453	4.33	4.5
	三向网格（两纵边支承）	0.578	3.39	
	三向网格（两端支承）	0.362	5.41	5.4
单层双曲椭圆抛物面网壳	正交单斜网格	0.420	4.67	5
	三向网格	0.386	5.08	

1.2　网壳结构的动力稳定性及其在强震下的失效机理

1.2.1　网壳结构动力稳定性的分析方法及判别准则

网壳结构具有极好的抗震性能，1994 年北岭地震和 1995 年阪神地震中许多网壳的优良表现给人以深刻印象；这种结构形式在强震下的抗震性能和失效机理问题也因此吸引了不少研究者的兴趣。单层网壳结构具有明显的几何非线性特征，无论在静力或动力作用下，都有易于失稳的倾向性；因而关于其失效机理的研究是从动力稳定性问题开始的。文献[5]对近十年来关于网壳动力稳定性的研究作了概括性的回顾，并明确提出，鉴于现在对网壳等复杂系统进行双重非线性动力响应分析已非难事，今后应该以直接基于响应分析的方法来研究网壳的动力稳定性，这样不仅使动力稳定性的判别准则更加直观，而且对于复杂的实际工程问题可以得出定量的可供实用的结果。具体操作方法是：针对某一特定的动力荷载（例如某一地震波），以逐步增大的荷载幅值为参数，对应每一荷载幅值做非线性动力时程分析，记录结构的某些特征响应（例如最大节点位移），绘制出这些特征响应随荷载幅值变化的完整曲线（可称之为该响应的荷载域全程曲线）；通过该曲线可全面了解结构随动力荷载幅值逐渐增大其动力性状不断变化乃至失稳的全过程。相应的动力失稳判别准则可表述为：当荷载幅值的微小增量导致网壳结构特征响应指标异常增大时，结构可视为动力失稳，所对应的荷载幅值定义为结构动力失稳临界荷载（或叫极限荷载）。此临界荷载值通常可根据上述荷载域全程曲线并参照有关计算点的时程响应变化来定量地确定。

作为说明例子，可参考下面算例一中图 6 所示的一 K8 型单层球面网壳（L=40m，f/L=1/3）在水平简谐荷载（频率 2.6Hz）作用下的全程分析结果。图 6b 表示不同荷载幅值（用加速度表示）下特征节点位移的时程曲线，可以看到，当荷载幅值为 97Gal 时，该特征节点仍在作平稳振动，而且最大位移仅为 0.06m；当荷载幅值增至 100Gal 时，特征节点的运动发散，出现动力失稳。反映在图 6a 的荷载幅值—特征节点位移全程曲线上，表现为荷载幅值由 97Gal 增至 100Gal 时，特征节点位移异常增大。图 6d 表示网壳的变形图，可以看到，荷载幅值为 97Gal 时，网壳变形还很小，而在 100Gal 时，突然发生严重的局部凹陷而倒塌。该系统的动力失稳临界荷载位于 97～100Gal 之间。

我们运用这一方法对单层球面网壳在突加荷载、简谐荷载和地震荷载作用下的动力稳定性进行了系统分析，得出了具有良好规律性的结果，其中部分成果在文献[5]中作了介绍。

1.2.2　网壳结构两种可能的失效机理

对网壳结构进行全过程动力响应分析的大量算例表明，随着动力荷载幅值的不断增大，绝大多数网壳结构最后都会发生倒塌破坏；但在有些算例中，网壳结构破坏时的位移并不大，内部塑性发展也不严重，结构破坏比较突然；而在另外一些算例中，结构要经历较长的塑性变形发展、结构刚度严重削弱以后才会发生倒塌破坏，结构临倒塌前节点振动平衡位置的偏移已相当大，即网壳形状已发生较大畸变，事实上应认为结构已达到了某种强度承载力的极限；也就是说，后一类算例可以认为属于强度破坏的范畴。

下面考察几个典型算例。在这些算例中，为了全面了解结构的塑性发展过程，按照文献[6]提出的综合考察多个响应指标变化规律的研究方法，在对网壳结构的动力响应进行荷载域的全程跟踪时，不仅记录下各特征节点的位移响应，而且全面记录下各杆件逐步进入塑性并且塑性变形不断发展的全过程；为此定义了最大相对应变、不同屈服程度的杆件比例、结构平均应变（所有杆件最大应变的平均值）等特征响应指标，并绘制了它们的荷载域全程曲线；还记录了结构在不同荷载幅值时的塑性分布范围。限于篇幅，下面各算例中只能给出部分结果来作说明。各算例均采用通用有限元分析软件 ANSYS 进行动力时程分析，杆件采用可实时输出应力应变的梁单元（Pipe20 单元），截面上共有 8 个积分点。为说明杆件截面的塑性发展程度，定义 1P 表示有 1 个积分点进入塑性，8P 则表示全截面进入塑性，余类推（见图 5）。

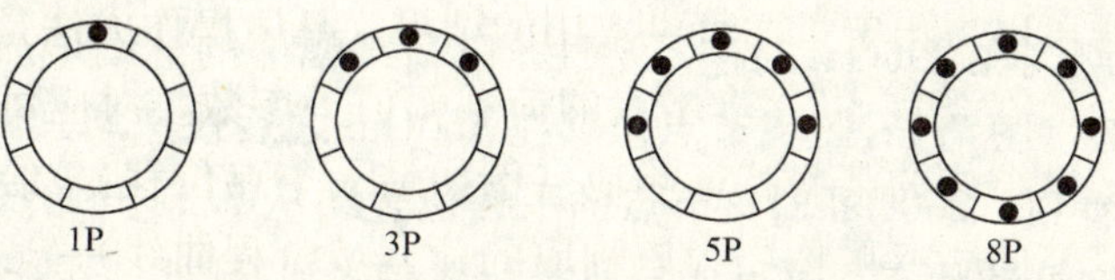

图 5　Pipe20 单元截面塑性发展定义

【算例一】 在水平简谐荷载（频率 2.6Hz）作用下的 K8 型单层球面网壳（$L=40$m，$f/L=1/3$）

此算例在前面已略作介绍，为了说明网壳内部的塑性发展过程，在图 6*c* 中还给出了

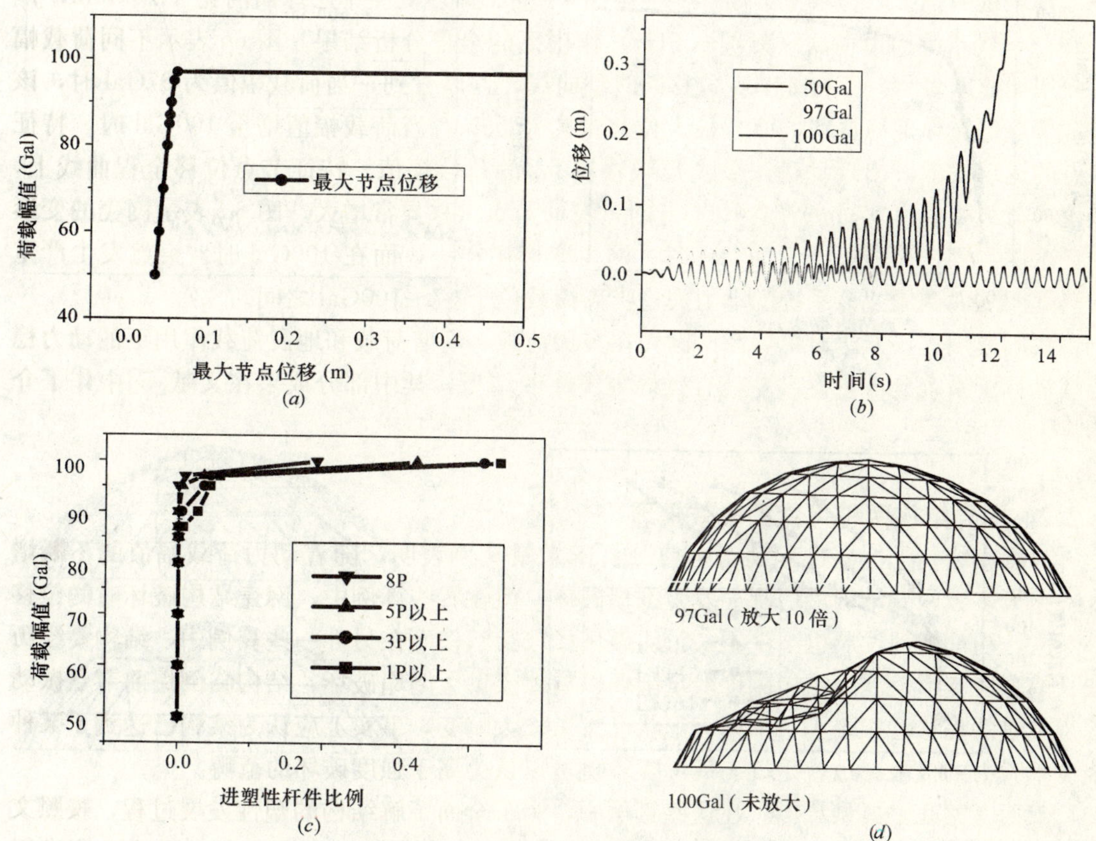

图 6　算例一动力全过程分析结果

(*a*) 荷载幅值—结构最大节点位移全过程曲线；(*b*) 最大节点位移时程曲线

(*c*) 荷载幅值—屈服杆件比例全过程曲线；(*d*) 网壳最大位移时刻变形图

表示不同屈服程度杆件的比例逐渐发展的全过程曲线。可以看到，当荷载幅值为97Gal时，屈服杆件数量还非常少，仅6%的杆件进入塑性；但当荷载幅值稍增大至100Gal时，位移和各项塑性响应指标均异常增大。这一算例中结构倒塌之前结构刚度无明显弱化，振动平衡位置未发生明显偏移，其破坏特征属于典型的动力失稳。

【算例二】 在El-Centro（1940）三维地震作用下的K8型单层球面网壳（$L=40m$，$f/L=1/3$）

在本算例中，网壳在296Gal时进入塑性，随荷载幅值的继续增加，屈服杆件比例基本呈直线变化的趋势（图7*c*）；最大节点位移在650Gal之后快速增加（图7*a*），从位移时程曲线（图7*b*）上看到，750Gal时该节点的平衡位置约从7.1s开始产生一较大偏移，但8s以后继续保持稳定振动状态，这可能与地震荷载在达到峰值以后随即迅速减弱有关；当荷载幅值达到800Gal，节点最大位移发散，网壳倒塌破坏。本例网壳倒塌临界荷载应在750～800Gal之间。对应750Gal地震作用，最大节点位移已达到0.67m，位移延性系数（即该最大位移与结构刚进入塑性时的最大弹性位移之比）16.9，1P以上杆件比例

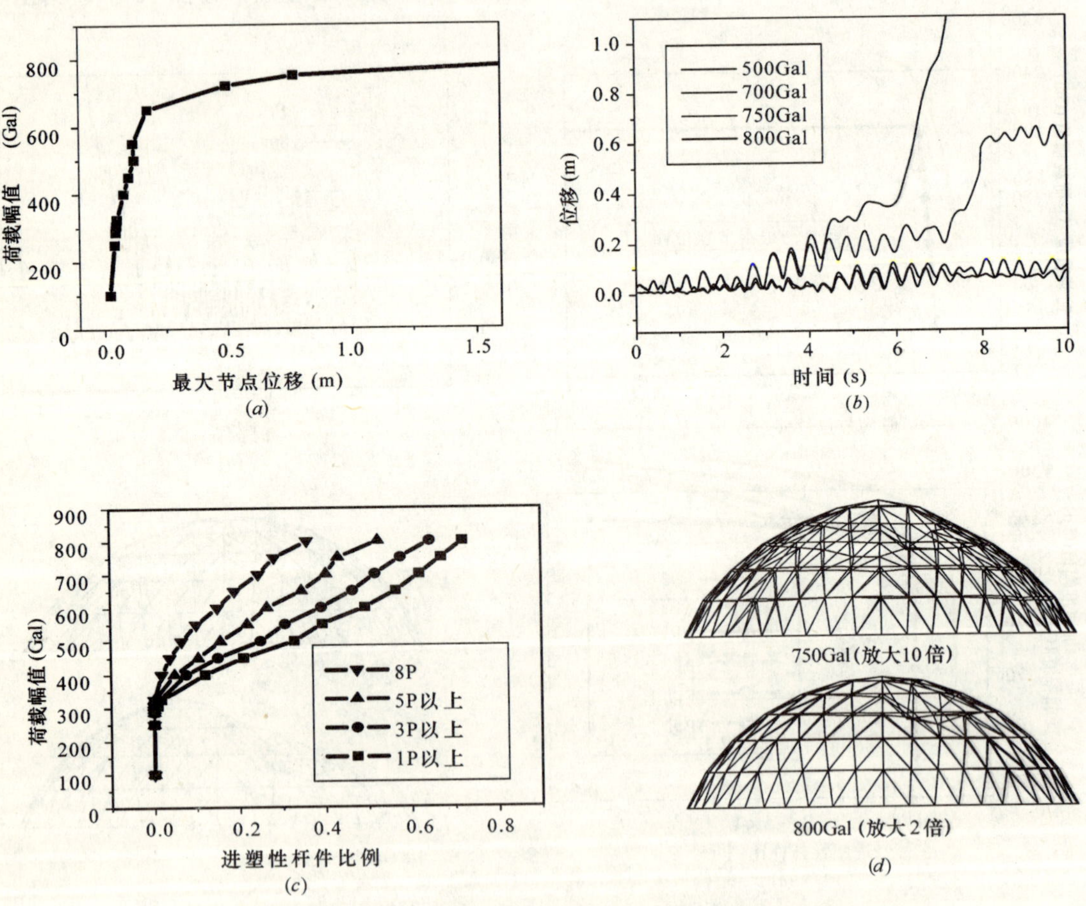

图7 算例二动力全过程分析结果

(*a*) 荷载幅值—结构最大节点位移全过程曲线；(*b*) 最大节点位移时程曲线

(*c*) 荷载幅值—屈服杆件比例全过程曲线；(*d*) 网壳最大位移时刻变形图

69.8%，8P杆件比例32.5%，最大相对应变（即最大应变与屈服点应变值之比）17.7；可见网壳在倒塌前塑性发展已相当深入，残余位移也很大；应认为结构的破坏是由于内部材料强度破坏累积到某种极限状态导致的。

【算例三】 在TAFT（1952）三维地震作用下的四边支承单层柱面网壳（$b=15$m，$f/b=1/2$，$L/b=1.8$）

网壳在1135Gal时开始进入塑性，然后随荷载幅值增加塑性变形持续发展（图8*c*），节点的振动平衡位置发生越来越大的偏移，但始终保持较平稳的振动（图8*b*）；当荷载幅值为3500Gal时，1P以上屈服杆件比例已高达85.7%，8P杆件比例也达到52.7%，最大位移达到2.35m，即网壳形状已严重畸变（图8*d*），但仍然保持平稳振动，荷载幅值—位移曲线仍然保持一定的上升趋势。这一算例的破坏特征显然应属于强度破坏的范畴。事实上，在3500Gal时，许多构件的最大应变已达到或接近材料强度极限，已经出现真正物理意义上的破坏。

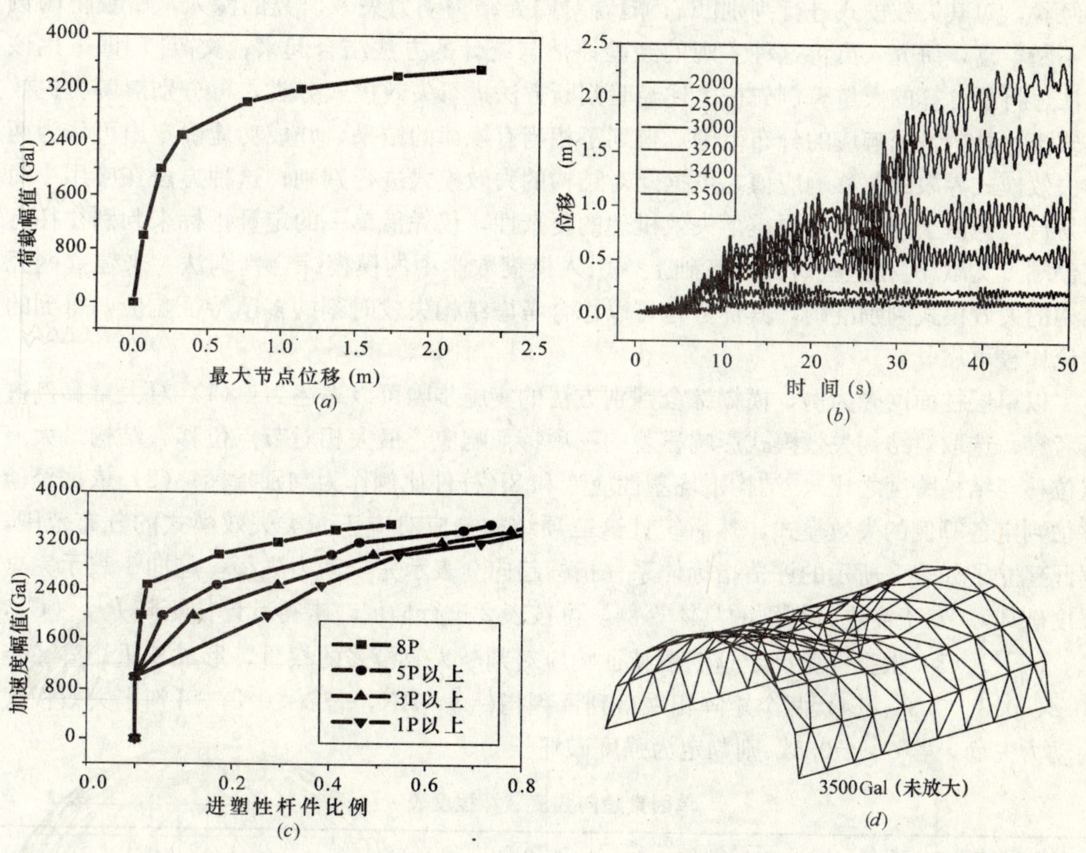

图8 算例三 动力全过程分析结果

（*a*）荷载幅值—结构最大节点位移全过程曲线；（*b*）最大节点位移时程曲线

（*c*）荷载幅值—屈服杆件比例全过程曲线；（*d*）网壳最大位移时刻变形图

由以上算例可以看到，网壳结构具有两种可能的动力破坏形式，即强度破坏与动力失稳。动力失稳具有较大的突然性；在这种破坏形式中，结构的几何非线性起到更为主要的作用，当各节点的振动平衡位置发生一定偏移后，结构即不能维持其总体平衡形状；失稳前内部塑性发展可能并不严重，结构位移也相对较小。强度破坏则表现为结构随荷载幅值增大要经受较大的反复塑性变形发展，结构刚度逐渐削弱，各节点的振动平衡位置发生越来越大的偏移，最后由于结构刚度的弱化再也不能维持其稳定振动状态而发生大塑性状态下的突然倒塌破坏，此时结构已达到了其强度破坏的极限。还有一些网壳甚至在形状已发生严重畸变，构件延性已消耗殆尽而仍不发生明显的突然倒塌现象，则更是表现为十分典型的强度破坏形式。结构的强度破坏形式表现出较好的延性和耗能能力。这里十分重要的是如何来区分这两类失效模式并确定强度破坏的极限荷载，这对工程应用具有重要意义。

1.2.3 网壳结构两类失效模式的判别方法

研究初期，主要是依据网壳结构在强震下破坏时某些特征响应的数值，结合研究人员的经验，对其失效模式进行判别[6]。随着对网壳结构动力失效研究的深入，所获得的数据不断丰富，研究人员将这种主观的经验评估与统计的方法结合起来；文献[7]即采用该方法，将所计算的大量算例先按上述经验判断方法进行失效模式分类，再分别对两类失效模式统计某些特征响应的分布范围，得到了相当有规律的结果，并成功地确定出可作为两类失效模式界限的位移响应值，可据此对结构的失效模式进行判别；这种方法在应用中简便易行，但由于网壳结构强震下失效机理的复杂性，仅凭借单一的定量指标来判别往往有失偏颇。文献[8]在这些研究的基础上，引入模糊数学中的模糊综合判别法，来建立网壳结构的失效模式判别准则，其优势是可以综合考虑结构失效时刻的多项特征响应，得到的结论比较可靠。

以单层柱面网壳为例，模糊综合判别方法的实施步骤可以表述为：(1) 对大量算例进行考察，选取对结构失效模式影响显著的三项特征响应：最大相对节点位移（结构最大节点位移与结构跨度之比）、结构平均塑性应变和 8P 杆件比例作为判别指标；(2) 依据经验评估判定各算例的失效模式，然后统计这三项结构响应对应于两类失效模式的分布范围，据此建立了如表 2 所示的评估指标体系（指标趋向 0 表示完全动力失稳，趋向 1 表示完全强度破坏）。对于需进行判别的具体算例，可按表 2 进行插值，求得其评估矩阵 $R=[C_m, C_a, C_p]^T$；(3) 根据经验确定这三项特征响应对判断失效模式的权重，形成权重矩阵 $A=[0.4、0.3、0.3]$；(4) 两个矩阵相乘得到评判指标 $b=AR$，若 $b<0.5$，可判定失效模式为动力失稳，若 $b>=0.5$，则判定为强度破坏。

单因素趋向强度破坏程度表 **表 2**

结构平均塑性应变	≤0.0003	{0.0015, 0.0023}	≥0.006
C_a	0	0.5	1.0
8P 杆件比例	≤0.05	{0.5, 0.55}	≥0.8
C_p	0	0.5	1.0
相对最大节点位移	≤0.0053	{0.055, 0.08}	≥0.133
C_m	0	0.5	1.0

1.2.4　网壳结构在强震作用下的强度破坏判别准则

网壳结构具有很好的延性，在强震作用下往往经历很大的塑性变形发展，产生很大的残余位移而不发生明显的物理意义上的破坏；但事实上，许多构件和节点连接部位已经由于经受反复塑性变形而产生不同程度的损伤，也不排除个别构件或连接部位由于经受过大的反复塑性变形作用而出现裂纹等实际破坏现象。所以十分重要的是需要建立某种综合的结构损伤模型，能较好地反映像网壳这样的复杂结构体系在强震下损伤不断累积乃至达到破坏的实际过程。

如前所述，对于动力稳定性问题，可以根据位移响应的全程曲线，按照动力稳定性判别准则获得结构动力失稳的临界荷载；而对于动力强度破坏的情况，目前在文献[9]中采取的方法是：在分析中充分考虑材料损伤累积及裂纹效应的影响，通过荷载域全程响应分析的方法，将网壳结构所能够承受的最大荷载强度定义为强度破坏极限荷载。通过对网壳结构动力强度破坏算例的考察发现，结构在发生动力强度破坏时，各项响应固定在一个范围，具有较好的规律性，为提出某种结构损伤模型提供了很好的条件。

文献[8、9]为此编制了考虑损伤累积及裂纹效应的材料接口程序，对网壳结构在强震下的响应进行了分析，通过对大量强度破坏算例破坏时特征结构响应的统计，拟合了网壳结构在强震下的损伤模型。以考虑初始缺陷的单层球面网壳为例，其损伤模型可以表述为下式的形式，该模型的误差分析也表明精度较好：

$$D_{SN}=(2.10\times L_k^2+5.57\times L_k+5.11)$$
$$\times\sqrt{\sqrt{f/L}\times\left\{4500\left(\frac{d_m-d_e}{L}\right)^2+m_k^{\frac{1}{3}}\left[100\left(\frac{\varepsilon_a}{\varepsilon_u}\right)^2+r_1^2+0.1r_8^2\right]\right\}}$$

式中 D_{SN} 为结构损伤因子；L 为球壳跨度；f 为矢高；ε_a 是结构平均塑性应变；ε_u 为钢材极限应变；d_m 为最大节点位移；d_e 是网壳材料出现塑性时刻的位移；r_1 是 1P 杆件比例；r_8 为 8P 杆件比例，L_k 为结构跨度（m）与标准结构跨度 40m 的比值，m_k 为屋面质量（kg/m^2）与标准屋面质量 $100kg/m^2$ 的比值。

根据上述网壳结构强震损伤模型，可提出网壳结构强度失效判别准则如下：对网壳结构采用荷载域全程响应分析时，当某荷载强度下对应的网壳结构损伤因子达到 1.0，即判定结构强度失效，对应的地震荷载幅值为结构的强度破坏极限荷载。

这种结构损伤模型不仅可以判定结构的强度破坏极限，而且能够评估在不同地震强度下结构的损伤程度，因此可以与结构的性态水准充分结合，作为网壳结构基于性态抗震设计方法研究的基础性。

1.3　考虑流固耦合效应的索膜结构风致动力响应研究

索膜结构属于典型的风敏感体系，其风致动力响应和气弹稳定性问题，由于涉及到多个研究对象，跨越多个学科，因此具有很大理论难度，至今尚未得到较好解决。文献[10]在总结前人工作的基础上，针对风—索膜结构的动力耦合特点，提出三种不同的研究思路：

(1) 解析方法：针对一些具有规则形状的简单膜结构，用解析方法来研究其风致动力效应，此时可将空气模拟为不可压缩的无粘性均匀流体，综合壳体的几何非线性无矩理论和流体力学的势流理论，建立结构与流场的耦合作用方程，从中提炼出一些反映膜结构气弹效应的基本概念或物理参数。

(2) 简化气弹模型法：从工程实用角度出发，对结构运动方程中的气动力项进行简化，将结构运动对流体的反馈作用以附加气动力项的形式表达出来；再借助风洞试验确定理论模型中一些必要的物理参数。

(3) CFD 数值模拟方法：综合运用计算流体力学和计算结构力学技术，建立适用于索膜结构流固耦合风振分析的数值计算模型；借助计算机有限元技术实现对结构及其周围流场运动过程的模拟。

1.3.1 索膜结构气弹稳定性研究的解析方法

文献[11]将壳体的无矩理论和流体的理想势流理论结合起来，提出一种用于判定膜结构气弹稳定性的解析方法，基本步骤是：

(1) 假设风为具有均匀流速的无粘性不可压缩理想势流，由此引入不可压缩理想流体的基本方程，并根据结构风工程的具体特点对基本方程进行了简化；

(2) 以小垂度封闭式薄膜结构为研究对象，应用扁壳的无矩理论并考虑膜结构的几何非线性，建立了膜结构的基本力学方程；

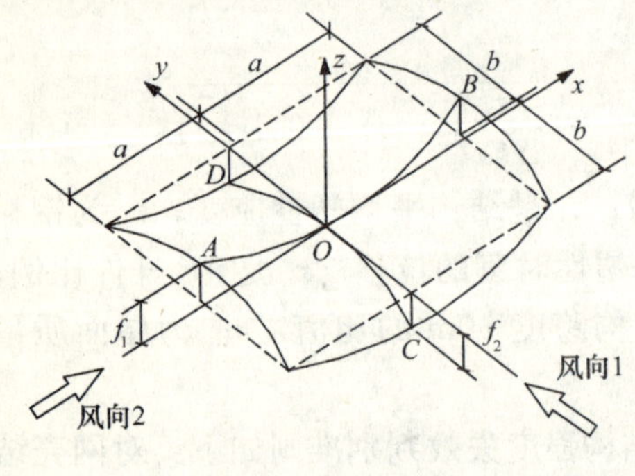

图 9 双曲抛物面薄膜结构

(3) 根据气流在结构前缘的分离情况分别建立了两种绕流模型，即均匀来流与连续点源层的叠加和均匀来流与连续旋涡层的叠加；在此基础上，借助经典空气动力学中的薄翼型理论确定作用于薄膜结构上的气动力，从而建立起风与薄膜结构的动力耦合作用方程；

(4) 针对上述两种绕流模型，采用 Bubnov-Galerkin 方法和 Routh-Hurwitz 稳定性准则推导了矩形平面双曲抛物面薄膜结构（图 9）的失稳临界风速公式为：

- 第一种绕流模型：

$$V_{cr}=\left(\frac{16(f_1/b-f_2/b)^2Eh/(1+\lambda^2)^2+\pi^2(N_{x0}+N_{y0}/\lambda^2)}{-\rho\lambda b\gamma_3}\right)^{1/2} \tag{5}$$

- 第二种绕流模型：

$$V_{cr}=\left(\frac{16(f_1/a-f_2/a)^2Eh/(1+\lambda_1^2)^2+\pi^2(N_{x0}+N_{y0}/\lambda_1^2)}{8\rho a^2\lambda_1\gamma_3}\right)^{1/2} \tag{6}$$

其中：N_{x0}、N_{y0} 为初始预张力，ρ 为空气密度，E 为薄膜的弹性模量，h 为薄膜的厚度，$\lambda=a/b$、$\lambda_1=b/a$ 为平面横顺比，γ_3 为平面横顺比的函数。应用上述方法，文献[17]对矩形平面双曲抛物面薄膜结构的气弹动力失稳临界风速进行了分析，并考察了预张力、矢跨比、顺风向跨度和平面横顺比等参数对失稳临界风速的影响。

应该指出的是，由于该解析理论采用了过多的简化假设，不应指望所取得的定量结果具有可供实用的精度；但从中提炼出来的一些概念或定性结论的确具有理论参考价值。例如，它揭示的增大膜单元的矢跨比能显著提高临界风速这一概念，对实际设计具有指导意义。此外，从解析理论的研究中有可能提供出一些对建立实用气弹模型有用的物理参数。

1.3.2 理论与实验相结合的简化气弹模型

结构在脉动风激励下的运动方程可以表述为：

$$M_s\ddot{x}(t)+C_s\dot{x}(t)+K_sx(t)=F(t,x(t),\dot{x}(t),\ddot{x} \tag{7}$$

上式左边各项分别代表结构自身的惯性力、阻尼力和弹性力项；等式右边代表气动力项，它是时间 t、结构位移 x 以及 x 导数的函数。也就是说，作用于结构上的风压不但与来流的脉动特性有关，还会受到结构自身运动的影响产生附加气动力（自激力）。

通过引入拟定常假定和强相关假定，对式（7）右端的气动力项解耦，整理后可得到考虑风与结构动力耦合作用的简化气弹模型方程为[10]：

$$(M_s+M_a)\ddot{x}(t)+(C_s+C_a)\dot{x}(t)+(K_s+K_a)x(t)=p(t) \tag{8}$$

式中：$p(t)$代表流体自身的脉动作用；M_a 可称为附加质量（added mass)，代表气动力中与结构加速度相关的部分，其物理意义可解释为随结构一起运动的那部分空气质量；C_a 可称为气动阻尼（aerodynamic damping)，代表气动力中与结构速度相关的部分，其物理意义可解释为结构周围空气与结构运动之间的能量转换；K_a 可称为气动刚度（aerodynamic stiffness)，代表气动力中与结构位移相关的部分，其物理意义可解释为由于结构瞬间移动在其上下（内外）表面形成的局部气压差，该参数仅对密封的膜结构（如气承式膜结构）有实际意义。

依据上述简化气弹模型，文献[13]进行了菱形平面鞍形索网结构和膜结构的刚性模型与气弹模型风洞实验研究。鉴于该实验的探索性质，作者设计了 4 个相对独立的子实验，目的是希望通过这一系列实验，来逐步揭示风与索膜结构的相互作用规律。具体包括：

（1）静风耦合实验：目的是确定结构自身的基频和阻尼比，同时可顺便研究振动结构与静止空气间的相互作用。

（2）刚性模型和弹性模型风洞测压试验：目的是通过对比考察结构振动对表面风压分布规律和频谱特性的影响。

（3）弹性模型风洞测振试验：目的是研究结构响应随风速、风向等参数的变化规律，并利用随机减量技术对气动阻尼和附加质量进行识别，从而考察了流固耦合作用影响的大小。

（4）弹性模型同步测振测压试验：目的是研究模型响应与脉动风压之间的相位关系，并利用气弹参数分析技术求解气动阻尼和附加质量，与（3）之结果相互印证。

实验取得了预期效果，即不但成功测出了简化气弹模型中的两个重要参数：附加质量和气动阻尼，而且对索膜结构的风振机理问题有了更深刻的认识。实验结果表明：结构在脉动风作用下的振动是一个宽带过程，其振动形态与风荷载的空间分布具有一定的相关性，表现出明显的受迫振动特征；气动阻尼对结构振动的影响较为显著，特别是对结构的低阶模态，气动阻尼比可达到 15％～20％左右；相比之下，附加质量对结构振动的影响较小，其量级仅为结构质量的 1～2 倍左右；另外，试验过程中未发现结构出现整体气弹失稳现象，但在某些情况下，在结构局部测点出现气动负阻尼。

1.3.3 考虑流固耦合效应的 CFD 数值模拟方法

薄膜结构在风荷载作用下的受激振动问题在理论上可归结为，不可压缩粘性流体与几何非线性弹性体之间的非定常耦联振动问题。因此可以考虑采用 CFD 数值模拟与结构动力有限元分析相结合的方法来模拟其响应全过程。相关研究主要从以下两个方面展开。

- 直接数值模拟方法

文献[14]基于分区求解的思想，提出了一种适用于索膜结构流固耦合风振分析的 CFD 数值模拟方法，并编制了相应的有限元计算程序。程序包含流体域、结构域和网格域三个计算模块。流体域采用 Taylor-Galerkin 有限元法和大涡模拟技术来求解不可压缩粘性流动；结构域的计算采用 Update Lagrangian 有限元列式和 Newmark 逐步积分技术，并考虑了几何非线性的影响；网格域采用拟弹性介质法来计算动态网格的更新。该程序目前还仅能用于二维问题的分析。

应用该程序对典型的单向张拉膜屋盖进行了刚性模型和弹性模型的绕流特性分析（图 10，图 11），还探讨了来流风速、屋面质量和初始预张力等参数对结构流固耦合响应特征的影响。在这些工作的基础上，关于索膜结构的风振机理问题得出了一些虽然是初步的、然而是十分重要的概念[15]：

图 10 平屋面刚性模型绕流图

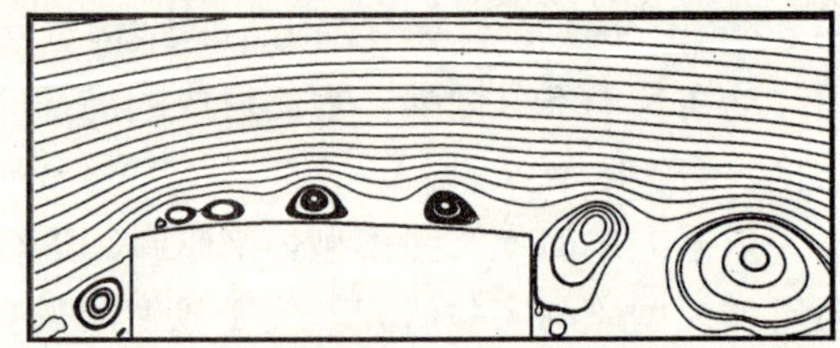

图 11 平屋面弹性模型绕流图

(1) 由屋盖前缘周期脱落的大尺度旋涡是诱导结构振动的主要原因，随着大尺度涡沿屋面向下游移动，其动能逐渐转化为结构变形能，使结构振动形态呈现明显的涡激振动特征。结构响应以背景响应为主，共振响应分量较小，呈现出宽带、受迫振动特征。由于背景响应具有拟静力性质，因此可以考虑将风与结构的流固耦合作用划分为稳态耦合和瞬态耦合两部分。在前一部分中，结构的几何非线性效应是主要的，而动力效应并不明显；在后一部分中，由于结构的瞬态振幅相对于稳态振幅要小的多，因此结构的动力效应是主要的，而几何非线性作用可以忽略。

(2) 索膜结构的流固耦合过程与桥梁和高层结构相比具有不同的特点。后者的惯性力和回复力通常较大，而且往往以单频振动为主；因此随着风速的增大，旋涡脱落频率加快，在理论上总是存在某一诱导结构出现气弹失稳的临界风速。但是对于索膜结构，由于其刚度与位形有关，且振型频谱密集；因此当结构的某一阶振型被激起后，随着结构振幅的加大，结构刚度和振动模态也会随之发生变化，从而使结构从某一共振频率中逃逸出来。因此索膜结构，也可以推论到具有负高斯曲率并且张紧的索膜结构，发生气弹失稳的可能性并不大；在结构与风荷载的相互作用问题中，气弹响应分析应是研究中的主要问题。

● 简化的数值模拟方法

上述的直接数值模拟方法能给出耦合振动的具体过程，从而得到此过程中的一些统计信息，如均值、方差等，但是这种方法的计算量十分巨大，在微型机上求解一道题往往需要数十天时间，无法在工程实践中进一步推广。从工程角度看，人们所关心的不是耦合振动的具体过程，而仅仅是此过程中的一些统计信息。因此不妨考虑换个思路，即放弃对流固耦合具体过程的模拟，转而采用某种更为简化的方法，仅通过少量计算来确定此过程的统计信息；当然，这样做的前提是要给出关于薄膜结构流固耦合效应机理的合理解释（或假定）。为此，文献[16]基于对薄膜结构流固耦合效应机理的探讨，提出了一种更具可操作性的简化数值模拟方法，并且对该方法中的若干具体问题进行了详细探讨。

根据 Davenport 提出的结构风振理论，可将结构风振响应分为：平均响应，背景响应和共振响应三部分，如图 12 所示。其中，平均响应$\bar{r}$是结构在平均风荷载作用下的响应；背景响应$\tilde{r}_B$是由那些低于结构基频的脉动风分量引起的，本质上是准静态的，不受结构动力特性的影响；共振响应$\tilde{r}_R$是由接近结构基频的脉动风分量引起的，具有明显的动力放大效应。这里假定薄膜结构的风振响应亦具有上述特征，则针对不同响应分量的性质，可以考虑采用不同的求解方法。

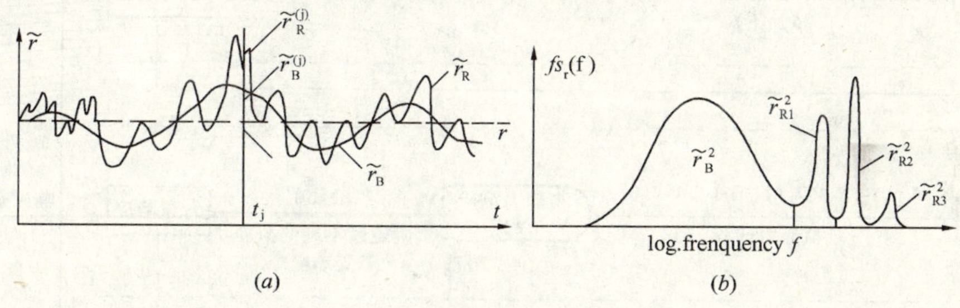

图 12　结构风振响应示意图

(*a*) response；(*b*) spectrum

在平均响应部分，主要考虑由于结构形状变化所导致的体型系数变化，这是一个静态过程；在背景响应部分，主要考虑脉动风中的低频部分与结构之间的耦合作用，体现了脉动风的空间相关性对结构整体振动的影响。这是一个慢变的平稳过程，可以看作是拟静态的；在共振响应部分，主要考虑脉动风中接近结构基频的部分与结构之间的动力耦合作用。

根据前面所述，可以将风与薄膜结构之间的流固耦合过程分为：静态耦合、拟静态耦合和瞬态耦合三部分（见图 13）。其中，平均响应$\bar{r}$属于静态耦合，背景响应$\tilde{r}_B$属于拟静态耦合，共振响应$\tilde{r}_R$属于瞬态耦合。对于静态耦合和拟静态耦合，需要根据结构的具体形状和绕流特点做个别研究，最适宜的方法是数值风洞法；对于瞬态耦合部分，可以考虑利用随机振动时程分析的方法来研究（见图 14）。由此得到结构的平

均风响应 $\overline{x}$，拟静力响应 $x_{1i}(t)$ 和相应的瞬态变形 $x_{2i}(t)$，则结构的极值响应可近似采用叠加原理得到：

$$x_{\max} = \overline{x} + \max(x_{1i}(t) + x_{2i}(t)) \tag{9}$$

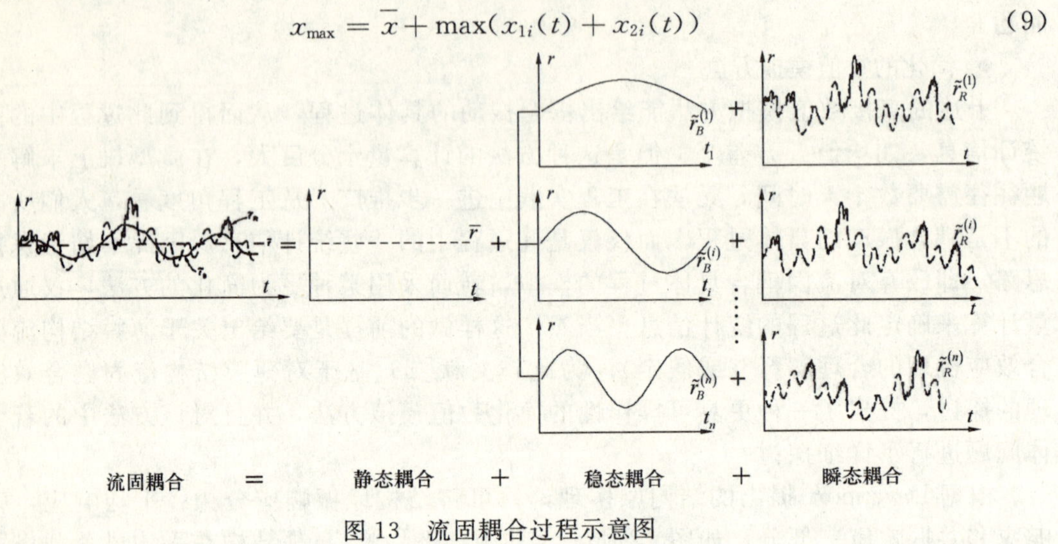

图 13 流固耦合过程示意图

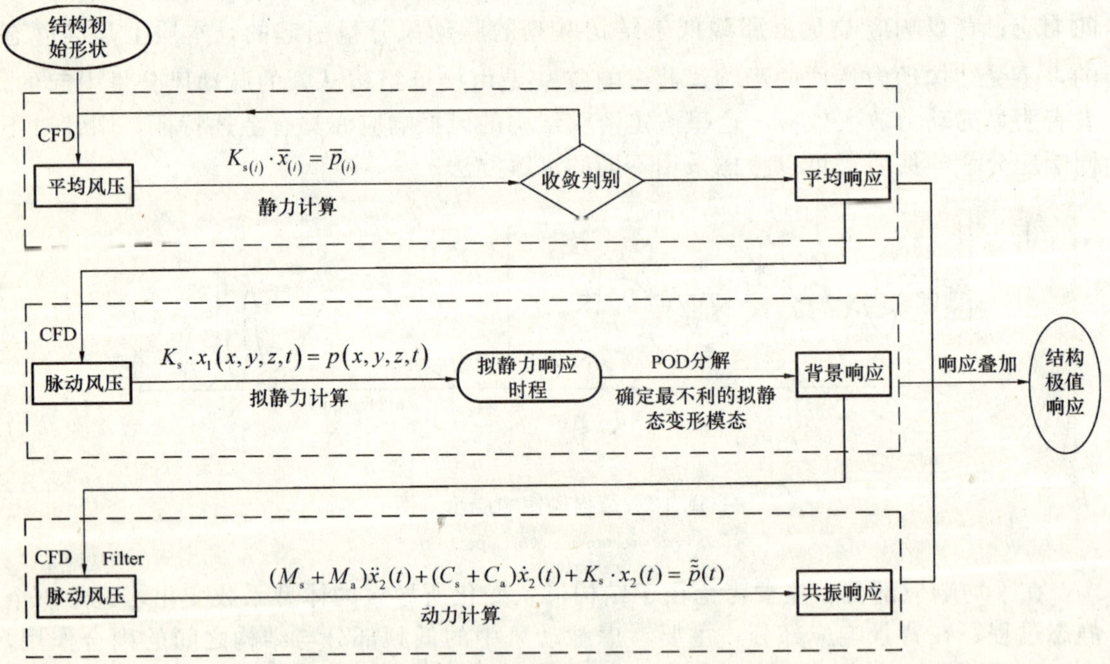

图 14 流固耦合数值模拟的基本框图

依照上述思想对单向柔性屋盖考虑流固耦合效应的风振响应进行数值模拟，分别从静态耦合、拟静态耦合和瞬态耦合三个过程对单向柔性屋盖的流固耦合效应进行了描述。图 15a 给出了用三种方法（直接数值模拟、简化数值模拟方法和随机振动时域分析方法）求得的平屋盖各点的最大位移，可以看出简化数值模拟方法和直接数值模拟方法的计算结果吻合较好，而以往采用的随机振动方法的计算结果偏大。此外，图 15b 和图 15c 还给出了拱形屋盖和悬挂屋盖的最大位移，也可得到相同的结论。此外，通过对这三种屋盖响应的流固耦合程度比较可以看出，拱形屋盖的流固耦合效应明显低于另外两种屋盖形式。这是由于流固耦合效应总是趋向于使屋盖接近流线形，因此可以推断，近似流线形结构的流固耦合效应要低于钝体结构。

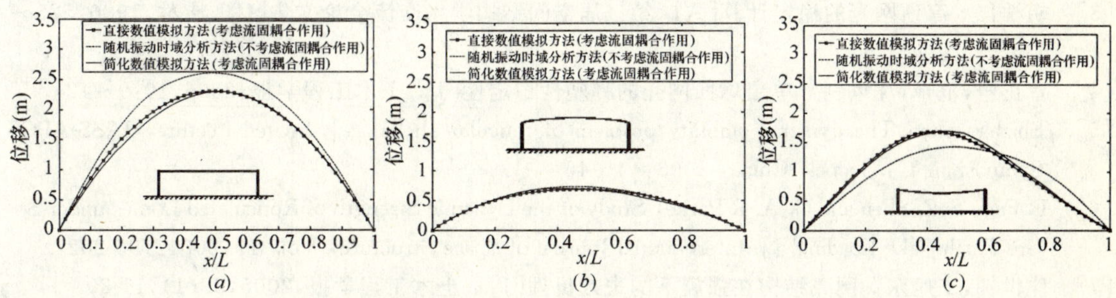

图 15　屋盖各点的最大位移

(a) 平屋盖；(b) 拱形屋盖；(c) 悬挂屋盖

文献[16]还采用简化数值模拟方法分析了菱形平面鞍形膜结构的风致流固耦合效应。图 16 给出了分别采用三种方法（直接数值模拟、简化数值模拟方法和随机振动时域分析方法）求得的屋盖最大位移分布图，可以看出简化数值模拟方法和直接数值模拟方法的计算结果吻合较好，而以往采用的随机振动方法计算结果偏小，可能会造成设计偏于不安全。

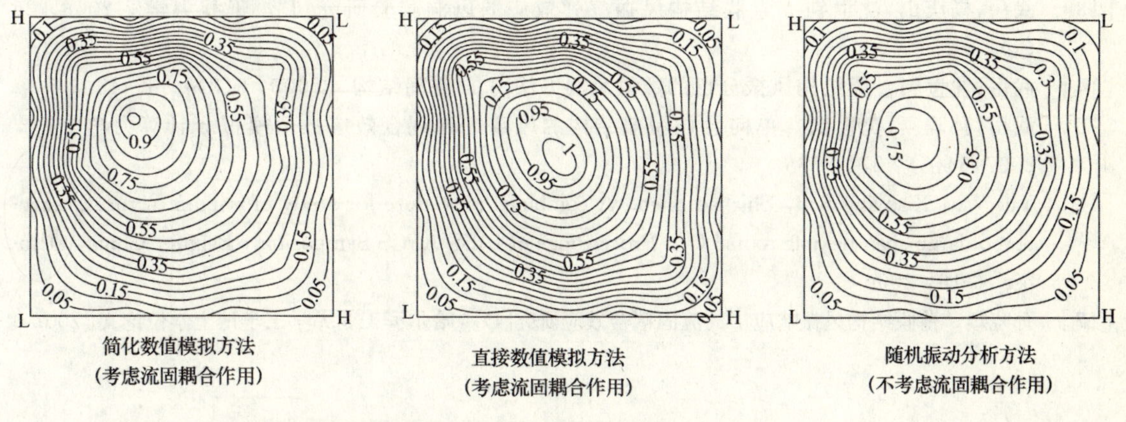

图 16　屋盖的最大位移（m）

综上所述，关于索膜结构风振机理的研究还处在探索阶段，尚未形成基本理论框架，因此需要从多个方面、不同角度进行理论和试验研究，积累素材，提炼概念，以逐步建立起较合理的能基本反映这种三维柔性体系流固耦合效应的气动力模型。从现有方法的比较来看，由于简化数值模拟方法引入了较先进的风振分析理念，因此在计算量方面大为降低，且完全可以满足工程精度要求；更为重要的是，其计算结果具有更加丰富的物理内涵，因此可以为进一步研究索膜结构的风振机理提供有效的途径。

* 本研究得到国家自然科学基金重点项目 50338010 的资助。

参考文献

[1] JGJ 61—2003. 网壳结构技术规程[S]. 北京：中国建筑工业出版社，2003

[2] 沈世钊，陈昕. 网壳结构稳定性[M]. 北京：科学出版社，1999

[3] 胡学仁. 穹顶网壳的稳定计算[A]. 第三届空间结构学术交流会论文集[C]. 吉林，1986：525～536

[4] 曹正罡，范峰，沈世钊. 单层球面网壳的弹塑性稳定性[J]. 土木工程学报，2006，10：9～12

[5] Shizhao Shen. The dynamic stability problem of reticular shells[A]. Invited Lecture, IASS-APCS Symposium[C]. Taibei, China, 2003：44～46

[6] F. Fan, S. Z. Shen and G. A. R Parke, Study of the Dynamic Strength of Reticulated Domes under Severe Earthquake Loading[J]. International Journal of Space Structure, Vol. 19，2004：195-202

[7] 沈世钊，支旭东. 网壳结构在强震下的失效机理[J]. 土木工程学报，2005，38(1)，11-20

[8] Zhi Xudong, Fan Feng, Shen Shizhao. Failure mechanism of single-layer reticulated domes subjected to earthquakes [J]. Journal of the International Association for Shell and Spatial Structures. 2007，(1)：29-44

[9] 支旭东. 网壳结构强震失效机理及支座减震研究[D]. 哈尔滨工业大学博士后出站报告，2008

[10] 武岳. 考虑流固耦合作用的索膜结构风致动力响应研究[D]. 哈尔滨工业大学工学博士学位论文. 2003

[11] 刘瑞霞. 薄膜结构气弹动力稳定性研究的解析方法[D]. 北京交通大学博士学位论文，2004

[12] 沈世钊，武岳. 膜结构风振响应中的流固耦合效应研究进展[J]. 建筑科学与工程学报. 2006，23(1)：1-9

[13] 武岳，杨庆山，沈世钊. 索膜结构风振气弹效应的风洞实验研究[J]. 工程力学. 2008，25(1)：8-15

[14] 武岳，沈世钊. 膜结构风振分析的数值风洞方法[J]. 空间结构. 2003，9(2)：38～43

[15] 武岳，孙晓颖，沈世钊. 单向悬挂屋盖结构的风致气弹耦合效应数值模拟. 计算力学学报. 2007，24(5)：571-578

[16] Yue Wu, Xiaoying Sun, Shizhao Shen. A combined procedure for study of wind-structure interaction of large-span flexible roofs [C]. Proceedings of IASS-APCS Symposium, Beijing, China, October 16-19, 2006

[17] 孙晓颖. 薄膜结构风振响应中的流固耦合效应研究[D]. 哈尔滨工业大学工学博士学位论文. 2007

第 2 章 Chapter 2

中国结构震动控制的研究与应用进展

Progress of Research & Application on Seismic Control for Structures in China

F. L. Zhou(周福霖), P. Tan(谭平),
J. Cui, Q. L. Xian, Y. Zhou,
X. Y. Huang, C. Y. Shen, L. S. Wei & D. Y. Huang
Earthquake Engineering Research Test Center, Guangzhou University, Guangzhou 510405, China

Abstract: This paper briefly introduces the recent research, testing analysis, design and application on seismic isolation, energy dissipation, tuned mass damper and active control for buildings and bridges in mainland China. Paper introduces some typical researches, testing and analysis, including the mechanical tests for bearings and control devices, and the shaking table tests for structural models with different control systems. Paper also introduces the Chinese design codes for structures with seismic isolation and energy dissipation. Paper describes the recent application status and typical examples, especially introduces the largest isolation buildings group in the world, and the using passive and semi active control for structures. Also the paper makes discussion some problems existed on passive and active control technique now and the tendency of development on seismic control in future

Keywords: Seismic Isolation, Energy Dissipation, Passive Control, Active Control

2.1 Development of techniques for seismic resistance of structures

During earthquake, the building structure which is fixed on the ground will respond gradually increasing from the building bottom (ground) to the building top, liking an "Amplifier". This will result in damage or collapse of structure or results in damage of non-structural components, decorations and equipment or facilities in buildings due to the large response of the structure. In order to reduce the response and avoid damage of structure, some techniques of seismic resistance for structures have been developed [1].

There are two kinds of techniques of seismic resistance for structures at present. One is the traditional Anti-seismic technique, another is the new technique——Seismic control of structures which will discussed in this paper. The traditional Anti-seismic technique uses the methods of strengthening the structure then rises the structural stiffness, which will induce to increase the structural response then to loss the safety of structure during earthquake. The new technique, Seismic control of structures, uses the methods only changing the structural dynamic characters (damping, frequency or tuned mass) to reduce the structural response in earthquake, instead of strengthening the structure. The new technique, Seismic control of structures, is very effective to control the structural response in earthquake or wind, also is able to ensure the structure to be safe during unexpected higher intensity earthquake. So the new technique may be safer, reliable, inexpensive and simple to be implemented in many cases, suitable to be widely used in general civil or important buildings, bridges, facilities or other structures in the regions of seismicity [2].

Some problems are existed in the traditional Anti-seismic technique for seismic resistance for structures:

(1) It is not very safe: It is difficult to control the structural damage level due to the inelastic deformation in earthquake, also it may be dangerous in severe unpredicted earthquake.

(2) It is limited to be used: The designing aim is only to protect the structure, but not protect the facilities inside the structure. So it is not able to be used in some important structures whose inside facilities could not be damaged in earthquakes, such as hospitals, City lifeline constructions, nuclear power plant, museum building, and the buildings with precise instruments in it.

(3) It is more expensive: The anti-seismic way of traditional technique is to strengthen the structures: making the columns, beams, shear walls or other structural elements bigger or stronger, increasing the stiffness of the structures. It will cause to increase the earthquake action(by increasing the stiffness), then, cause again to strengthen the structures. Finally, it will rise the cost of structure, but the safety level of anti-seismic is still unsure. The cost of building using traditional anti-seismic technique will rise as bellow comparing with the cost of no-anti-seismic building depending on the present design code in some countries [3]:

- For moderate earthquake(Ground 0.125g), rise the cost 3% ~ 10%.
- For major earthquake(Ground 0.250g), rise the cost 10% ~ 15%
- For strong earthquake(Ground 0.400g), rise the cost 15% ~ 25%
- For worst earthquake(Ground > 0.500g), rise the cost 30% ~ 60 %, also the design is difficult.

Because the traditional method and system are not satisfied in many cases for earthquake resistance, so it is limited to be developed in future although it is used very common at present. Many experts have paid more attention to find out some new systems for earthquake resistant structure: Seismic Isolation, Energy Dissipation and Structural Controlling

System. Some available results of research have been obtained, and the some systems have been used successfully in engineering application in China and many countries [4,5].

The range of application and technical maturity of Seismic Control of Structures are shown in Table 1 [2].

The range of application and technical maturity of Seismic control of structures Table 1

Name of control	Application Ranges	Technical maturity
Seismic isolation	2~50 stories buildings (New design or existed) Bridges, subway Equipment or facilities	Mature technique Have rich theoretical and testing results Widely application Successfully experienced the Strong earthquake
Energy dissipation	High rise building Long span bridges High tower or skeleton	Mature technique Have rich theoretical and testing results Application for seismic or winds
Passive control (TMD, TLD, etc)	High rise buildings Long span bridges High tower or skeleton	Basically mature technique Have some research and testing results Application for seismic or winds
Active control	High rise buildings Long span bridges High tower or skeleton	Does not become a mature technique Have some research and testing results Application for seismic or winds

2.2 Seismic isolation

There are over 500 buildings with isolation rubber bearings built in China until 2005. These buildings include houses (about 70%), office, school, museum, library, and hospital. The story of buildings is 3~19 stories. The most of structural types of buildings are concrete frame or shear wall-frame and brick wall structures. Some railway bridges and highway bridges with seismic isolation have been built also in China. It has become a very strong tendency to widely using seismic isolation rubber bearings system in China now.

2.2.1 Many Significant Advantages of Structures with Seismic Isolation System (Table 2)

Safe in Strong Earthquake

Comparing the seismic isolation structures with the traditional anti-seismic structures, the response of isolating structures can be reduce to 1/2~1/8 of the response of traditional structures, according the testing results and the records in real earthquake also. It is very effective to reduce the response of structures in earthquake and is able to prevent the structure from damage or collapse in earthquake. So it can ensure the structures with isolation system to be safe in strong earthquakes.

Save the Structure Cost

Comparing the seismic isolation structures with the traditional anti-seismic structures, the building cost of isolation structures can be saved 3%～15% of the general building cost in some cases, because re-designing the supper structure which seismic response is very small, according the final statistics results of 30 buildings with rubber bearings completed in southern, western and northern China [6,7].

Wide Ranges of Application

The seismic isolation rubber bearings system can be used in, both new design structures and existed structures, both important buildings and civil buildings especially for house buildings, both for protecting the building structures and for protecting the facilities inside the buildings [3].

Free Architectural Design

The seismic isolation system can be used in the buildings with irregularities configuration, by putting the isolation layer on the suitable vertical level and, by arranging the isolators with different stiffness and damping in plan of isolation layer. But it is impossible to be done for the traditional anti-seismic buildings that must be regular configuration very strictly [8].

Technical and economical comparing of seismic isolation and traditional buildings　Table 2

	Traditional anti-seismic buildings	Seismic isolation buildings
Acceleration response	1.00	1/2～1/8
Working state of structure During earthquake	inelastic	basically elastic
Building cost	1.00	0.85～1.05
Building configuration Requirements	regular	irregular

2.2.2　Testing and Design of seismic isolation system

There are five kinds of material have been used for isolators in China, including Sand layer, Graphite lime mortar layer, Slide friction layer, Roller and Rubber bearing (Fig. 1,2)——This is the laminated steel sheet rubber bearing with or without lead core.

Many tests have been finished and whole sets of computation theory of seismic isolation rubber bearings system have been established in China now. The tests include two kinds of work:

(1)Test of mechanical characteristics for isolator, includes compression tests (capacities, stiffness), compression with shear cycle loading tests (stiffness, damping radio and maximum horizontal displacement) (Fig. 3).

Fig. 1 Design of Rubber Bearing

Fig. 2 Compression. -Shear Test for Bearing

Fig. 3 Tests of Creep

(2) Tests of Durability for isolator, includes low cycle fatigue failure tests, creep tests and ozone aging tests

(3) Test of structural system, includes shaking table tests for large scale structural model, including one 6 stories steel frame models with different location of isolation bearings layers are tested on the shaking table (Fig. 4). The testing results show that the acceleration responses on each stories of structural model are nearly the same. It means that the elements and joints of structure with isolation rubber bearings nearly work within elastic range only. The acceleration response on structure with isolation is only (1/3～1/10) response of structure fixed on shaking table. It means the isolation structure is more effective to attenuate the structural response in earthquake than any other methods.

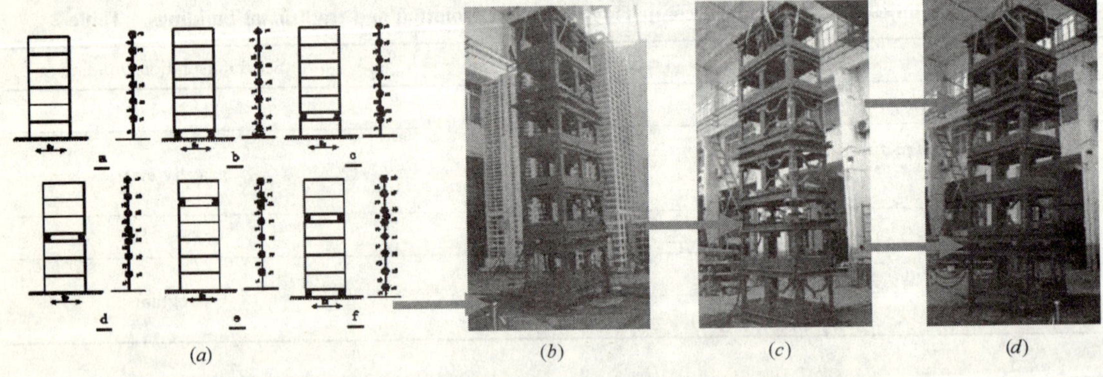

Fig. 4 Shaking Table Test for Different Location of Isolation Bearings Layers

(*a*)Different location of isolation layers; (*b*)On the base; (*c*)On the Story; (*d*)Multi-layers

2.2.3 Different Isolation Structural Systems

There are five kinds of locations of isolation layer with rubber bearings in China (Fig. 4, 5)

- *Base isolation*. Isolation layer is located on the base of building.
- *Basement isolation*. Isolation layer is located on the certain story of the basement (Fig. 5*a*).
- *Story isolation*. Isolation layer is located on the top of the first story (Fig. 5*b*) or

certain story of supper structure (Fig. 5*c*).

● *Top isolation*. Isolation layer is located on the top of building (Fig. 5*d*), like TMD, is always used to add 1-2 stories on the top of existed building for seismic retrofit.

● *Over bridge linking isolation*. Isolation layer is located at the linking joints between over bridge and buildings (Fig. 5*e*) to decouple the different model shapes of buildings linked by over bridge

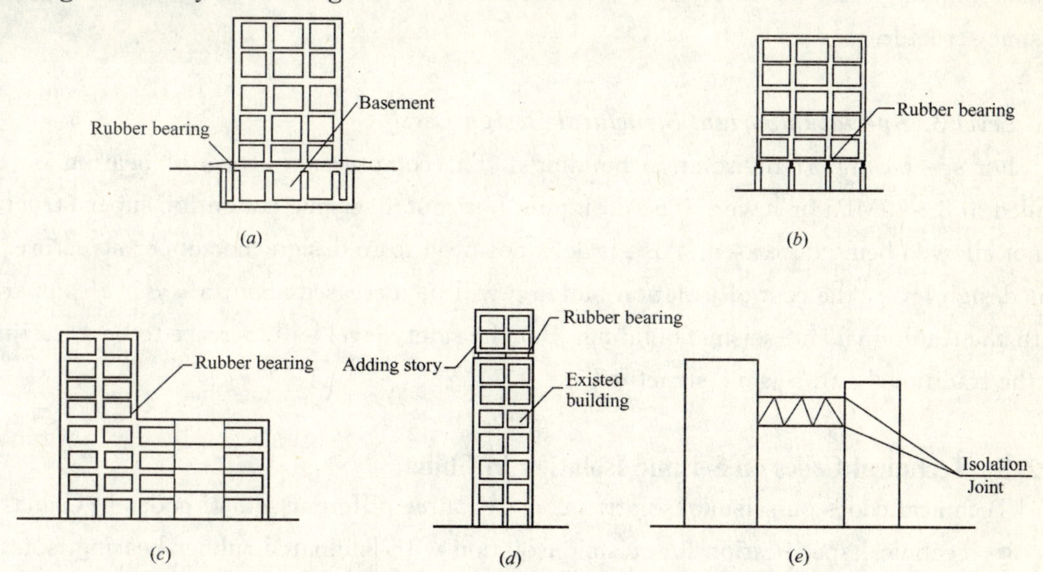

Fig. 5　Different Locations of Isolation Layer in Structure

(*a*) Basement isolation; (*b*) Story isolation with isolators on the top of the first story

(*c*) Story isolation with isolators on certain story of supper structure;

(*d*) Top isolation; (*e*) Over bridge linking isolation

2. 2. 4　Different Designing Levels for Isolation Buildings

Level 1. General Structural Design Level

For common isolation civil buildings: The compression stress of bearing is controlled in 12～15MPa. The designing horizontal seismic load for super structural is allowed being decreased to be 1/2～1/8 of traditional anti-seismic structure, and is allowed to re-design the supper structure for decreasing the section or reinforces of structural elements. In this design level, the cost of isolation building will be saved about 3%～15% comparing with the traditional anti-seismic building. And the safety level will increase to be 2～4 times of the traditional anti-seismic structure.

Level 2. Important Structural Design Level

For important isolation buildings: The compression stress of bearing is controlled in

10～12MPa. The designing horizontal seismic action for super structure is allowed being decreased to be 1/2～1/4 of traditional anti-seismic structure, and is allowed to re-design the supper structure for a little decreasing the section or reinforces of structural elements, or does not need to re-design the supper structure. In this design level, the cost of isolation building will be balanced or increased about 3%～5% comparing with the traditional anti-seismic building. But the safety level will increase to be 3～6 times of the traditional anti-seismic structure.

Level 3. Special Important Structural Design Level

For special important isolation buildings: The compression stress of bearing is controlled in 8～10MPa or lower. The designing horizontal seismic action for super structure is not allowed being decreased. Also it does not need to re-design the supper structure. In this design level, the cost of isolation building will be increased about 5%～7% comparing with the traditional anti-seismic building. But the safety level will increase to be 4～8 times of the traditional anti-seismic structure.

2.2.5 Technical Codes on Seismic Isolation in China

Technical codes on seismic isolation consists three different sets of codes in China:

- Technical specification for seismic isolation with laminated rubber bearing isolators (CECS 2000). This is the national code for design and construction of buildings and bridges with seismic isolation in China.
- Standard of laminated rubber bearing isolators (JG 118—2000). This is the national standard of isolators for laminated rubber bearing in China
- Seismic isolation and energy dissipation for building design (Chapter 12 in code for seismic design of buildings, GB 50011—2001). This is a part of national code in China for seismic design of buildings, in which is the chapter 12 [9].

Some main introductions for all these three codes (standards) on seismic isolation in China are described as below:

- Provide the design methods of seismic isolation for buildings, bridges, special structures and industry facilities.
- Provide the design methods of seismic isolation for new design structures also for retrofit of existed structures.
- Allow to fellow three design level depending the importance of structures and requirements of owners in the areas with different economic situation in China.

Level 1, for general structures, using isolation will save building cost about 3%～15%.

Level 2, for important structures, using isolation will increase building cost 3%～5%.

Level 3, for special important structures, using isolation will increase building cost 3%～5%. But the isolation buildings designed by any level will increase the seismic safety about 2～8 times comparing the traditional anti-seismic buildings.

● Provide two methods of structural analysis for seismic isolation of structures:

Equivalent shear method, it is the static analysis methods for structures that are not higher than 40m or 10 stories, regular configuration and with shear deformations predominantly.

Time-history analysis, it can be used for all structures.

● Allow reducing the seismic shear load for designing super structure for saving the building cost for general civil buildings or for some poor economic areas.

● Allow choosing the different compression stress level for isolators.

σ=12～15MPa, for general civil buildings or for some poor economic areas.

σ=10～12MPa, for important buildings or for general areas.

σ=8～10MPa or $\sigma \leqslant$8MPa, for special important buildings or for rich areas.

● To control the maximum horizontal shear displacement Dmax of isolation layer.

Dmax shall not be larger than 0.55 times of diameter of bearings and 300% shear strain deformation of bearings. Dmax shall be the total displacement including both translation and torsion of structural system.

● Require high quality of rubber bearing to be proved by complete testing [9].

2.2.6 Examples of Seismic Isolation Building

Example 1. RC Frame Multi-Stories House Building with Base Isolation

Some 7～8 stories RC frame house building - one of the most popular isolation building types in China built from 1991 in China. Rubber bearings layer is located on the base (Fig. 6). The design level is level 1. The compression stress of bearing is nearly 15MPa. The designing horizontal seismic action for super structural was decreased to be 1/4 of traditional anti-seismic structure. The supper structure was re-designed for decreasing the section or reinforces of structural elements. The cost of isolation building is saved about 7% comparing with the traditional anti-seismic building. And the safety level increases to be over 3 times of the traditional anti-seismic structure. There are about 180 buildings of this type have been built in China.

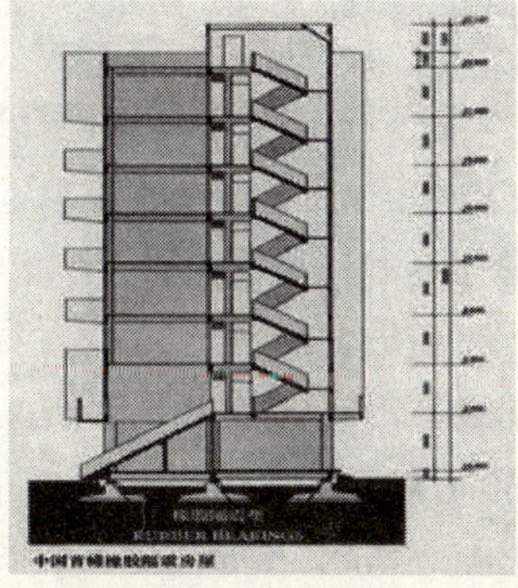

Fig. 6　RC Frame 7 Stories House Building with Story Isolation

Example 2. Masonry Multi-Stories house Building with Basement Isolation

Some buildings group with basement isolation consists of over 60 buildings with 6～7 stories masonry house building - one of the most popular isolation building type in China built from 1996 in western China (Fig. 7). The rubber bearings layer is located on the top of the basement. The design level is level 1. The compression stress of bearing is nearly 15MPa. The designing horizontal seismic action for super structural was decreased to be 1/6 of traditional anti-seismic structure. The supper structure was re-designed the cost of isolation building is saved about 8% comparing with the traditional anti-seismic building. And the safety level increases to be 4 times of the traditional anti-seismic structure. There are about 300 buildings of this type have been built in China.

Fig. 7 Group of Masonry Multi-Stories House Building with Basement Isolation

Example 3. RC Frame-shear 13 Stories Museum with Story Isolation.

One 13 stories RC Frame-shear museum is built in 1996 with 28,000m^2 in southern China (Fig. 8). The rubber bearings layer is located on the top of the column in the first story because the building without basement. The design level is level 2. The compression stress of bearing is nearly 12 MPa. The designing horizontal seismic load for super structural is allowed to be decreased to be 1/4 of traditional anti-seismic structure. The cost of isolation building is increased about 2% comparing with the traditional anti-seismic building. But the safety level increases to be 4 times of the traditional anti-seismic structure. It satisfies to protect not only the structure, but also the history relic inside the building.

Fig. 8 RC Frame-Shear 13 Stories Museum with Story Isolation

Example 4. RC Frame 2 Stories Platform + 9 Stories House with Story Isolation

The Seismically isolated artificial ground which is the largest area in the world (Fig. 9, 10). There is a very large platform (2 stories RC Frame) with 1500m wide and 2000m long to cover a railway area in Beijing City. There are 50 isolation buildings (7～9 stories RC frame) built on the top floor of the platforms The rubber bearings layer is located on the top floor of the platform to isolate the seismic motion also to isolate the railway vibration.

2.2.7 The Isolated Buildings which is the Largest Area in the World

The Background of using Isolation for This Large Area of Buildings

● The very large area of subway and railway communication hub on the ground is located nearly the Beijing City center, where the land is very expensive. The City Government urgently wants to effectively use the land in this large area as house buildings, also want to solve the seismic and environmental problems of railway vibration and noise in this city center area. This project is named Isolation House Buildings on Subway Hub (IHBSH).

● There is a very large platform of two stories RC frame, which is used to put all equipment and facilities for railway hub in it, and cover the noise from the railway trains. The size of platform is 1500m wide and 2000m long. There are 50 house buildings (7～9 stories RC Frame) built on the top floor of the platforms. The floor area of all isolation house buildings is approximately 480,000m^2 which is the largest area using seismic isolation in the world. From the results of analysis and testing, using Stories Isolation is the best way for seismic design (Design ground motion 200～300Gal). The rubber bearings layer is located on the top floor of the platform. (Fig. 9, 10)

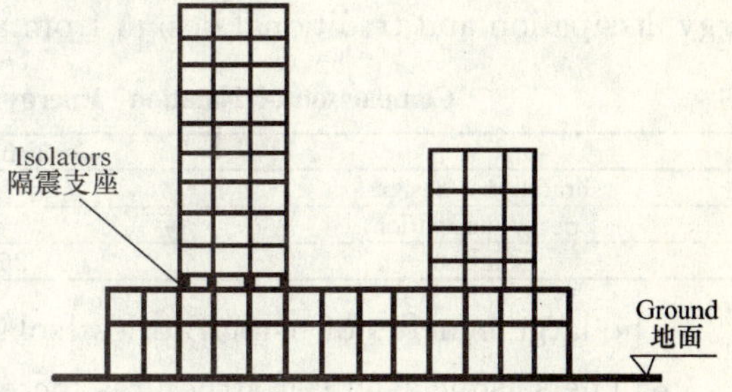

Fig. 9　Stories Isolation used in IHBSH

Fig. 10　A Part of View of IHBSH

Design and Shaking Table Tests

● The design level is level 2. The compression stress of bearing is about 10 MPa. The size of rubber bearings mainly is ϕ 700mm.

● The shaking table test shown that the horizontal seismic loads were decreased to be 1/4 for super structural and 1/2 for platform structure comparing with the traditional anti-

Fig. 11 Shaking Table Tests for Structural Model of IHBSH

seismic structure (Fig. 11). During the strong earthquake (400 Gal) input, the structure is perfect with no any damage for putting isolators between the platform and the building, but the structure is severe damaged and nearly collapse for fixing joint between the platform and the building.

Technical and Economical Comparison of Isolation with No-isolation

The comparison results of isolation, energy dissipation and traditional design from analysis are listed in Table 3 [10].

Comparison of Isolation, Energy Dissipation and Traditional Design **Table 3**

	Seismic shear force	Construction cost
Traditional design	100%	100%
Energy dissipation	80%	95%
Isolation	25%~35%	75%

The large benefits of using Stories Isolation are described as below:

- The seismic safety of structures increases to 4 times.
- The construction cost could save 25%.
- The number of building stories could rise from 6 stories to 9 stories, therefore, The building floor area increase from 380000m^2 to 480000m^2, adding 100000m^2
- The whole pure profit of real estate reaches RMB 240 million—￥600000000 Calculated from pure profit RMB ￥6000/m^2 × 100000m^2 (whole pure profit USD 75 million—$75000000)
- The environmental problems of railway vibration and noise in the city center area could be solved

2.3 Energy dissipation

There are over 30 buildings with Energy dissipation dampers until 2005. Energy dissipation system is formed by adding some energy dissipaters into the structure. The energy dissipaters provide the structure with large amounts of damping which will dissipate most vibration energy from vibration sources previous to the structural response reaching the limitation, then ensure the structure to be safe in earthquake or to satisfy the using requirement in wind. The energy dissipaters may be set on the bracing, walls, joints, connection parts, nonstructural elements or any suitable spaces in structures, which may reduce 40%~60% of the structural response comparing the traditional structure without energy dissipaters. This technique is very reliable and simple, suitable to be used for general or important new or existed buildings or facilities in seismic regions[11, 12].

2.3.1 Research, Testing and Application for Energy Dissipation Dampers in China

Three kinds of dampers have been used in China now, including Metallic Damper, Friction Damper and Viscous Fluid Damper.

Metallic Damper

The X shape of steel plate is chosen (Fig. 12) as damper[13, 14], to avoid stress concentrations at the clamped boundaries or at the middle during bending. When this flat plate of mild steel is loaded, the bending stresses become uniform along the whole length, then allowing yielding to develop over the full length of steel plates. A series of bending cycles fatigue failure tests of X shape of steel plate have been carried out (Fig. 13, 14), then the relationship between Strain and the number of cycles to failure have been established for structural design (Fig. 15).

Another Square Frame Steel Damper has been developed in 1978. A full scale test of Square Frame Steel Damper have been finished (Fig. 16) and the 1st industry structure have been constructed in northern China in 1979 (Fig. 17).

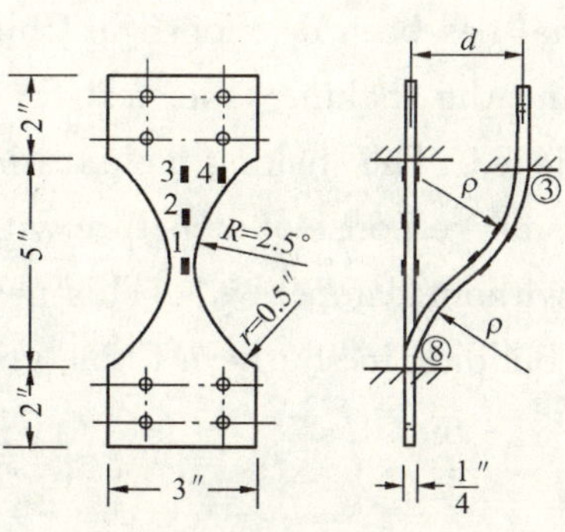

Fig. 12　X Shape of Steel Plate

Fig. 13　Bending Cycles

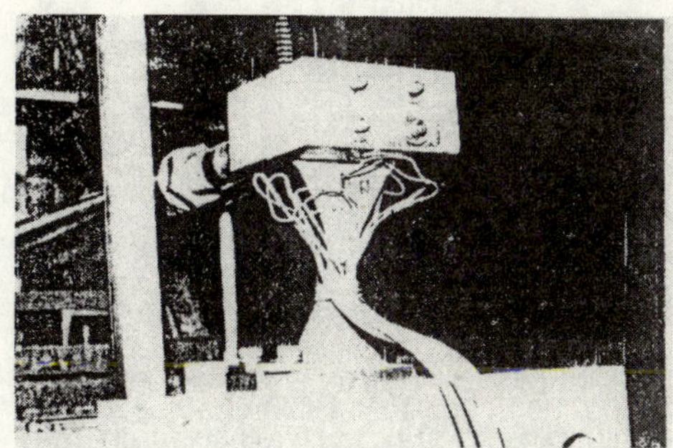

Fig. 14　Strain Records

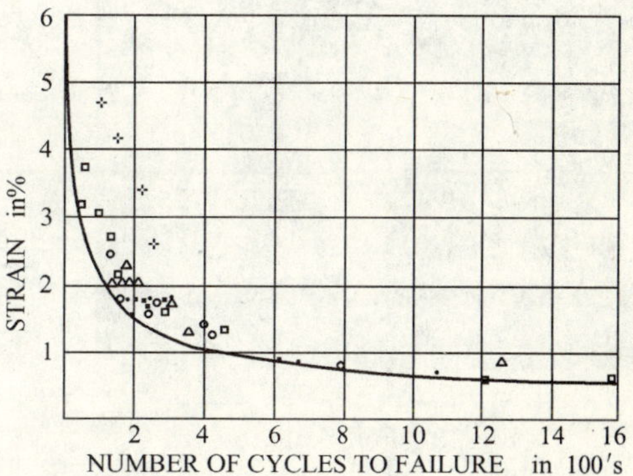

Fig. 15　Relationship Between Strain and the Number of Cycles to Failure

Fig. 16 Full Scale Test of Metallic Damper

Fig. 17 The 1st Structure with Damper in China

Friction Damper **(*by Asso. Prof. Q. L. Xian*)**

A damper with new Composite Friction material has been developed in China. A full scale test of this new Friction Damper (Fig. 18) and the shaking table test for structural model with this new Friction Damper have been finished. The shaking table testing of large models with energy dissipaters shown that, the seismic response of structure with dampers is decreased 30%～40% comparing the structure without dampers[2, 4]. This new Friction Damper have been used in the design for high rise building in southern China (Fig. 19).

Fig. 18 Full Scale test of Friction Damper

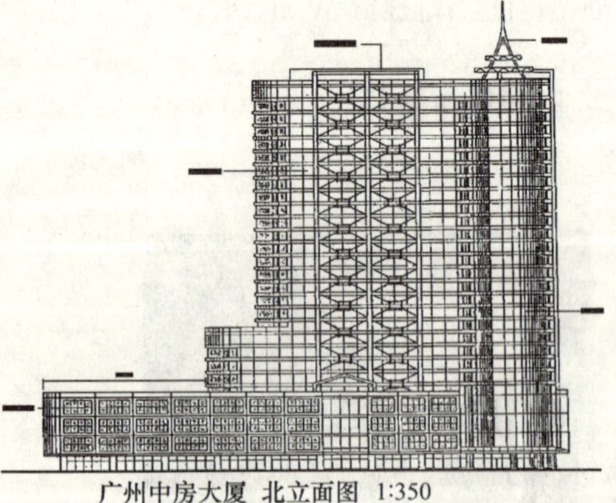

Fig. 19 High Rise Building with Friction Damper

Viscous Fluid Damper

A series of Viscous Fluid Dampers are produced in China and widely been used, the damping force from 20 ton to 200 ton. The tests of characteristics for dampers have been carried out (Fig. 20). Nearly 30 buildings used these dampers for new design or retrofit of existed building (Fig. 21, by Prof. W. Q. Liu). This Viscous Fluid Dampers also been combined used with the isolation systems.

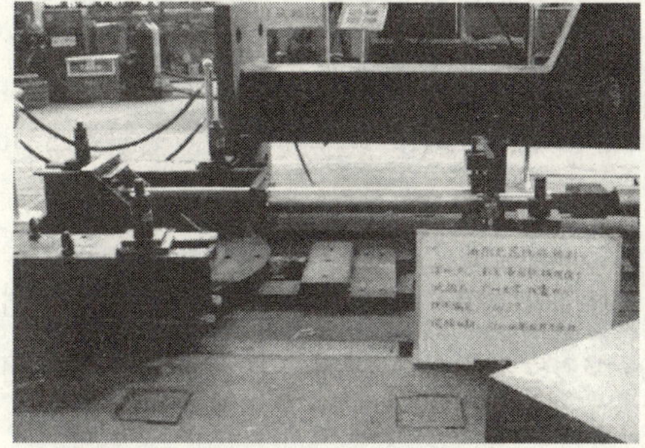

Fig. 20　Test for Viscous Fluid Damper

Fig. 21　The Building with Viscous Fluid Damper

2.3.2　Design and Application of Energy Dissipation System

There are three kinds of energy dissipation system being used now in China:

(1) Energy Dissipation Bracings (Fig. 22): Energy dissipaters put on the bracing, setting along whole structural height for regular building to rise the capacity of anti-seismic, or only setting in some spaces (or stories) of structure for irregular building to avoid torsion damage during earthquake.

(2) Energy Dissipation Wall (Fig. 23): Energy dissipaters put on the walls in weak stories of structure to improve the structural working state, avoiding dangerous damage of weak stories during earthquake.

(3) Energy dissipation Connection (Fig. 24): Energy dissipaters put on the gaps between adjacent buildings to avoid impact damage during earthquake.

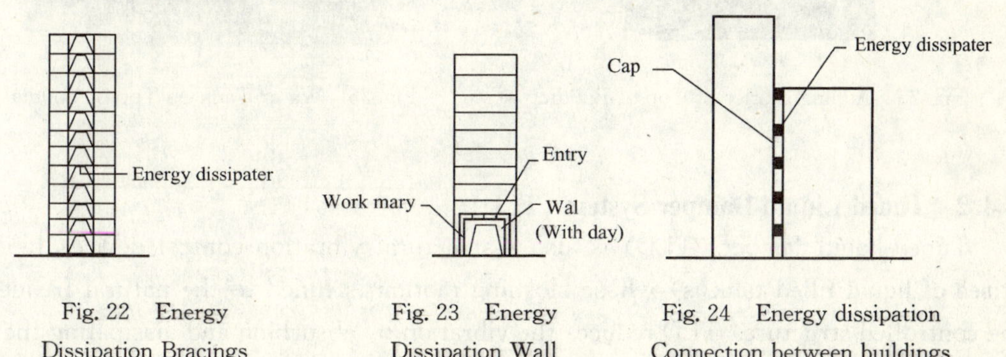

Fig. 22　Energy Dissipation Bracings

Fig. 23　Energy Dissipation Wall

Fig. 24　Energy dissipation Connection between buildings

2.4 Passive control

2.4.1 Tuned Mass Damper System (TMD)

The system is, on certain position (such as on roof) in new or existed structure (main structure), places an additional Filial Structure (possesses mass, stiffness and damping) which natural period is nearly equal to the natural period of main structure. The dynamic characteristics of structure system are changed. During the earthquake, the Filial Structure will move against the direction of main structures vibration then reduce the response of main structure.

In China, this Filial Structure may be formed by adding one or more stories supported by rubber bearings on the roof of main building structure (Fig. 25), or adding a certain mass, such as water tank, supported by rubber bearings on the roof or other floors in main building structure (Fig. 26).

The shaking table tests show that, the response of main structure adding filial structure is reduced 30%~40% of response of structure without Filial structure [2, 9].

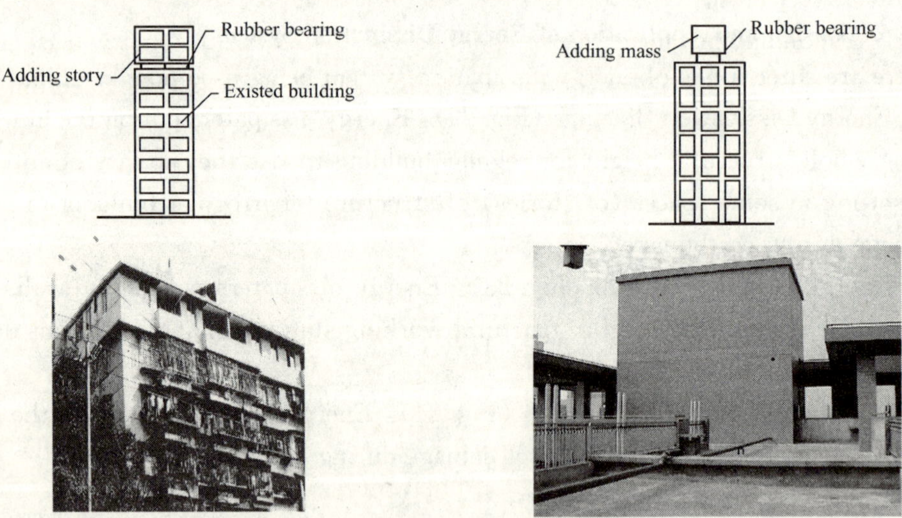

Fig. 25 Adding Stories of Top of Building

Fig. 26 Water Tank on Top of Building

2.4.2 Tuned Liquid Damper System (TLD)

Tuned liquid damper (TLD) is also a structural vibration control device. It is comprised of liquid filled tank(s) whose sloshing motion is tuned to the natural frequency of the controlled structure. TLD reduces the vibration by absorbing and dissipating the external excitation energy through fluid motions. It can increase the equivalent damping of the controlled structure. Compared with tuned mass damper (TMD), TLD has many favourable properties: 1) economic; 2) multifunctional; 3) easy to run; 4) the absence of mobile mechanical components due to less maintenance costs and more reliability; 5) useful in both short-term and long-term; 6) can operate in all directions in the horizontal plane; 7) the needlessness of additional stiffness and damping makes it easily be used in a controllable hybrid system. Thus, TLD is widely used all over the word[15, 16].

2.5 Semi-active control system

Semi-active control system is formed by adding some simple devices on the bracing or walls in structure (Fig. 27), which devices provide the structure with variable damping and stiffness to reduce the seismic response of structure in earthquake. The devices will work as energy dissipaters during moderate earthquake or wind, and will work as active controller during unexpected strong earthquake by inputting very small power. This semi-active control system is much more reliable and simple than general active control system, and more effective to reduce the structural response than other passive control system. This technique is suitable to be used in some important buildings or facilities.

Semi-active control system—Active Variable Stiffness/Damper AVSD (by Dr. P. Tan), corresponding advanced device and algorithm based on the energy dissipation passive control, active control and other semi-control systems[8]. AVSD combines Active Variable stiffness (AVS) and Active Variable Damping (AVD) to become one system, which possesses the all advantages of AVS and AVD. So AVSD is more effective to control the structural response in very wide range of vibration input, more saving energy, more stable and reliable in deferent working state, more simple, economical and practical in application.

The theoretical model of AVSD is shown in Fig. 28, which shows that j th AVSD device located on i th story of structure. Where m_i is the mass of i th story of structure, k_i and c_i are the original stiffness and damping of i th story of structure, k_{0j} and c_{0j} are the stiffness and damping of AVSD system. When the system switch on the k_{0j}, the AVSD works as AVS system, while switch on the c_{0j}, the AVSD works as AVD system. How to switch on, is controlled by the control algorithm, according to the sign of value of response of the structure, based on analysis of mechanism of AVSD within every sample period in vibration of structure.

AVSD system could provide two kinds of working state to control the vibration of structure: working as AVS system provided from closing the valve, working as AVD system from opening the valve. Also, this AVSD could amplify the story drift for damper, which makes the system to dissipate more energy input from ground motion (Fig. 28).

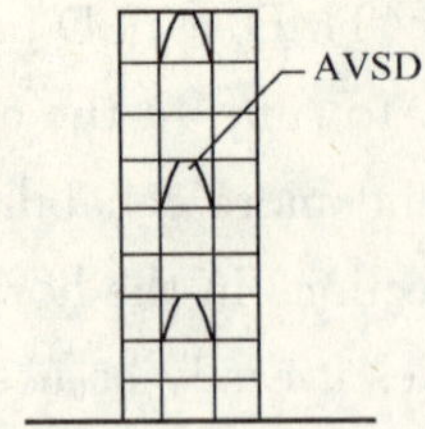

Fig. 27 AVSD on Certain Stories

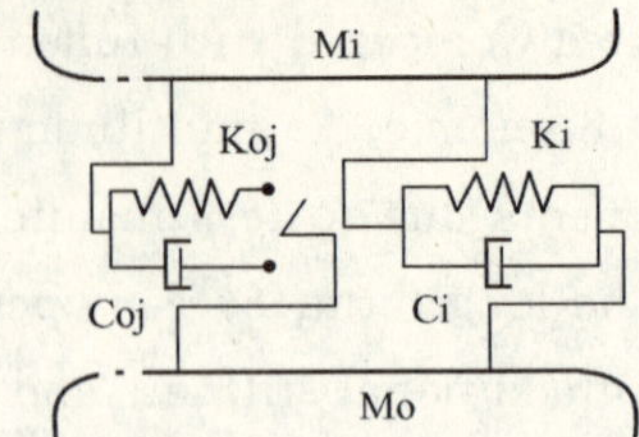

Fig. 28 Mathematic Model of AVSD

The results of shaking table testing (Fig. 29) and analysis show that, AVSD is more effective than other control system[8]. Comparing AVS, the AVSD is working in that state the original structure is replaced by an energy dissipation system, which is more effect and reliable to control and reduce the structural response (Fig. 30). The control algorithm suggested by authors is reasonable. It can reach the optimal state to control the structural response in earthquake or wind load.

Fig. 29 Full Scale AVSD

Fig. 30 Shaking Table test for AVSD System

MR Semi-Active Control is used at Ton-Tin Lake Bridge (Fig. 32), and it is very effective to reduce the vibration in the wind load. It is the first application of MR in the world (by Prof. Z. Q. Chen).

Hybrid control have been used in Guangzhou TV Tower with height 650M (Fig. 33, 34) for reduce the horizontal response about 40% from wind load (by Prof. F. L. ZHOU,

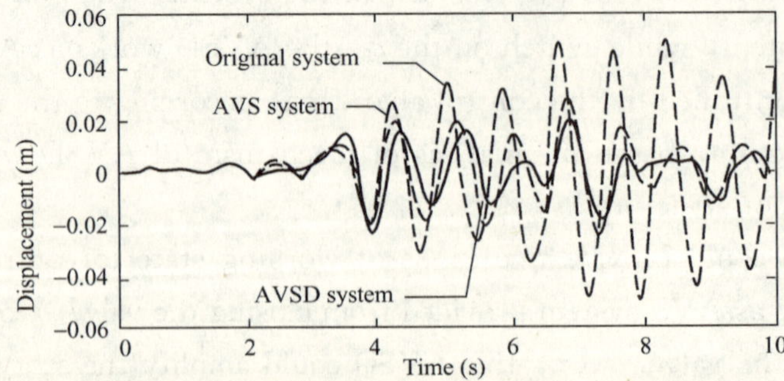

Fig. 31 Time History of Response of AVSD System from Shaking Table Tests

(J. P. Ou, P. Tan ad J. Teng et al).

Fig. 32　MR is used in Ton-Tin Lake Bridge

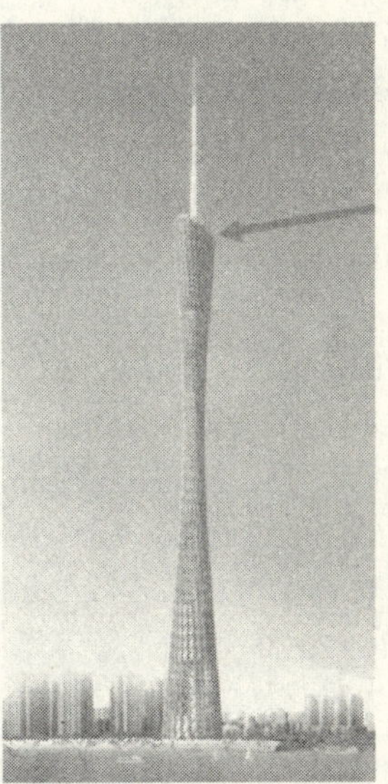

Fig. 33　Hybrid Control(HC) for Guangzhou TV tower with height 650m

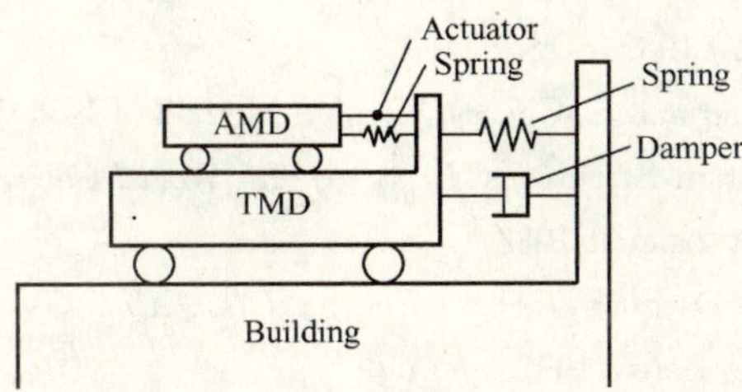

Fig. 34　Model of Hybrid Control (HC)

References

[1]　Z. M. Sheng, *Aseismic Engineering*, Beijing: China Architecture Engineering Publishing House, 2000.

[2]　F. L. Zhou, *Seismic Control of Structures*, Beijing: Chinese Seismic Publishing House, 1997.

[3]　F. L. Zhou, "Technical Report on Mission to Santiago, Chile" as Consultant to attend the *International Post-SMiRT conference Seminar on Seismic Isolation Passive Energy Dissipation and Active Control of Vibrations and Structures*, 1995. 8.

[4]　F. L. Zhou, "Development and Application for Technique of Isolation and Energy Dissipation and Control of Structural Response (1st Part)", *Journal of International Earthquake Engineering*, Vol. 4 (1989).

[5]　F. L. Zhou, "Development and Application for Technique of Isolation and Energy Dissipation and Control of Structure Response (2nd Part)", *Journal of International Earthquake Engineering*, Vol. 1

(1990).

[6] F. L. Zhou, S. F. Stiemer and S. Cherry, "A New Isolation and Energy Dissipating System for Earthquake Resistant Structures" *Proc. of 9th European Conference on Earthquake Engineering*, Moscow, Russia, September 1990.

[7] F. L. Zhou, J. M. Kelly, K. N. G. Fuller, T. C. Pan, et al. , "Recent Research Development and Application on Seismic Isolation of Buildings in P R China" *Proc. of International Workshop IWADBI*, Shantou, ,China, 1994.

[8] F. L. Zhou, L. L. Xie, W. M. Yan and P. Tan, "Optimum semi-Active Control system AVSD with Advanced Device and Algorithm" *Proc. of* 2 *WCSC*, *Kyoto*, Japan, July 1998.

[9] F. L. Zhou, "Discussion on Compiling Chinese Design Code of Seismic Isolation" *Proc. of The PRC-USA Bilateral Workshop on Seismic Codes*, *Guangzhou*, China, December 1996.

[10] X. Y. Huang, "Earthquake Engineering Research & Test Center. Technical Reports" Guangzhou University, 2003.

[11] F. L. Zhou, "New System of Earthquake Resistant Structures in Seismic Zone" *Recent Development and Future Trends of Computational Mechanics in structural Engineering*. Published by ELSEVIER APPLIED SCIENCE PUBLISHERS LTD, London & New York. (1996).

[12] F. L. Zhou, et al. , "Recent Research, Application, and development of Seismic Control for Structures" *Proc. of 9th World Conference on Earthquake Engineering*, Wuhan, China, 1992.

[13] F. L. Zhou and G. H. Yu, "Testing and Design for Energy Dissipation Bracing", *Report of Earthquake Engineering*, Vol. 1 (1982).

[14] F. L. Zhou, S. F. Stiemer and S. Cherry, "Design Method of Isolating And Energy Dissipating System for Earthquake Resistant Structures" *Proc. of 9th World Conference on Earthquake Engineering*, Tokyo-Kyoto, Japan, August 1988.

[15] T. T. Soong and G. F. Dargush, *Passive Energy Dissipation System in Structural Engineering*, New York: John Wiley & Sons, 1997.

[16] H. N. Li, *Structure Vibration and Control*, Beijing: China Architectural and Industrial Press, 2005.

第 3 章 Chapter 3

结构健康监测集成系统及其在基础设施中的应用

Structural Heath Monitoring Integrated Systems and Their Implementation In Infrastructures

J. P. Ou[1] (欧进萍) and H. Li[2] (李 惠)

[1] School of Civil and Hydraulic Engineering, Dalian University of Technology, Dalian, 116024, China

[2] School of Civil Engineering, Harbin Institute of Technology, Harbin, 150090, China

Abstract: Structural health monitoring (SHM) provides a useful tool for ensuring integrity and safety, detecting the evolution of damage, and estimating performance deterioration of civil infrastructures. A number of civil infrastructures under construction have greatly motivated and promoted development and application of SHM in mainland China. In the past decade, Chinese researchers have made great progress in all aspects of SHM, including advanced smart sensors, wireless sensors and sensor networks; data acquisition systems, data communication systems, signal processing systems, data management systems, and system integrated techniques; approaches to damage detection, model updating and safety evaluation; design and implementation of SHM systems for practical civil infrastructure; and data interpretation and utilization. Additionally, almost 100 bridges, offshore platforms, tall buildings, spatial structures, underground infrastructures and pavement have been implemented with SHM systems in mainland China at present. This chapter summarizes the state-of-the-art application of SHM for civil infrastructures in mainland China, in particular sensors, sensor technology, and applications of SHM systems.

Keywords: Fiber optic sensing technology; wireless sensors; cement-based strain sensors; damage detection; structural health monitoring; implementation

3.1 Fiber optic sensing technology

Fiber optic sensing technology is one of most critical achievements in the SHM field.

Great efforts have been made to promote the study and application of this sensing technology due to the advantages of electro-magnetic shielding, small size, resistance to corrosion, and convenience for multiplexing a large number of sensors along a single fiber. Four kinds of fiber optic sensors have been developed for infrastructures: OTDR (BOTDR), F-P sensors, white-light interferometers and optical fiber Bragg-grating (OFBG) sensor (Ou, 2003). However, only the FRP-packaged optical fiber Bragg-grating (OFBG) sensors and FRP-packaged optical fiber sensors based on BOTAD (R) technology are practical.

3.1.1 FRP-packaged optical fiber Bragg-grating sensors

Due to the fragility of the OFBG, naked OFBG without any package is not appropriate for direct use in civil infrastructures. Additionally, the performance of the OFBG sensors packaged by glue suffers from short life due to creep and aging of the glue. To solve this problem, the first author has proposed embedding OFBG sensors into FRP when the FRP is fabricated (Ou, 2003). Consequently, FRP can protect the OFBG sensors against breakage and short life. On the other hand, the FRP products embedded with OFBG sensors can sense their own strain. The fabrication procedure of FRP-OFBG rebar is shown in Fig. 1. The FRP-OFBG rebar is further processed to form several kinds of FRP-OFBG strain and temperature sensors for monitoring steel or reinforced concrete (RC) structures, as shown in Fig. 2.

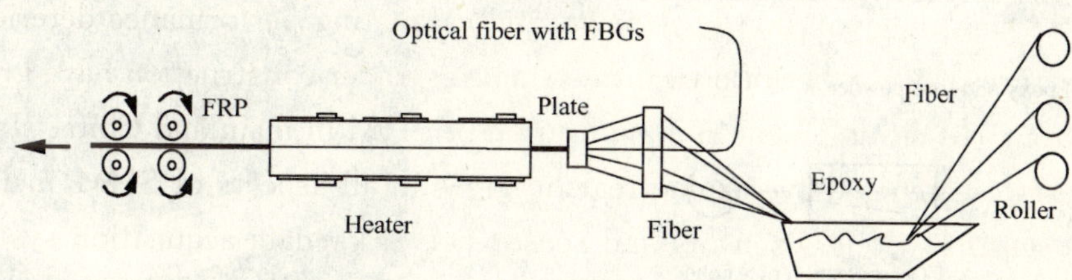

Fig. 1　Fabrication procedure of FRP-OFBG rebars

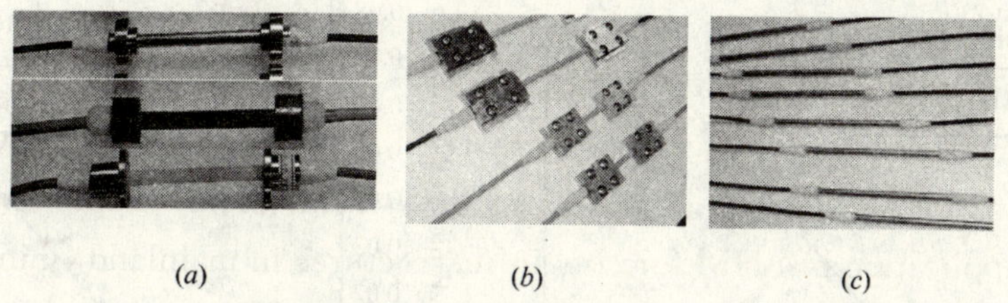

(a)　(b)　(c)

Fig. 2　FRP-OBFG sensors

(a) Strain sensors embedded into RC structures; (b) Strain sensors welded with steel structures or to surface of RC structures; (c) Temperature sensors

The verification and calibration of FRP-OBFG sensors have been conducted both by laboratory and field tests. The performance, including sensitivity, resolution, discrimination, range, linearity, accuracy, repeatability, stability, environmental constraints, fatigue resistance and corrosion resistance of the FRP-OFBG sensors, has been systematically and experimentally investigated. The results indicate that the measurement range of the FPR-OFBG sensors is about 5000$\mu\varepsilon$ to 10000$\mu\varepsilon$; accuracy is about 1～2$\mu\varepsilon$ depending on the interrogator; repeatability error is less than 0.5%; linearity error is less than 0.8%; sensitivity coefficient is about 7.8×10^{-7}; hysteresis error is less than 0.5%; fatigue life is higher than 1000000 times at 1000$\mu\varepsilon$ level; and the FRP-OFBG sensors can be used for at least 20 years.

3.1.2 FRP-packaged optical fiber sensors based on BOTDA (R) technology

In the last two years, to develope a practical BOTDA or BOTDR solution for long-term SHM of large-scale infrastructures, a series of novel, low-cost and highly reliable BOTD sensors using FRP-optical fiber (OF) rebar, as shown in Fig. 3, has been developed which can be manufactured as long as several kilometers. For the same reason mentioned above, the FRP also protects the OF against breakage, which makes the OF appropriate for civil infrastructure. The performance of the FRP-OFs at various gauge lengths were experimentally investigated under different spatial and readout resolutions using commercial BOTDA, as shown in Fig. 3. It can be observed that the FRP-OFs have good sensing properties. Therefore, the FRP-OFs can be used in long-term SHM for civil infrastructures.

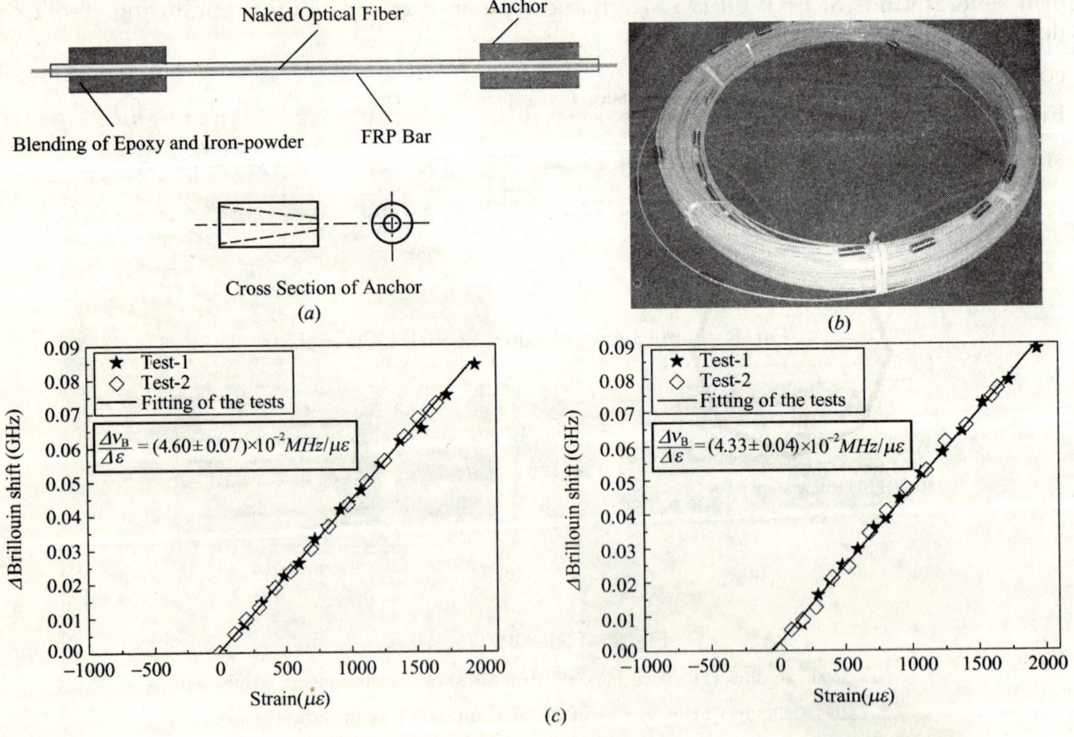

Fig. 3 BOTDA(R) FRP-OF sensors:

(a) Configuration of the FRP-OF sensors; (b) Photo of the FRP-OF sensors;

(c) Strain sensing properties at different spatial and readout resolution

3.1.3 FRP-OFBG-based smart products

As mentioned above, FRP-OFBG rebar can sense its own strain, which means that it can play the role of a strain sensor. At the same time, the FRP-OFBG rebar has the same strength as FRP rebar and thus can play a role of rebar. The FRP-OFBG smart rebar has been commercially produced by the Harbin Tider Co, as shown in Fig. 4.

Fig. 4 Commercial FRP-OFBG-based smart rebars

Cables and suspenders are very important components of civil infrastructures. However, they suffer from corrosion, fatigue and sever coupled effects in long-term service. Monitoring technology for cables is very critical, though it has been a challenging issue so far. A novel FRP-OFBG-based smart cable was developed by HIT group. For fabrication of FRP-OFBG-based smart cable, several FRP-OFBG smart rebars were placed into hollow cable and anchored as other steel rebars are, as shown in Fig. 5. Consequently, FRP-OFBG-based smart cable can sense its own strain and also bear force as common steel cable does. Performance verification and calibration of FRP-OFBG-based smart cables have been comprehensively carried out by filed tests, as shown in Fig. 5. The strain versus load in Fig. 5 is linear and the small error between different sensors is attributed to different stress states of the rebars in cable.

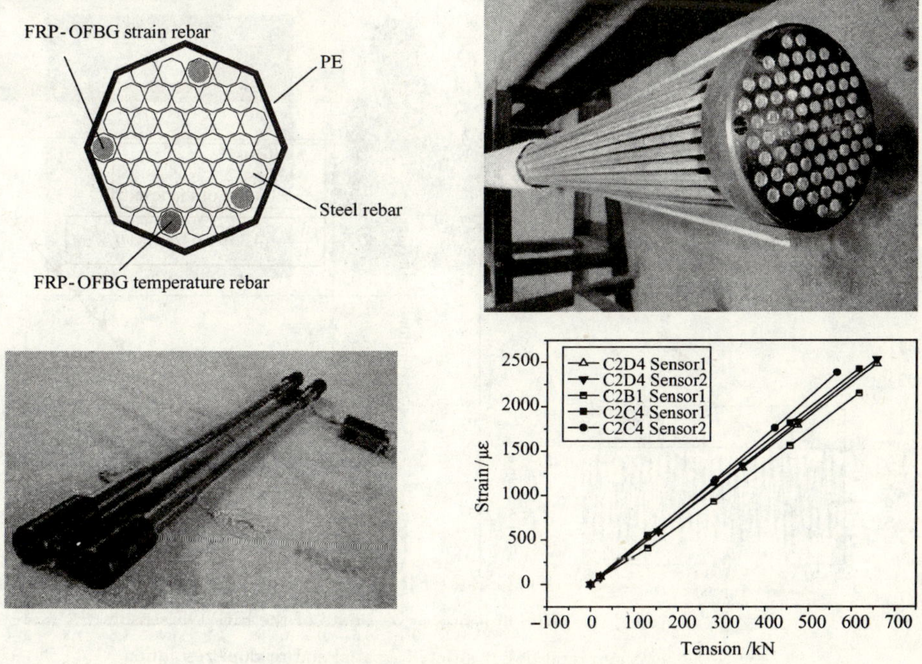

Fig. 5 FRP-OFBG-based smart cables and performance

3.1.4 FBG demodulators

Three types of simple and low-cost wavelength demodulators with high resolution for multi-channel dynamic FBG measurements were developed at HIT, as shown in Fig. 6. The TFBGD-210 is a portable double-channel FBG demodulator with a strain resolution better than 1pm at a frequency band of 1-100Hz. The TFBGD-700 is a single-channel FBG demodulator with a strain resolution of 0.5pm at a frequency of 1Hz. The TFBGD-9000 is a 16～64-channel FBG demodulator with a strain resolution of 1pm at a frequency of 300Hz. The performance of these three demodulators types was verified and calibrated by laboratory and field tests.

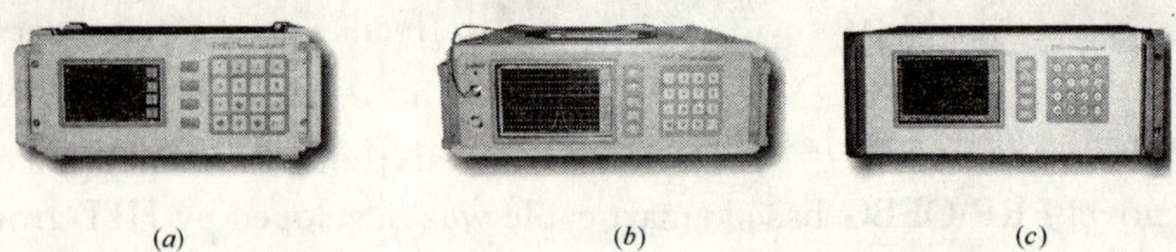

Fig. 6 FBG demodulators:
(*a*) TFBGD-210; (*b*) TFBGD-700; (*c*) TFBGD-9000

3.2 Wireless sensors and sensor networks

MEMS-based wireless sensing technology is another important achievement in the SHM field. Great efforts have been made in MEMS-based wireless sensors and sensor networks in mainland China. Ou (2003) has comprehensively studied wireless accelerometers and their networks,

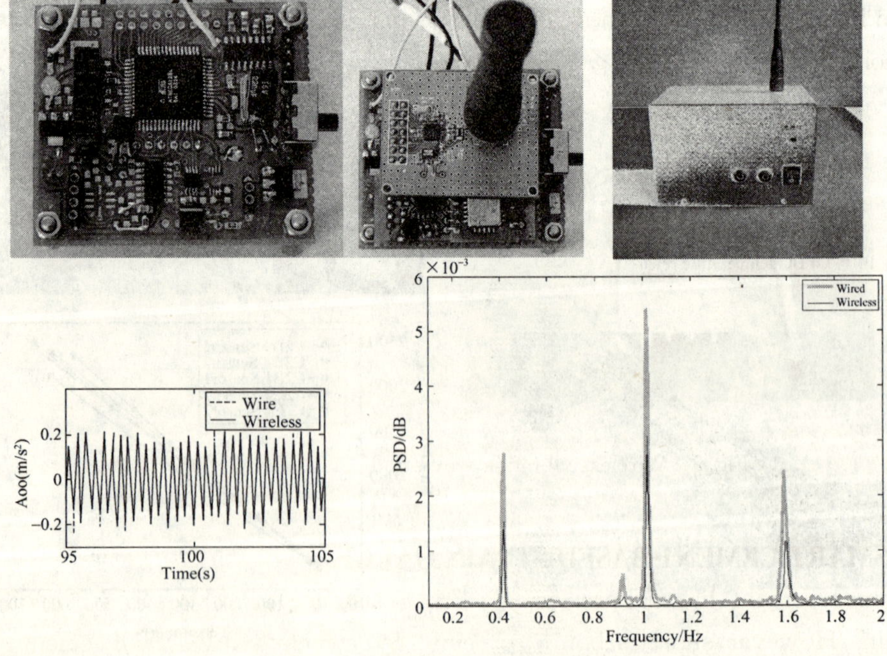

Fig. 7 Wireless accelerometers and performance

as shown in Fig. 7. The performance of the wireless accelerometers and their networks has been investigated through shaking table tests of a small-scale frame model and a small-scale offshore platform model. The results indicate that the wireless accelerometers can record the time-history of acceleration as well as the wired accelerometers. The frequency response functions are almost identical to each other. Five wireless accelerometers have been attached on the Shenzhen Diwang Building to measure wind-induced vibration under typhoons, as shown in Fig. 8.

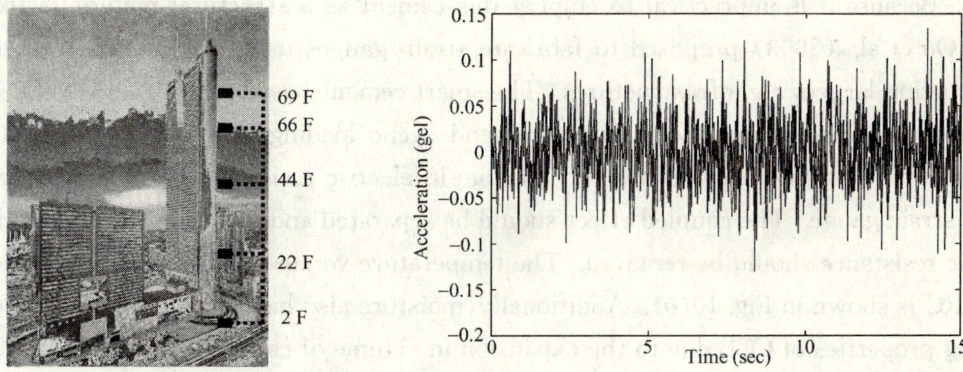

Fig. 8　Application of wireless accelerometers in the Shenzhen Diwang Building

The vibrating wire strain gauge is an effective sensor for measurement of static strain. However, the cables for signal transmission increase labor and cost, and decrease reliability of SHM systems. A wireless vibrating wire strain gauge, as shown in Fig. 9, was thoroughly developed by the HIT group. The performance of the wireless vibrating wire strain gauge was verified through static loading tests on a reinforced concrete beam and the results are shown in Fig. 9. It can be seen from Fig. 9 that the strain measured by the wireless sensors agrees well with that measured by common vibrating wire strain gauges.

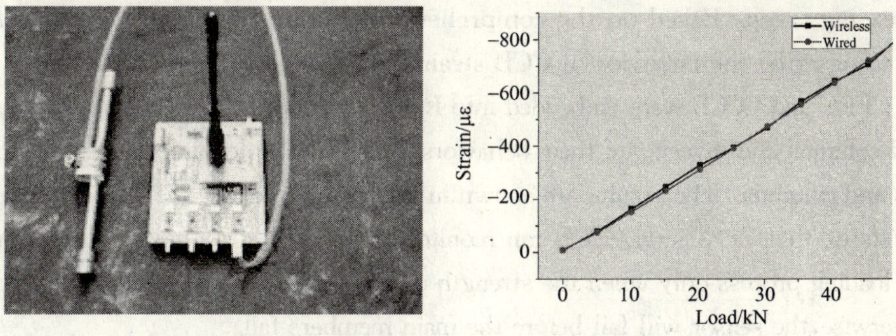

Fig. 9　Wireless vibrating wire strain gauge and performance

3.2.1　SMART CEMENT-BASED STRAIN GAUGE

Common cement has no pressure-electric resistance properties and cannot sense its own strain. However, short carbon fiber reinforced cement (CFRC) and cement containing carbon black (CCB) have good pressure-electric resistance properties and can play the role of strain gauge, namely smart cement-based strain gauges (Ou, 2005). The change in resistance for CFRC may be attributed to the interface between carbon fiber and cement (contact resistance), whereas, for CCB, the change in resistance may be attributed to the tunnel effect of carbon black. The fabrication procedure, mixture proportion, and CD circuit for measurement of electric resistance were systematically studied for these two smart cements. Because it is impractical to employ this cement as a structural material at present stage, Ou et al. (2003) proposed to fabricate strain gauges using this cement rather than constructing large-scale infrastructures. The smart cement-based strain gauge and its pressure-resistance properties under monotonic and cyclic loadings are shown in Fig. 10(*a*). Both strain and temperature can induce change in electric resistance of the smart cement-based strain gauge. The coupled effect should be separated and the effect of temperature on electric resistance should be removed. The temperature versus change in electric resistance of CFRC is shown in Fig. 10(*b*). Additionally, moisture also has a significant impact on the sensing properties of CCB due to the expansion in volume of carbon black after the absorption of water. A CCB strain gauge wrapped with epoxy resin is presented to prevent water from seeping into the gauge. The change in electric resistance versus strain of the CCB strain gauge under water for three months was investigated and the test results are shown in Fig. 10(*c*). It can be seen that the epoxy resin can prevent water from seeping into CCB and that the sensing properties remain stable even when the CCB strain gauge was dipped into water (100% moisture). Creep is also an important factor for cement materials and the change in electric resistance versus strain of CCB was also investigated under sustaining constant stress for a half year. The results are shown in Fig. 10(*d*). It can be seen that the change in electric resistance decreases dramatically at the early stage and then converges with decreasing creep. Based on the comprehensive study, an electric-mechanical model is proposed to describe the behavior of CCB strain gauge.

Both CFRC and CCCB were embedded into RC beams and columns to measure strain of the beams and columns and investigate their behaviors during whole loading process up to failure of the beams and columns. The results are shown in Fig. 10(*e*) and (*f*). The results in Fig. 10(*e*) and (*f*) indicate that CFRC and CCCB can monitor strain of the RC beams and columns during the entire loading process only when the strength of the sensor is larger than the beams and columns; Otherwise, the sensor will fail before the main members fail.

In the last two years, smart cement-based sensors have been embedded into RC beams of a practical bridge and inner wall of a practical tunnel. The measurement of strain of the tunnel is shown in Fig. 11 and it can be observed that the smart cement-based strain gauge can measure the strain as well as the vibrating-wire strain gauge does. However, the smart cement-based strain gauge has the same life expectancy as the monitored RC structure and is compatible with the RC structure.

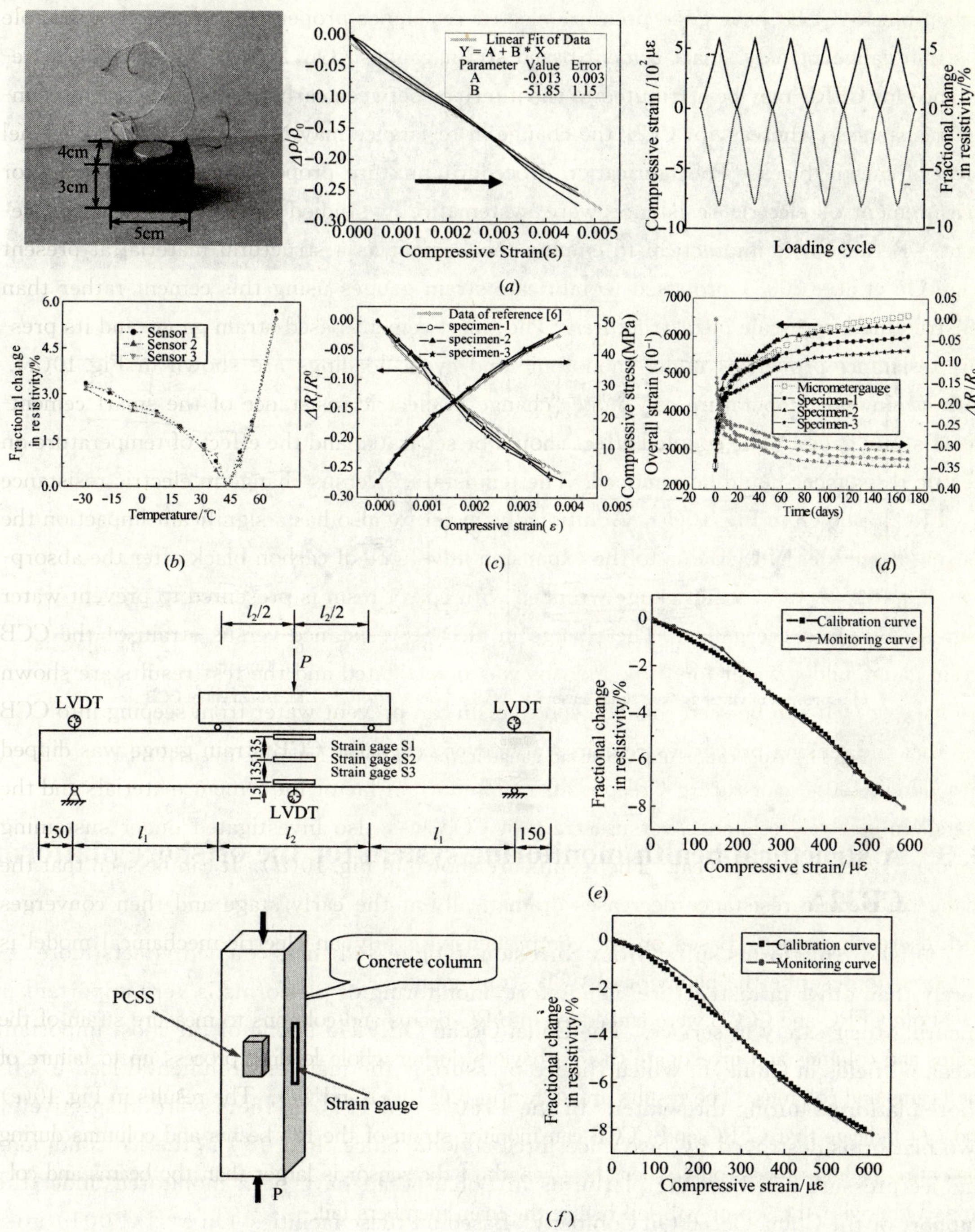

Fig. 10　Cement-based smart strain gauge:

(*a*) CFRC and CCB strain gauges and their pressure-resistance properties under monotonic and cyclic loadings; (*b*) Change in resistance versus temperature; (*c*) Pressure-resistance properties of CCB with water-proof measure; (*d*) Influence of creep on pressure-resistance properties; (*e*) Pressure-resistance behavior of CFRC in RC beam; (*f*) Pressure-resistance behavior of CFRC in RC column

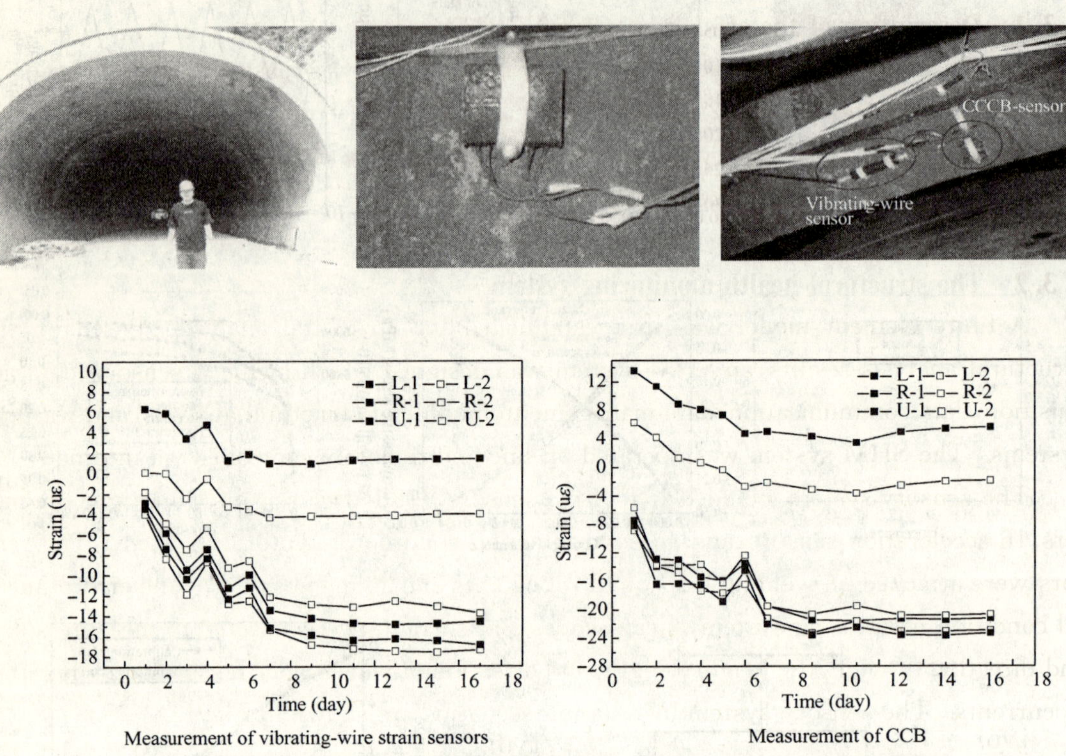

Fig. 11 Application of CCB strain gauge in the Chuankai Tunnel, Zhejiang, China

3.3 A structural health monitoring system for the offshore platform CB32A

Offshore platforms suffer from corrosion, fatigue and their coupled effects more severely than other infrastructure. Therefore, monitoring of platforms is very important in ensuring their safety in service. The Bohai Ocean Oil Field is one of the most important ocean oil fields in China, in which the ice pressure is the main environmental load to offshore platforms during the winter. In the 1960′s and 1970′s, there were, respectively, two platforms destroyed by heavy ice force action. Since the 1980′s, the ice conditions and ice pressure acted on the platforms in Bohai ocean have been monitored under the support of the China Ocean Oil Company. Based on these facilities, Ou et al (2001) integrated an on-line structural health monitoring system for steel jacket platform JZ20-2MUQ (a typical platform structure in this area), which has been operating since 1999. In recent years, Ou et al (2005) has implemented an intelligent structural health monitoring systems into the offshore platform CB32A under the support of the Ministry of Science and Technology, China.

3.3.1　Description of the offshore platform CB32A

An offshore platform CB32A with a jacket height of 24.7m was constructed in 2003 and located in water with a depth of 18.2m, as shown in Fig. 12. The offshore platform consists of a jacket, pier and platform. The jacket consists of four legs with a slope angle of 1 : 10. The altitudes of the foot and top floor of the jacket are respectively -19.7m and 5.0m.

3.3.2　The structural health monitoring system

A finite element model was first established for structural analysis. Based on the structural analysis results, an SHM system was designed which included sensory, data acquisition, data communication, data management, module of structural analysis, and warning systems. The SHM system was operated on-line and remotely controlled via internet.

The sensory system included 259 OFBG sensors, 178 PVDF sensors, 56 fatigue life meters, 16 acceleration sensors, and an environmental condition monitoring system. The sensors were attached on welding lines of 12 tube joints and 20 members. For the environmental condition monitoring system, anemoscopes and currents were selected to measure speed and direction of wind, the height, length, and direction of waves, and the speed and direction of currents. The sensory system installation scenarios are shown in Fig. 12.

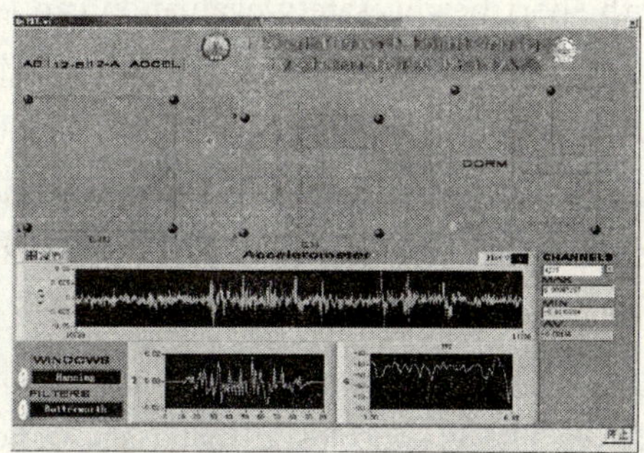

Fig. 12　The Offshore Platform CB32A and its SHM system

For the data acquisition system, the PXI bus technique was employed and the sampling rates were 1000 Hz for accelerometers and 50 Hz for other sensors. For the data communication system, a wireless communication system was employed to transmit the collected data on site from the CB32A to the administrative office, as shown in Fig. 12. Oracle database was employed to manage collected data and structural analysis results. The collected data can be automatically processed in real time and warnings automatically emitted.

The SHM system was then integrated and has been operating online in real time since October 2005. Users can download data via the internet.

3.3.3 SHM-based safety evaluation methods

Safety evaluation of the CB32A can be directly done by comparison of the peak of measured stress with the strength of the steel material or by comparison of the peak of measured loads with the specified limits for design. The fatigue accumulated damage can also be calculated through rainfall analysis based on time-history of measured stress. Warnings are immediately emitted in the following three cases: (i) A critical change in stress, loads and/or fatigue accumulated damage occurs; (ii) Stress, loads and/or fatigue accumulated damage equal to or exceed the specified limits; and/or (iii) Damage is detected using various damage detection strategies. However, all these methods are for member-level safety evaluation, not for structure-level safety evaluation.

Ou et al. (2003) proposed a method of structure-level safety evaluation for jacket offshore platforms. Considering that the base shear is the control variable for the safety of jacket offshore platforms with short height, safety evaluation can be conducted by comparison of the peak of the measured base shear (obtained using the mass and measured acceleration response of the structure) with the load capability along various orientations under various loadings, as shown in Fig. 13. Ou et al. (2003) further proposed that the load capacity of a jacket offshore platform can be calculated using a pushover method based on the commercial finite element model software package SACS. The typical load capacity curves of a jacket offshore platform under ice action are shown in Fig. 14. It can be observed that the base shear versus the top floor displacement of the platform behaves like elasto-plastic materials. The first and second characteristic points correspond to yield and ultimate capacities of the platform, which are listed in Tables 1 for wave-current loading and ice loading.

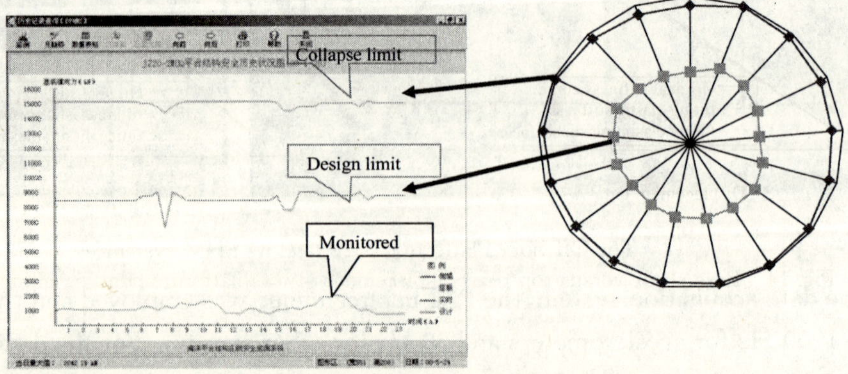

Fig. 13 Safety evaluation method in real time for jacket offshore platforms

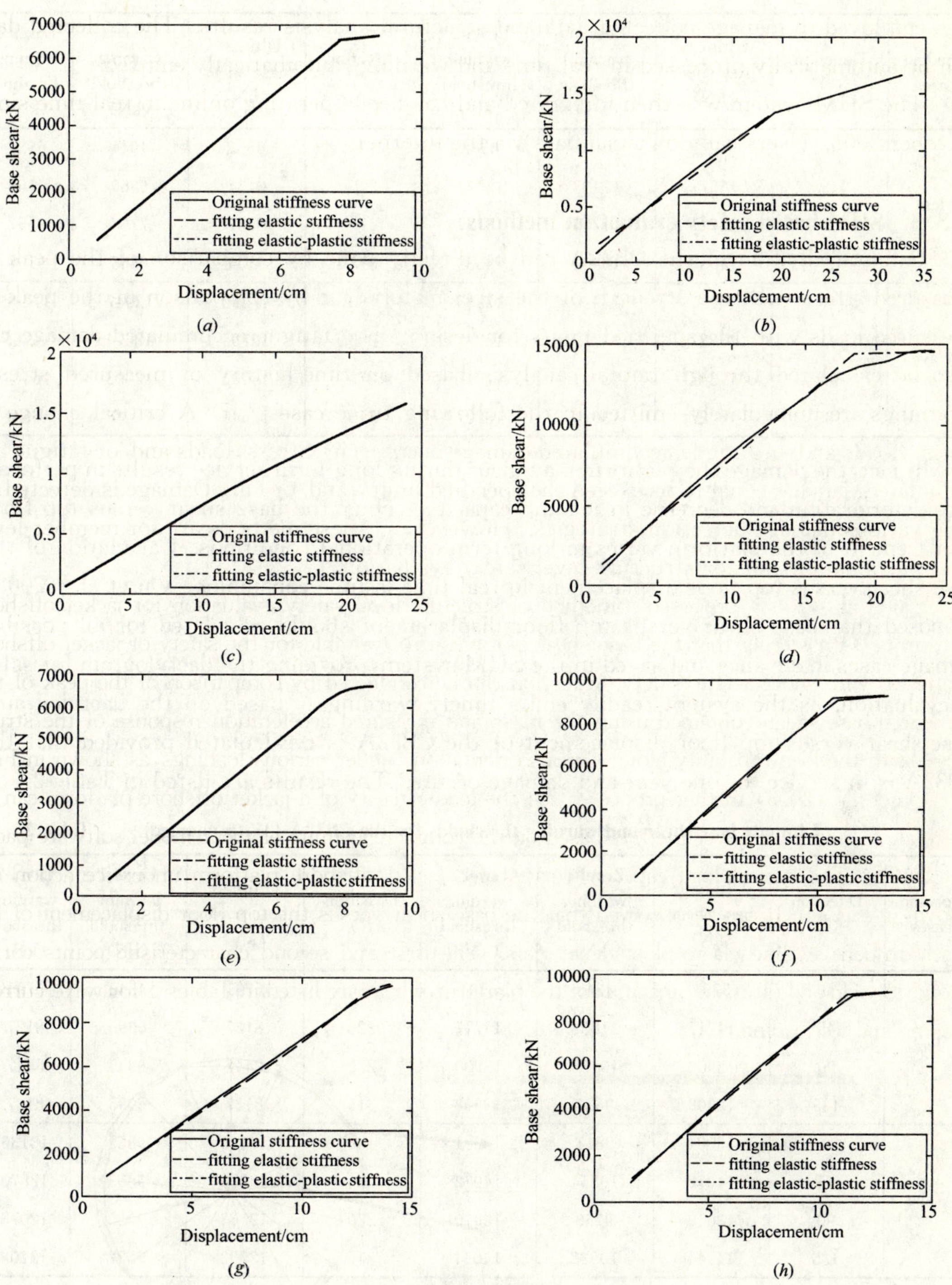

Fig. 14　Base shear versus top floor displacement of the platform under ice action

(*a*)0°;(*b*)45°;(*c*) 90°;(*d*) 135°;

(*e*) 180°;(*f*) 225°;(*g*) 270°;(*h*) 315°

Ultimate base shear and warning thresholds for the CB32A (without damage) Table 1

Loading cases	Ditection (°)	Ultimate base shear (kN)	Level 1 warning threshold (kN)	Level 2 warning threshold (kN)	Direction (°)	Ultimate base shear (kN)	Level 1 warning threshold (kN)	Level 2 warning threshold (kN)
Ice	0	6552	4914	6552	180	6552	4914	6552
	45	16577	12433	16577	225	9154	6865	9154
	90	15727	11795	15727	270	9557	7168	9557
	135	14539	10904	14539	315	9154	6865	9154
Wave	0	17235	12926	17235	180	17262	12947	17262
	45	18287	13715	18287	225	16765	12574	16765
	90	17241	12931	17241	270	16004	12003	16004
	135	18227	13670	18227	315	16646	12485	16646

In fact, the damage the platform may incur during long-term service results in performance deterioration and decrease in loading capacity. Thus, the base shear versus top floor displacement of the platform varies in long-term operational conditions. Calculation of the base shear versus top floor displacement in real time is time-consuming. Ou et al. (2003) proposed that base shear versus top floor displacement should calculated for all possible damage cases in advance and saved in the SHM system, providing the dactylogram for safety evaluation. is the system readily emits timely warnings based on the dactylogram. Base shear versus top floor displacement of the CB32A is recalculated provided that the CB32A is in service for one-year and damage occurs. The results are listed in Table 2.

Ultimate base shear and warning thresholds for the CB32A (With damage) Table 2

Loading cases	Ditection (°)	Ultimate base shear (kN)	Level 1 warning threshold (kN)	Level 2 warning threshold (kN)	Direction (°)	Ultimate base shear (kN)	Level 1 warning threshold (kN)	Level 2 warning threshold (kN)
Ice	0	5791	4343	5791	180	5791	4343	5791
	45	14717	11038	14717	225	8127	6095	8127
	90	13900	10425	13900	270	8447	6335	8447
	135	12908	9681	12908	315	8127	6095	8127
Wave	0	13081	9811	13081	180	13136	9852	13136
	45	14072	10554	14072	225	12870	9653	12870
	90	13010	9758	13010	270	12073	9055	12073
	135	14044	10533	14044	315	12703	9527	12703

3.3.4 Warning thresholds

The Offshore Platform Design Code is formulated based on the limit-state concept and offers performance criteria for various limit states. The two principle limit states are defined as (i) serviceability limit state; and (ii) ultimate limit state. Two warning levels corresponding to these two limit states were proposed by Ou et al (2003). Consider when the offshore platform and its accessory facilities are out of order when the deformation of platform is too large, which means that the stiffness of the offshore platform is not permitted to decrease dramatically. Therefore, a Level 1 for warning should be emitted to notify that the structure behaves in the elastic stage and the base shear cannot exceed the yield point. Consequently, the yield point is the Level 1 warning threshold. For level 2, it refers to ultimate loading capacity.

For offshore platform CB32A, the methods of safety evaluation based on SHM are shown in Fig. 15.

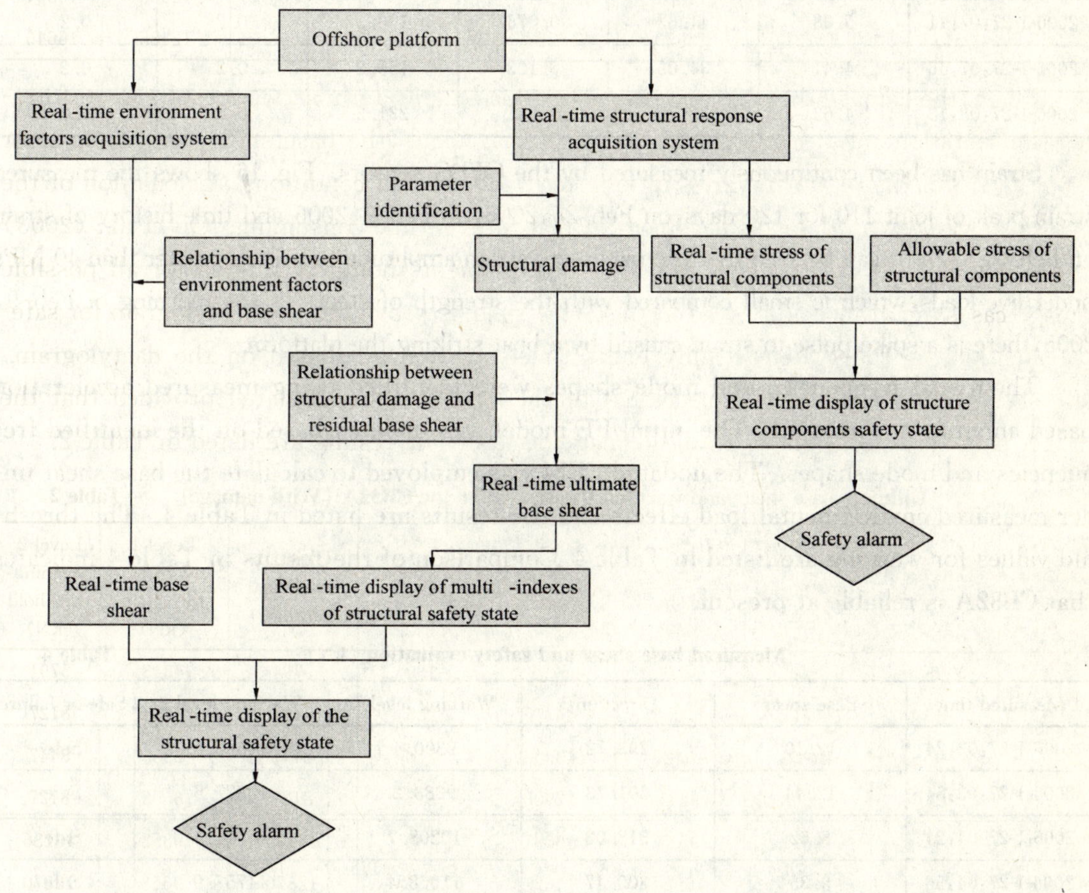

Fig. 15　Methods of safety evaluation based on SHM for jacket offshore platforms

3.3.5　Overview of findings

The wind, wave and current are measured by the SHM system and listed in Table 3. It can be seen that the environmental load effects are quite small compared to the design limits.

Measured environmental load effects **Table 3**

Measured time	Wind speed/m/s	Wind direction/℃	Current speed/m/s	Current direction/℃	Wave height/m	Wave direction/℃
2006-1-27:03:24	5.13	293.72	0.199	134	0.23	0.2
2006-1-27:03:54	5.3	301.73	0.185	134.1	0.23	0.2
2006-1-27:04:24	3.85	318.02	0.221	125.4	0.23	0.2
2006-1-27:04:56	3.9	309.47	0.247	146.2	0.23	0.2
2006-1-27:05:32	3.86	296.65	0.239	139.2	0.18	0.1
2006-1-27:06:02	3.27	316.96	0.216	134.4	0.2	0.2
2006-1-27:06:38	3.98	332.44	0.135	161.4	0.19	0.2
2006-1-27:07:11	5.08	41.62	0.076	160.3	0.2	0.2
2006-1-27:07:46	4.41	34.05	0.103	155.2	0.2	0.2
2006-1-27:08:18	4.62	32.49	0.061	231.2	0.2	0.2

Strain has been continuously measured by the OFBG sensors. Fig. 16 shows the measured strain peak of joint 110 for 120 days, on Feb. 25, 2006, on Feb. 9, 2006 and time-history of strain on Feb. 9, 2006. It can be seen that the maximum strain amplitude can reach greater than 40 MPa under live loads, which is small compared with the strength of steel. In the morning of Feb. 9, 2005, there is a spike pulse in strain caused by a boat striking the platform.

The modal frequencies and mode shapes were identified using measured acceleration based an ambient vibration. The initial FE model was updated based on the identified frequencies and mode shapes. The updated FEM was employed to calculate the base shear under measured environmental load effects and the results are listed in Table 4. The threshold values for warning are listed in Table 4. Comparison of the results in Table 4 indicates that CB32A is reliable at present.

Measured base shear and safety evaluation (kN) **Table 4**

Measured time	Base shear	Direction	Warning level 1	Warning level 2	Safe or failure
2006-1-27:03:24	12.10	293.72	9390.4	12520.6	Safe
2006-1-27:03:54	12.44	301.73	9283.2	12377.6	Safe
2006-1-27:04:24	8.62	318.02	10208.7	13611.6	Safe
2006-1-27:04:56	9.35	309.47	11068.4	14757.9	Safe
2006-1-27:05:32	8.60	296.65	9351.2	12468.3	Safe
2006-1-27:06:02	6.81	316.96	10201.2	13601.6	Safe
2006-1-27:06:38	7.08	332.44	10311.0	13748.0	Safe
2006-1-27:07:11	9.97	41.62	11211.3	14948.4	Safe
2006-1-27:07:46	8.03	34.05	11083.1	14777.5	Safe
2006-1-27:08:18	8.35	32.49	11056.7	14742.2	Safe

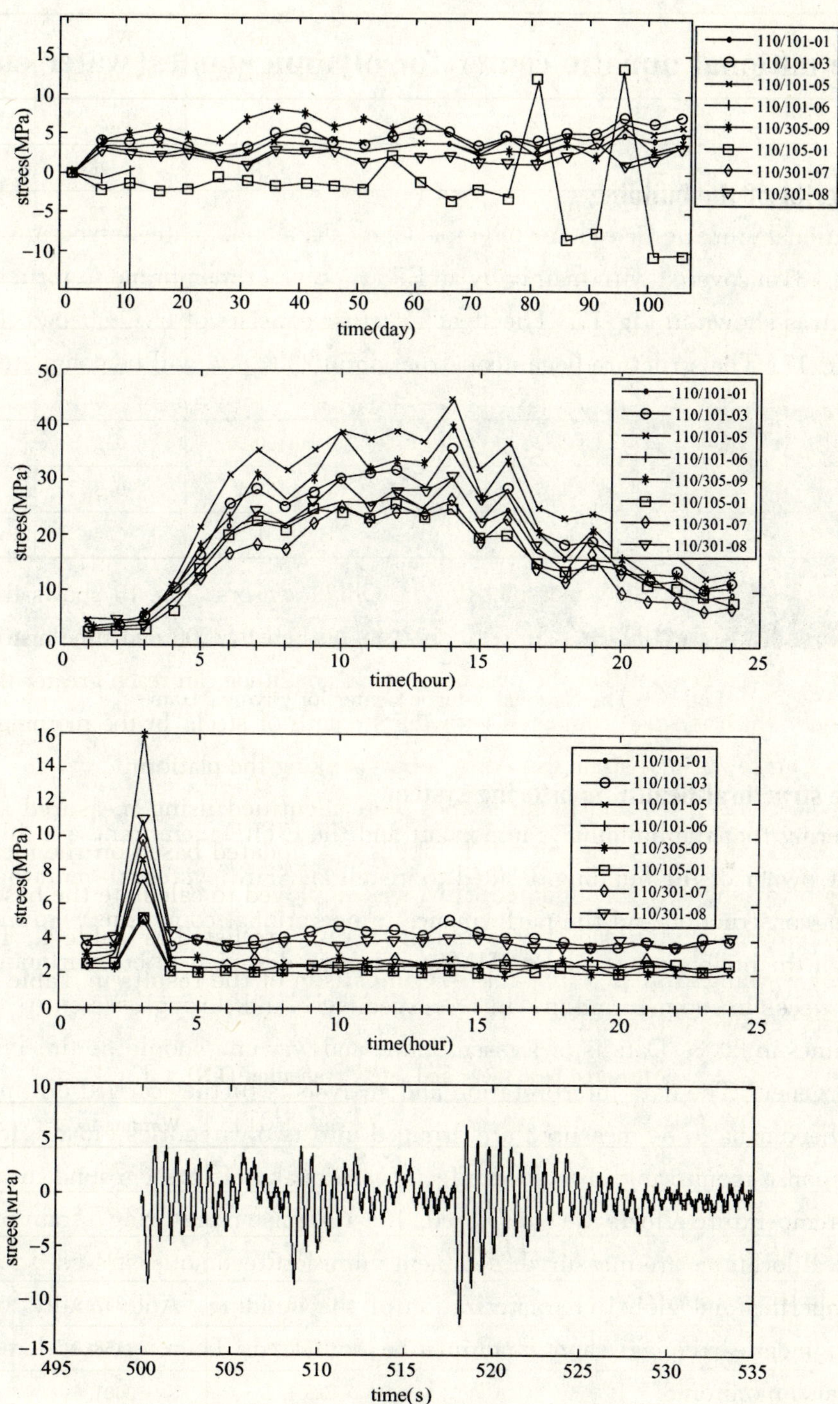

Fig. 16　Measured strain segment of joint 110

3.4 The national aquatic center for olympic games (water cube)

3.4.1 Descript of the building

The National Aquatic Center for Olympic Games is a cuboid steel structure with a size of 171×171×31m covered with mainly by an EFTE polymer membrane to form a sandwich configuration, as shown in Fig. 17. The steel structure consists of H_2O-shaped elements, as shown in Fig. 17. The structure began construction in 2004 and will be completed in 2007.

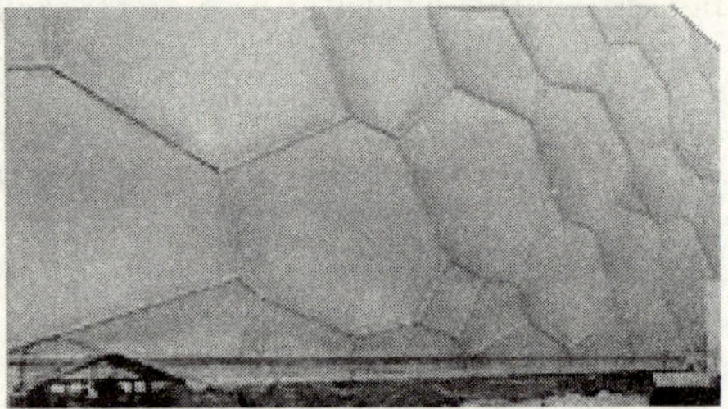

Fig. 17 The National Aquatic Center for Olympic Games

3.4.2 The structural health monitoring system

Considering that the building is important and the ETFE membrane was first adopted in China, the owner of the building decided to install an SHM system to monitor the stress states during construction and the performance in operational conditions. To meet the requirements of the building owner, the SHM system was designed to be continuously operated for two weeks in summer and in winter, respectively, and through the entire duration of Olympic Games in 2008. Data is processed online and warning should be timely emitted in case of emergency. The data interpretation and analysis is further carried out in detail offline. The phenomena to be measured are grouped into two categories: load-effect monitoring and response monitoring. For load effects, earthquake effects (ground motion), wind effects and temperature effects are considered. For response monitoring, strain of the members at critical locations are measured. Ambient vibration techniques are used to determine dynamic properties and global characterization of the building. Additionally, vibration of the building under earthquakes and wind must be monitored. Therefore, acceleration of this building is also monitored.

To determine the types, characterization, quantities and locations of sensors, a prior ANSYS FE model to best represent the existing knowledge of the building was established, as shown in Fig. 18. Additionally, an FE model was also established for simulation of the change in stress and deformation when the support jacks are removed at the end of construction. The response of the building was calculated using the FE model under 140 loading cases and the sensory system of the SHM system was then designed according to the computational results.

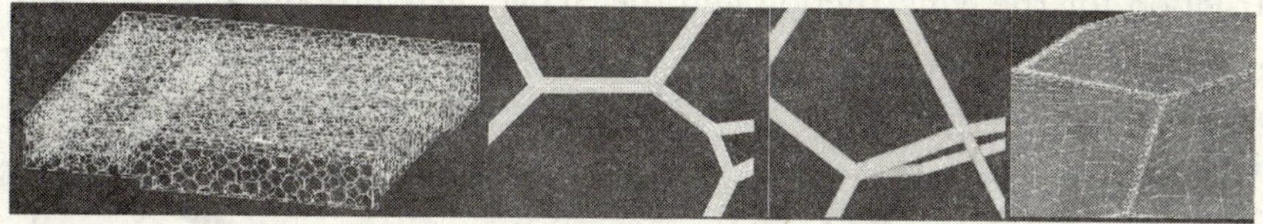

Fig. 18　FE model of the structure

For wind load effect monitoring, an ultrasonic anemoscope was installed near the building to measure wind speed, direction, attack and ambient temperature. The sampling rate for wind load was 32 Hz. Additionally, the wind pressure distribution on the membrane was critical in verifying design assumptions, design parameters and analytical processes used in the design and construction with an aim to improve computational and assessment wind load models for safety evaluation. Therefore, 29 wind pressure meters were pasted over a corner of the roof, as shown in Fig. 19. 30 OFBG temperature sensors were welded at some elements to the measure temperature lapse rate for temperature compensation. For response monitoring, 230 OFBG strain sensors were welded at 230 critical elements. 1 tri-axial accelerometer was attached at a foot of a column for earthquake ground motion monitoring, 1 tri-axial accelerometer was attached on the midpoint of the roof for measurement of vibration and another 27 axial accelerometers were dispersedly attached at the roof for measurement of vibration and extracting global dynamic properties based on ambient vibration technique. It should be noted that the temperature of the roof between two membranes was in the high 70's℃ and all sensors were selected to meet the requirement of the temperature. The schematic of the integrated SHM system is shown in Fig. 19.

Fig. 19　The structural health monitoring system of the building

3.4.3 Overview of findings

For the dome structure, stress monitoring is one of greatest issues when the support jackets are removed during the end of construction. The SHM system monitors the stress of critical elements during the removal of support jackets and the variation of the stress during this process is shown in Fig. 20. At the same time, the calculation results using the

FE model is also depicted in Fig. 20 for comparison. It can be observed that stresses vary dramatically, even to about 450 MPa of web-member 3620, during this process and the calculated results agree well with the monitored stresses. The measured results provide very critical instruction for the construction and the SHM system plays an important role in ensuring structural safety during construction.

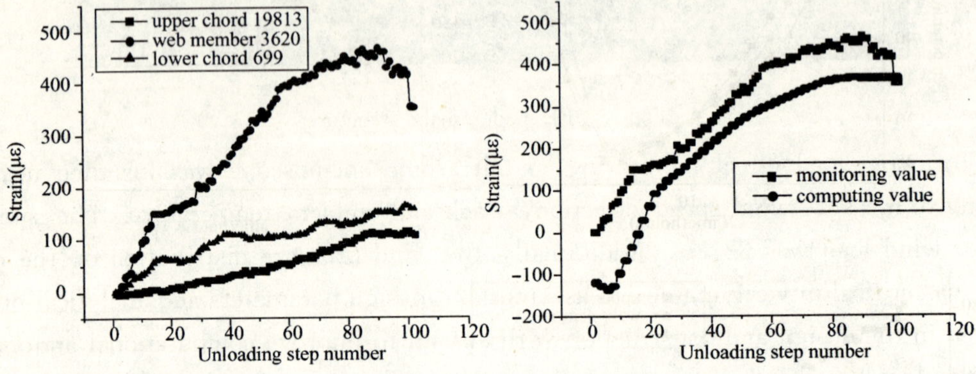

Fig. 20 Monitoring and computational stress during removing support jackets

The National Aquatic Center is redundant. Therefore, variation of temperature generated significant change in intrinsic forces. The temperature and the change in stress of elements were measured in the winter of 2006 and summer of 2007. The measured temperatures and strains are shown in Fig. 21. It can be seen that the temperature variation generates great change in the strain of elements (the maximum variation in strain reaches to about 400MPa). The strain increases with decreasing temperature, and vice-versa, decreases with rising temperature. For the lapse rate, the upper chord temperature was lowest and the temperature of the web layer and lower chord is almost the same, though slightly higher in winter. However, in summer, the temperature along the height of the roof is uniform and in the high 60's℃. Because this building is still under construction and without damage, all measured data can serve as a baseline for evaluating future changes in the condition, performance and health of the building.

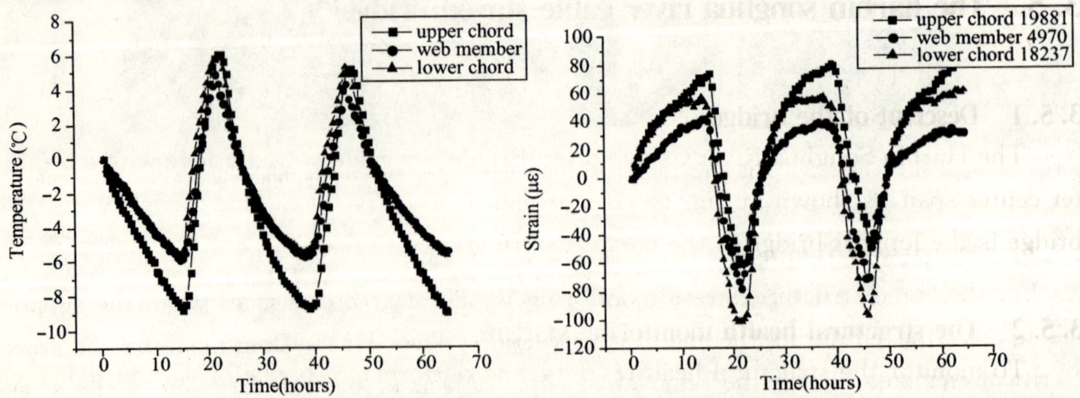

Fig. 21 Monitoring temperature and strain in 2006 and 2007(—)

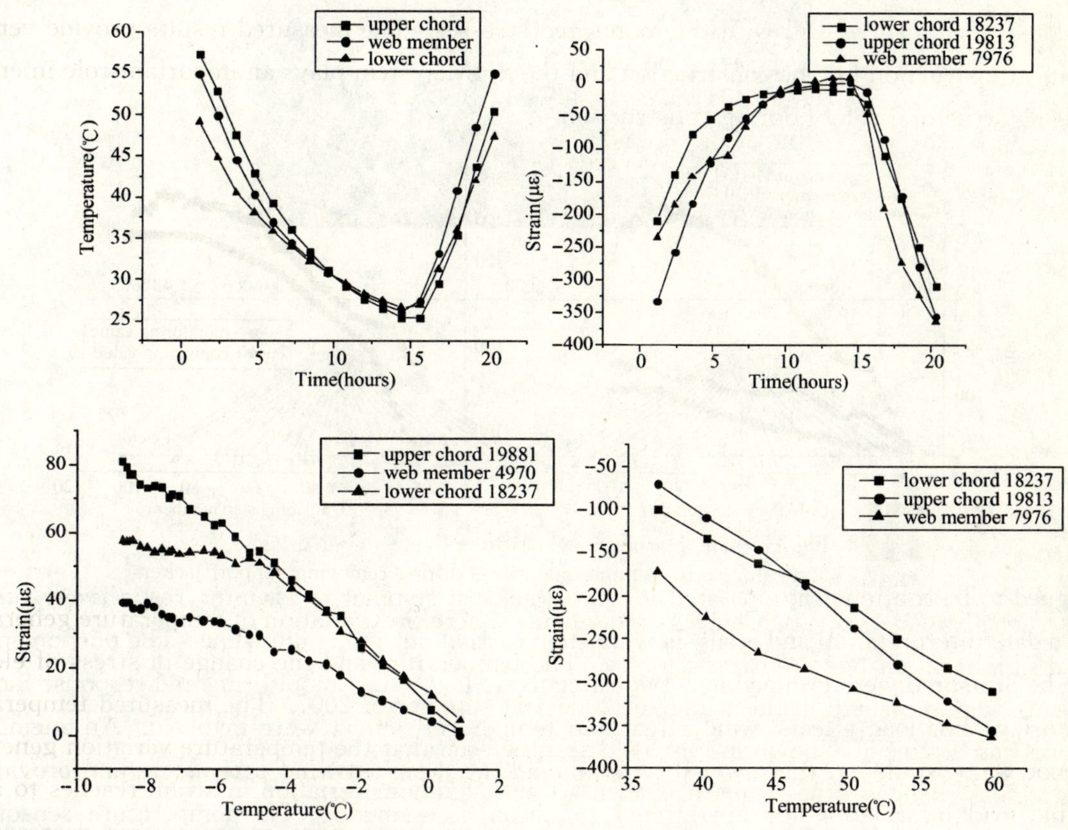

Fig. 21 Monitoring temperature and strain in 2006 and 2007(二)

There is no snow in late winter and the wind load effect is very small; therefore, the temperature variation is the main load effect at this stage. Because the structure is redundant, vibration is very small under ambient excitation. No available acceleration related data has been measured so far.

3.5 The harbin songhua river cable-stayed bridge

3.5.1 Descript of the bridge

The Harbin Songhua River Cable-stayed Bridge is a cable-stayed bridge with a 360 meter center span, as shown in Fig. 22. It is designed to carry two 2-lane carriageways. The bridge is the longest bridge in the north-eastern area of China.

3.5.2 The structural health monitoring system

To monitor the structural health, safety, and performance of the bridge, an SHM system was designed. To meet the requirement of the bridge owner, the SHM system was de-

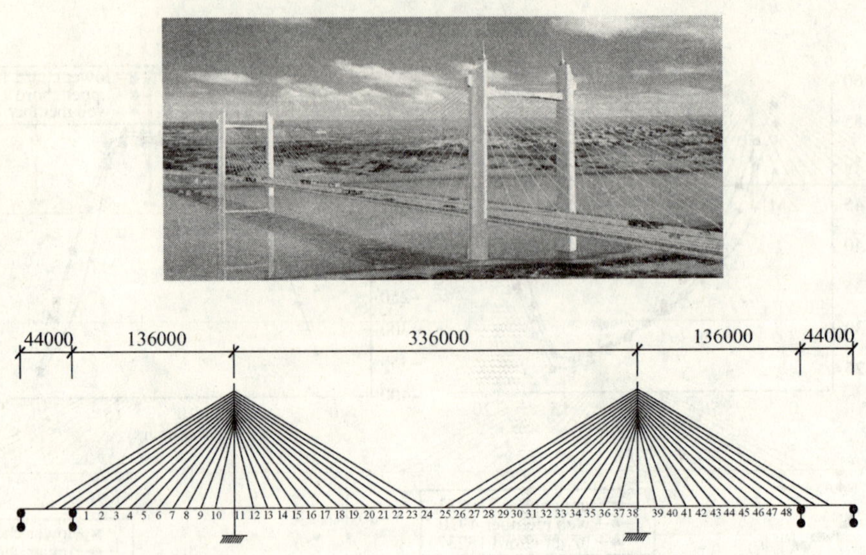

Fig. 22 The Harbin Songhua River Cable-stayed Bridge

signed to be continuously operated for two weeks in summer and winter, respectively, and the data interpretation and analysis were then carried out in detail offline. The phenomena to be measured were grouped into two categories: load-effect monitoring and response monitoring. For load effects, wind effect and temperature effect were involved. An anemoscope was installed at the tower top to validate the designed wind parameters and provide wind load information for operational condition assessments. The temperature sensors were installed to measure the air temperature above and below the girder levels. The temperature sensors simultaneously play the role of temperature compensation. For response monitoring, strain of the girder at critical cross-sections and cable forces were measured. 24 OFBG sensors were attached at critical locations for measurement of girder strain and temperature. The location of the OFBG sensors is shown in Fig. 23. Ambient vibration techniques were used to determine dynamic properties from which the cable tension forces were calculated. Therefore, 52 accelerometers were attached to cables for cable force measurement. An additional 26 accelerometers were symmetrically installed on the deck. The location of the 26 accelerometers was determined based on the sensor placement optimization procedure proposed by the authors, as shown in Fig. 23. The global dynamic properties were determined to calibrate the bridge theoretical model and form basic references of subsequent structural damage detection and safety evaluation.

3.5.3 Overview of findings

On August 26, 2004, the proof load tests are carried out to capture the response of all of critical elements before the bridge opened to traffic. Since that day, the SHM system has been operated. The monitored data indicated that the measured load-effects and response of

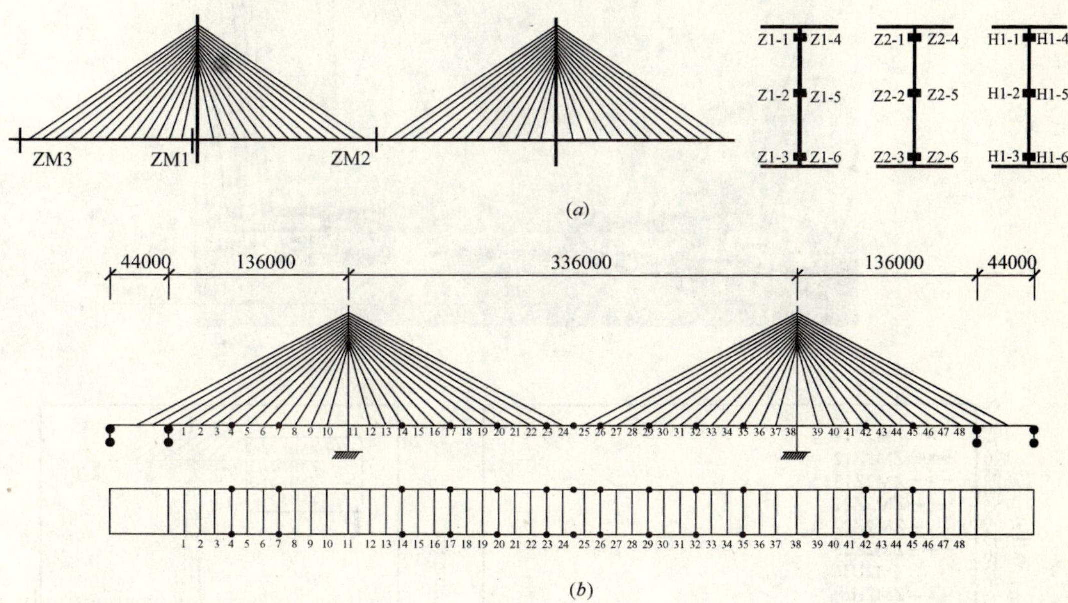

Fig. 23　Locations of sensors:

(*a*) OFBG strain and temperature sensors; (*b*) Accelerometers

the bridge were all less than their corresponding design values. More information from the monitored data is summarized as follows:

(i) Controlled load tests

The proof load tests were conducted at the bridge to obtain response of all of the critical elements under known truck loads for open public traffic. Twenty-four 30-ton trucks were used for loading, as shown in Fig. 24. A total of ten loading position lay-outs were arranged in the test. During the tests, the SHM system recorded the response. Fig. 24 shows the girder strain at the midspan cross section measured by OFBG sensors when the twenty-four trucks were symmetrically distributed over the center span of the bridge. The maximum measured strain reaches about 400$\mu\varepsilon$, which is about 25% of the yield strain of the steel girder.

Crawl tests with two loaded trucks were also performed. Two 30-ton trucks on two of four lanes crawled throughout the bridge in various speeds. Strain at critical locations were measured by OFBG sensor at a sampling rate of 250 Hz. Fig. 24 shows the girder strain at the midspan cross section measured by OFBG sensors. It can be observed that the strain reaches its peak value (about 20$\mu\varepsilon$) at the moment the trucks cross the midspan.

The frequencies were extracted after post-processing of the ambient monitoring acceleration and are shown in Table 5, providing an insight to the global properties of the bridge.

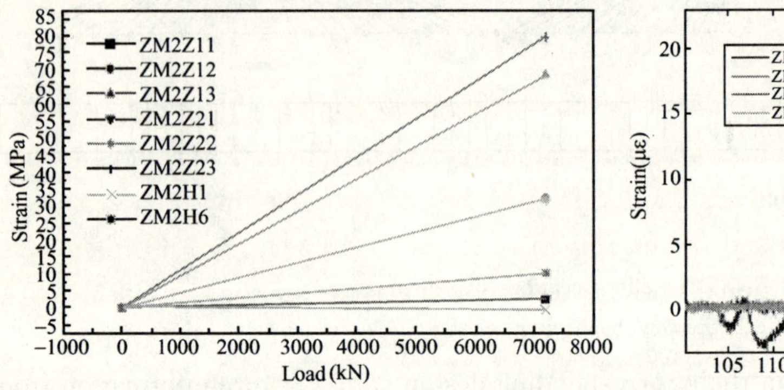

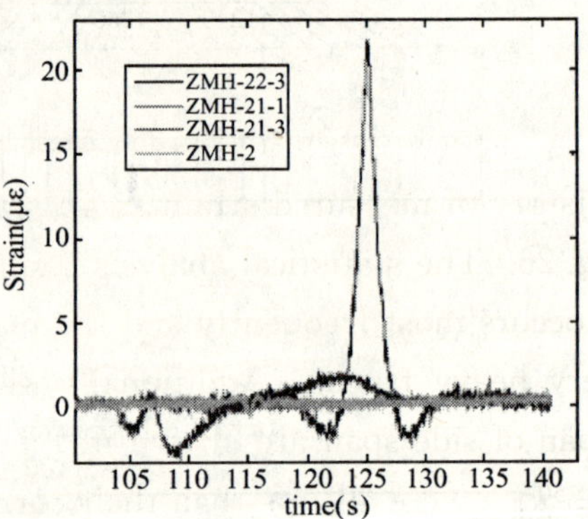

Fig. 24 Proof load tests and measured strain of the girder

Identified frequencies of the deck (Hz) **Table 5**

	1st	2nd	3rd	4th	5th	6th
Longitudinal	0. 031					
Vertical	0. 361	0. 449	0. 826	1. 168	1. 181	1. 342
Torsion	0. 593	0. 777	1. 015			
Transverse	0. 703					

Additionally, wind effects were also monitored during the controlled load tests. The maximum wind speed was about 7m/s, within the design range of 28m/s. The wind usually blows in the N-S direction.

(ii) Operational condition monitoring

A comprehensive measurement by this SHM system was performed in May 2005. The girder strain was continuously collected for two weeks at a sampling rate of 62. 5Hz and two typical segments of the girder strain at midspan cross section are shown in Fig. 25. The peaks in the time-history curve were generated by trucks when they crossed through the measured strain section. Therefore, the volume of trucks crossing over the bridge can be accounted through the time history of measured strain. The peaks were picked up from

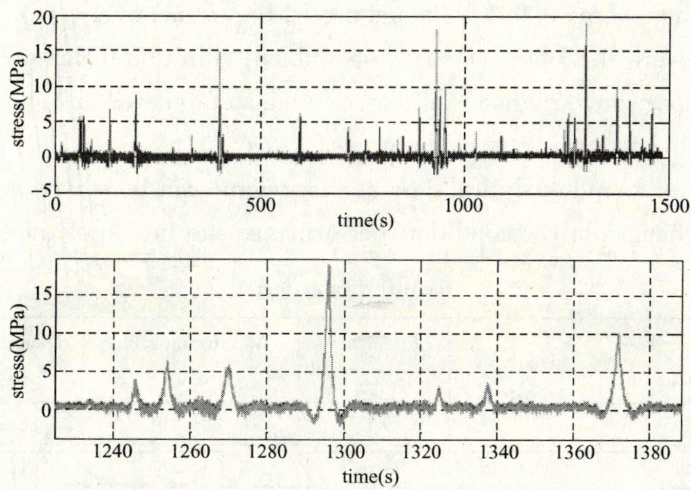

Fig. 25　Measured Strain of the bridge girder

the history of measured strain and a statistical analysis carried out. The results are shown in Fig. 26. The statistical analysis results indicate that the stress within the range of 30-40 MPa occurs most frequently and the maximum stress is about 50 MPa, which is generated by very heavy trucks. Additionally, the stresses of the cross-sections at tower-foot and midspan of side-span are shown in Fig. 26. The maximum stresses at the tower-foot and

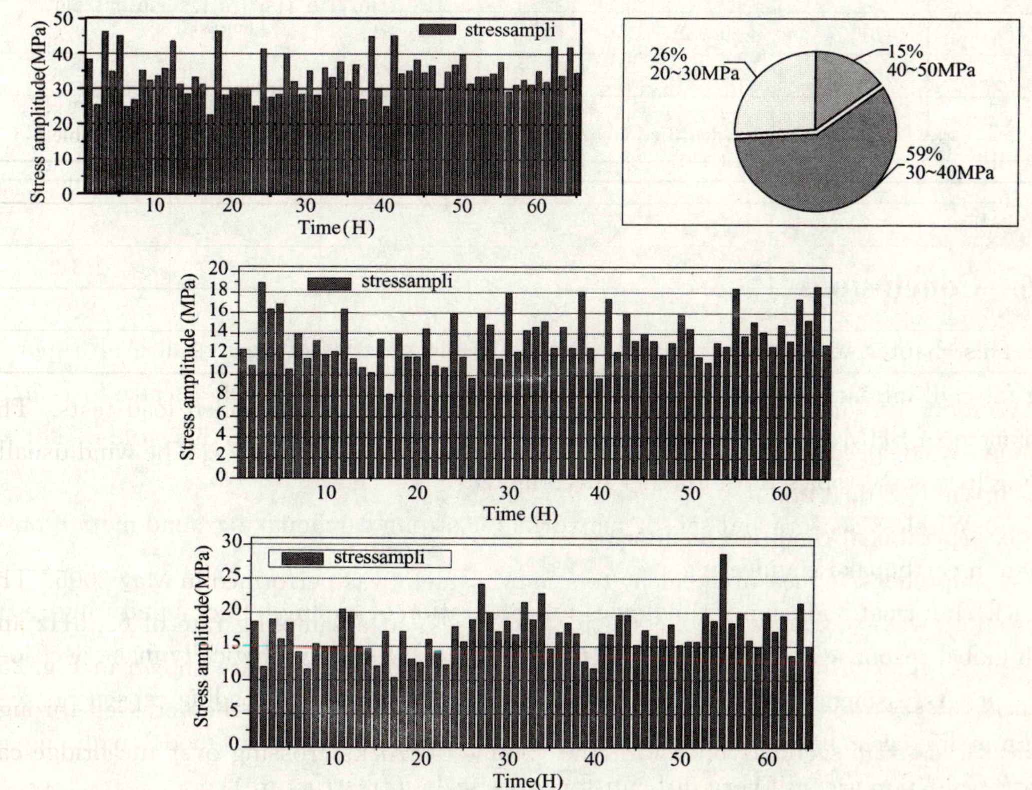

Fig. 26　Monitoring stress and statistic analysis

midspan of side-span were 18. 962 MPa and 28. 7 MPa, respectively.

Cable forces were also obtained based on ambient vibration techniques and are listed in Table 6. For comparison, designed cable forces are also tabulated in Table 6. It can be seen that there is no obvious change in cable forces.

All the monitored and analytical data at this regime can be served as a baseline for evaluating future changes in the condition, performance and health of the bridge.

Identified cable forces **Table 6**

Cable No	Mass per unit/kg/m	Length/m	Frequency /Hz	Monitoring cable force /kN	Design cable force/kN	Error (%)
C13	120. 48	177. 9728	0. 75256	8644. 9	8869. 6	2. 53
C12	109. 89	166. 4528	0. 7813	7433. 8	8144. 8	8. 73
C11	99. 11	155. 0362	0. 8793	7367. 3	7274. 8	−1. 27
C10	87. 23	143. 7388	0. 9206	6110. 1	6404. 4	4. 75
C9	78. 68	132. 5915	1. 0375	5956. 0	5824. 4	−2. 60
C8	78. 68	121. 6321	1. 1344	5991. 3	5824. 4	−2. 87
C7	78. 68	110. 9106	1. 1797	5387. 7	5824. 4	7. 50
C6	78. 68	100. 4935	1. 3231	5563. 9	5824. 4	4. 47
C5	61. 37	90. 4672	1. 4706	4344. 7	4519. 6	3. 87
C4	61. 37	80. 9376	1. 6379	4314. 2	4519. 6	4. 54
C3	54. 08	72. 0057	1. 8117	3681. 4	3936. 6	6. 46
C2	54. 08	63. 7567	2. 0882	3834. 4	3936. 6	2. 60
C1	54. 08	54. 9513	2. 4306	3858. 9	3936. 6	1. 97

3. 6 Conclusions

This chapter summarizes the state-of-the-art and practice of structural health monitoring for civil infrastructures. Although Chinese researchers have made great efforts in development of SHM, a number of challenging issues are in need of further study. The following lists some topics to be further investigated:

(i) Wireless sensors and sensor networks for strong earthquake ground motion observation in earthquake engineering.

(ii) Information fusion technologies for integration of loading-effect monitoring results with global response and local response monitoring results to synthetically make decisions.

(iii) Assessment and evaluation methods of performance and loading-capacity deterioration in life-cycle based on SHM technologies (Damage detection, model updating and safety evaluation are still very difficult for large-scale infrastructure).

(iv) Extension of structural health monitoring technology to disaster monitoring, including strong wind, earthquake, fire, accidents and so on, and the integration of accumulated and evolving damage monitoring with disaster monitoring.

ACKNOWLEDGMENTS

The authors gratefully acknowledge the financial support provided by the NSFC with grant No. s: 50538020, 50525823 and the Ministry of Science and Technology with grant No. 2006BAJ03B05.

References

[1] J. P. Ou (2003) *Some recent advances of intelligent health monitoring systems for infrastructures in the Mainland of China*, Proceedings of 1st International Conference on Structural Health Monitoring and Intelligent Infrastructures, Tokyo, Japan(Keynote Lecture)

[2] J. P. Ou, H Li and Z D Duan (2005), *Structural health monitoring and intelligent infrastructure*, Taylor & Francis (Editors)

[3] J. P. Ou, Z. D. Duan and Y. Q. Xiao (2003), *Safety evaluatin for offshore platform : Inpscetion, Monitoring and Maintaince*, *Press of Science*, *China*

第 4 章　Chapter 4

混凝土随机损伤力学
——背景、意义与研究进展

李　杰　同济大学土木工程学院，上海四平路 1239 号，200092

摘　要：混凝土结构非线性行为研究的核心和灵魂是关于混凝土本构关系的研究。20 世纪 80 年代中期以来，混凝土损伤力学的研究为混凝土本构关系的研究开辟了新的道路。然而，经典连续介质损伤力学在本质上属于唯象学的研究框架，对于带有根本性的损伤演化法则，在逻辑上只能采用经验归纳或理论假设的方式给出，而难以说明损伤演化的内在物理机制。缘于此，从对混凝土材料构成性质的随机性和受力行为非线性的考察入手，我们从随机损伤及其演化的角度对混凝土本构关系的形成机理进行了系列的研究，初步形成了混凝土随机损伤力学的新型研究框架。本文对这一方面的研究进展做出了阶段性总结。

研究表明：损伤演化的非线性源于细观层次断裂应变分布的随机性。利用作者提出的细观随机断裂—滑移模型，可以建立细观损伤与宏观损伤之间的物理桥梁、合理反映基本的损伤演化规律。在此基础上，通过引入能量等效应变，可以将随机损伤演化法则推广于多维受力状态，从而建立起应用于结构分析的混凝土随机损伤本构模型。通过引入非线性有限元法与概率密度演化分析原理，可以实现结构随机非线性反应的分析。在上述理论框架中，确定性损伤本构关系与随机损伤本构关系可以互相转化；在细观—宏观、本构—结构等不同分析尺度上，随机要素的传递与演化、演化过程中随机涨落，也可以得到合理的反映。

上述理论框架的提出，为合理地实现基于可靠性的混凝土结构设计与控制提供了基础。

关键词：混凝土结构；混凝土随机损伤力学；细观随机断裂—滑移模型

Stochastic Damage Mechanics of Concrete Structures

J. Li

School of Civil Engineering, Tongji University,

1239 Siping Road, Shanghai 200092, China

Email: lijie@mail. tongji. edu. cn

Abstract: Research on the damage mechanics is a modern development trend of concrete mechanics. However, despite the celebrated works of early researchers and the substantial research efforts, secrete of damage evaluation remains somewhat challenging. Starting from the analysis of constitutive structures of concrete materials, two micro stochastic damage models are proposed in our works. Based on the irreversible thermodynamics, a class of elastoplastic damage model of concrete is developed. Then the concept of energy equivalent strain is derived to bridge the gap between micro stochastic damage model and continuum damage model. On the basis of traditional finite element method and the general probability density evaluation equation developed by author in recent years, a complete frame for stochastic nonlinear response analysis of concrete structures is established. Several numerical simulations are presented whose results allow for validating the capability of the proposed theory for reproducing the typical nonlinear stochastic performance of concrete materials and structures.

Keywords: Concrete Structure; Concrete Stochastic Damage Mechanics; Micro Stochastic Rupture-slippage Model;

4.1 研究背景

在我国工程建设中，混凝土结构是应用最为广泛的一类结构。2006 年，我国水泥产量再次刷新世界记录，达到 9.4 亿吨，占世界水泥产量的 45%。根据经验折算，2006 年我国混凝土工程用量已达到 45～52 亿吨，相当于每人每年 4 吨的水平。

与工程建设繁荣兴旺的背景相适应，我国在混凝土科学方面的研究也得到长足进步。据初步统计，1997 至 2007 年我国在土木工程领域发表刊物研究论文约 66100 篇，其中，涉及混凝土材料、混凝土结构和混凝土力学的研究达 29600 篇，占上述研究论文总数的 45%。

混凝土结构研究的终极目的，是建立合理的、可以客观反映结构受力机理与破坏机制、有效控制结构性态的工程结构设计理论。在这一意义上，混凝土结构研究的核心和灵魂是关于结构非线性性态和演化机理的研究，其基础，则是混凝土本构关系的研究。令人遗憾的是，在国际范围内，在长达 40 余年的时间里，尽管先后引入了非线性弹性理论、经典塑性力学理论、断裂力学等，试图为混凝土本构关系的研究建立合理的理论框架，但由于问题的高度复杂性，仍然没有形成可以较好地反映混凝土材料在多维空间中非线性受力机制的基础理论。20 世纪 80 年代中期以来，这种局面开始出现转机。伴随着损伤力学登上混凝土研究的历史舞台、非线性随机系统研究的重要进步，混凝土本构关系研究与结构非线性分析研究均开始出现一系列观念上的变化。

一般认为，最早将损伤力学的基本概念应用于描述混凝土非线性特性的是 Dougill (1976)。但是，混凝土损伤力学理论研究中第一个具有开创意义的工作是 Ladeveze 和 Mazars 作出的（Ladeveze 1983，Mazars 1984，1986）。众所周知：混凝土最典型的性质是其在拉应力和压应力作用下表现出迥异的强度与刚度特性。基于此，1983 年，Ladeveze 首先引入应力张量的正负分解，并假设正应力引起受拉损伤，负应力引起受压损伤，在复杂加载时，损伤为受拉损伤和受压损伤的组合。这种损伤变量的表达形式，后来被证明普遍适用于混凝土材料。1986年，Mazars基于应力张量分解的思路，引入弹性损伤能释放率建立损伤准则，为混凝土损伤力学的进一步发展奠定了热力学基础。Ladeveze-Mazars损伤本构模型能够很好地模拟混凝土材料在低周反复荷载作用下的刚度退化，分析结果与单轴试验结果的吻合程度也令人满意，但是，它不能很好地反映双轴受压应力状态下混凝土强度和延性的提高。

此后，不少研究者试图在弹性损伤的框架内修正和完善Ladeveze-Mazars模型，其中代表性的工作包括Mazars & Pijaudier-Cabot（1989）、Lubarda et al（1994）以及Comi & Perego（2001）等。由于弹性损伤框架的局限性，上述模型并没有在Ladeveze-Mazars模型的基础上取得实质性进展，所建立的模型均不能很好地反映混凝土多轴非线性行为的机理和特性，尤其是在多轴受压区，模型预测结果与试验结果存在显著差距。

弹性损伤模型的根本缺陷在于没有反映混凝土受力非线性发展过程中部分存在塑性变形这一事实。缘于此，自20世纪80年代中期以来，一批研究者试图突破弹性损伤的理论框架，将塑性应变及其演化规律引入到损伤本构关系的建模过程中，以期能够反映混凝土材料的残余变形。在此方向，研究工作大体上沿着两个侧面展开。其一是在Cauchy应力空间建立塑性应变的演化方程，如Ortiz（1985）、Simo-Ju（1987）、Lubliner et al（1989）、Abu-Lebdeh & Voyiadjis（1993）、Prisco-Mazars（1995）、Carol et al（2001）以及Ananiev & Ozbolt（2004）等。Cauchy应力表征宏观水平的表观应力，材料进入软化段后，宏观应力会产生下降，基于Cauchy应力空间建立的弹塑性损伤模型必然涉及屈服面收缩问题，因此会出现数值收敛和稳定性等一系列问题。在另一侧面，在有效应力空间建立塑性应变演化方程，则不会出现这些问题。事实上，在加载过程中，有效应力空间内的屈服面一直处于膨胀状态而不存在收缩，因此可以避免软化段的复杂处理问题。在此侧面的典型研究见于Ju（1989）、Lee-Fenvas（1998）和Jason（2006）等。考虑到直接运用塑性力学方法在屈服状态判断时需要进行迭代，出于大型结构非线性分析时的计算效率考虑，也有不少学者采用经验表达的方法来考虑塑性变形，如Resende（1987）、Faria et al（1998，2004）、Valliappan et al（1999）和Hatzigeorgioiu et al（2001）等。

在上述背景下，基于对混凝土材料构成性质的随机性和受力行为非线性的考察，我们从随机损伤及其演化的角度对混凝土本构关系的形成机理进行了较为系列的研究，初步形成了混凝土随机损伤力学的新型研究框架。本文，试图对此方面的研究进展做一阶段性的总结。

4.2 混凝土的非线性与随机性

众所周知，混凝土是具有高度非线性力学行为的材料。图1是典型的混凝土单轴受拉应力—应变曲线。一般认为：造成这种非线性特征的本质原因，在于混凝土内部存在初始微裂缝与微孔洞。在外力作用下，在这些内部缺陷附近将产生应力集中，从而导致裂缝进一步扩展。这种细观的非均匀受力及其演化过程是形成非线性的本质原因。

在另一方面，混凝土是一类多相复合材料。其构成组分及各组分强度都具有随机分布特征。初始微缺陷或损伤的随机性，势必导致后续的损伤演化路径具有随机性。在本质上，这种随机性与前述非线性是天然地耦合在一起的。由此导致混凝土本构关系及强度表现具有不可避免的随机性（李杰，2004）。图 2 是作者所带领研究梯队的一组单轴受力混凝土应力—应变曲线试验结果。在材料基本配比、制作工艺和试验方法完全一致的条件下，出现这样的离散性在定量上是正常的，在概念上也是合理的：随机的初始损伤必然导致损伤的随机演化，随机的损伤演化，必然导致随机的强度表现和随机的本构关系。

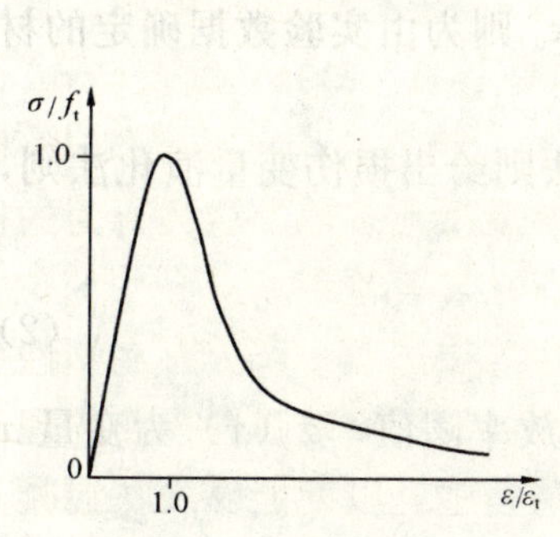

图 1　混凝土单轴受拉应力—应变曲线

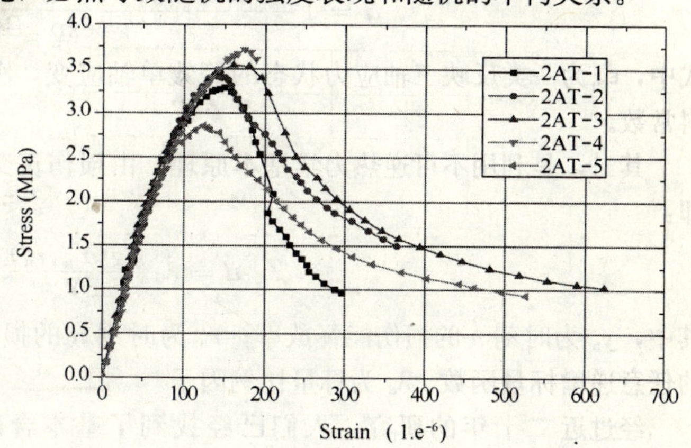

图 2　混凝土单轴受拉应力—应变试验曲线

与非线性弹性理论、经典塑性理论相比较，损伤力学对混凝土非线性形成机制的解释更为符合于真实背景：非线性源于损伤的逐步累积。1976 年，Dougill 首先将损伤力学的基本概念应用于描述混凝土的非线性。他认为：混凝土材料的非线性是由渐进断裂（损伤发展）所引起的刚度衰减造成的。尽管 Dougill 仅研究了弹性断裂问题，完全忽略了塑性应变对非线性的贡献，但将损伤力学引入到混凝土力学研究中，是 Dougill 的一个重要贡献，从此，混凝土力学研究进入到了一个新纪元。

图 3 的简化模型可用之考察混凝土单轴受力的非线性力学行为。当图 3*a* 因受力致使其中某根弹簧断裂形成图 3*b* 时，因弹簧束中内力重分布形成新的内力分配格局。与之相适应，应力—应变曲线出现刚度折减，表现出非线性的特征（图 4*a*）。显然，当弹簧个数趋于无穷时，这一断裂—内力重分布—刚度折减的过程将形成光滑的非线性应力—应变曲线（图 4*b*）。

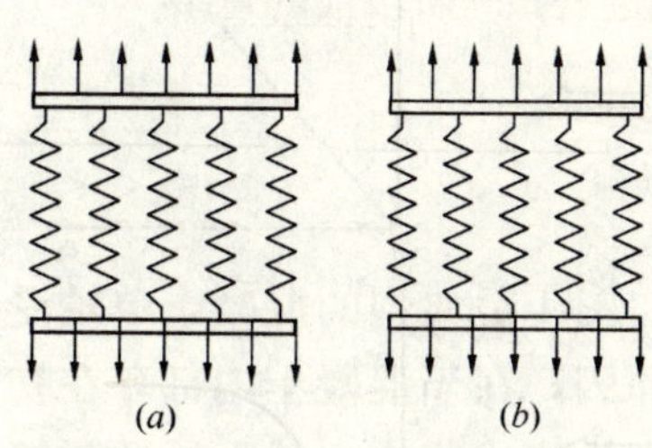

图 3　弹簧模型及其断裂

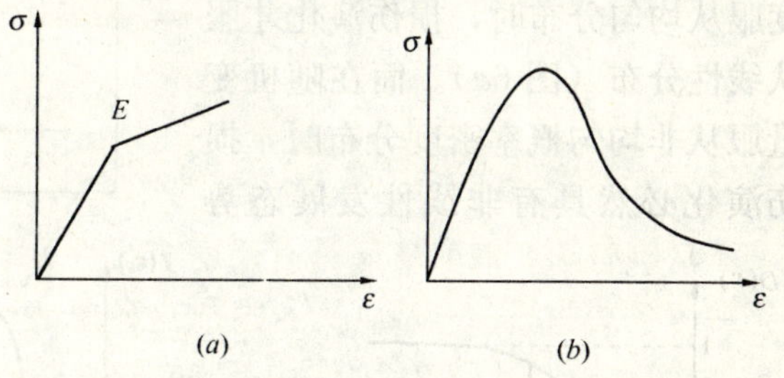

图 4　损伤导致非线性

经典损伤力学对于非线性形成机理的上述出色解释，使人们有理由相信损伤力学是研究混凝土本构关系的合理工具。但由于损伤力学在本质上是具有内变量的理论，随着研究工作的深入，人们发现：关于损伤变量演化规律的研究构成了研究进程中的主要障碍。在

20 世纪 90 年代中期之前，一般从两条途径反映损伤变量的演化规律。

其一，是通过经验总结的方式给出损伤变量演化规律，其典型表述形式是（Mazars，1984）

$$d=1-\frac{(1-A)\varepsilon_0}{\varepsilon_{eq}}-\frac{A}{\exp\left[B(\varepsilon_{eq}-\varepsilon_0)\right]} \tag{1}$$

式中，ε_{eq} 为一类反映三轴应力状态的等效单轴应变，A、B、ε_0 则为由实验数据确定的材料常数。

其二，是利用不可逆热力学基本原理，由损伤正交流动法则给出损伤变量演化法则，即：

$$\dot{d}=\dot{\lambda}_d\frac{\partial g(Y_n,r_n)}{\partial Y_n} \tag{2}$$

其中，y_n 为时刻 n 的损伤能释放率；r_n 为时刻 n 的损伤能释放率阈值，g（x）为变量 x 的任意递增标量函数，λ_d 为标量比例因子。

经过近二十年的研究，人们已经找到了基本合理的损伤能释放率表达形式（Ju，1989，李杰，吴建营，2005）。但是，决定上述演化准则的核心—g（·）的具体形式—仍然不得不通过理论假设给出。事实上，经典连续介质损伤力学在本质上属于唯象学的研究框架。因此，对于带有根本性的、具有物理内涵的损伤演化法则，在逻辑上只能采用经验归纳或理论假设（猜想）的方式给出。换句话说，在唯象学的研究框架里，是难以说明损伤演化的内在物理机制的。

然而，人们研究实践的总结也揭示了一个十分有趣的现象：不同的研究者，无论从经验归纳角度、还是理论假设的途径，其损伤演化规律都殊途同归：大体具有类似于图 5 所示的基本形式。从这一图示中，我们自然可以提出这样的问题：损伤为何不是线性发展、而是具有某种非线性特征呢？换句话说，在损伤发展过程中，为什么同样的应变增量会导致不同的损伤增量？在不同的损伤阶段，为什么会出现能量的非均匀耗散？

在本文作者看来：是随机性导致了损伤发展的非线性。

仍以图 3 所示模型为例，若将细观弹簧的断裂应变视为随机变量，则不难推断：在均值意义上，仅当随机变量的概率密度服从均匀分布时，损伤演化才服从线性分布（图 6*a*）。而在随机变量服从非均匀概率密度分布时，损伤演化必然具有非线性发展态势

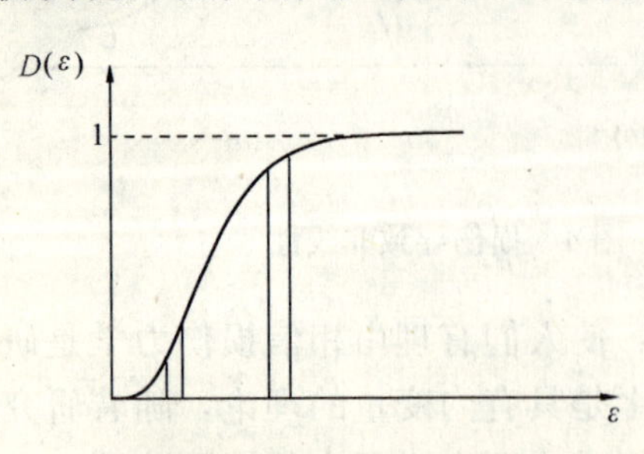

图 5 典型损伤演化曲线

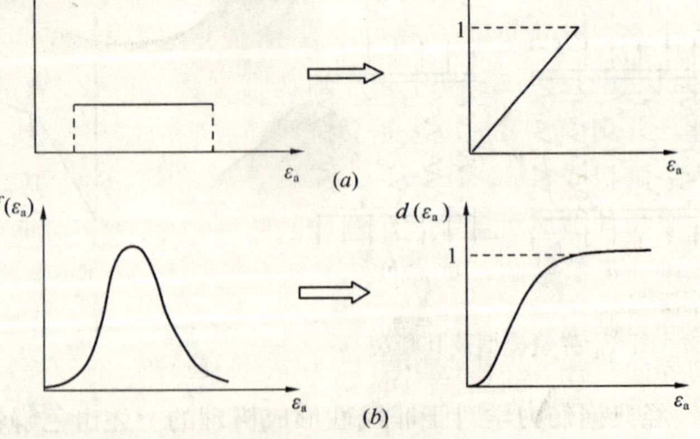

图 6 两类损伤演化

（*a*）均匀分布时的损伤演化；（*b*）非均匀分布时的损伤演化

(图 6*b*)。一个单轴拉伸试件，可以视为一个关于细观断裂应变变化的集合样本，根据作者在文献[34]中阐述的随机建模原理，可以完整地解释经验的损伤演化规律赖以产生的物理原因。事实上，作为一个集合样本，其中断裂应变集合具有随机变量的经验分布性质。若这一分布具有非均匀分布性质，则在受力断裂过程中，不同时刻的相同应变增量必然导致不同的细观单元断裂数量，从而导致非线性的损伤演化特征。

上述分析告诉我们：损伤演化的非线性源于细观层次断裂应变分布的随机性，宏观的损伤演化规律，应该在细观层次的物理分析中寻求其内在机理与建模途径。细观物理分析与宏观唯象分析的联系途径，可以采用基于自洽理论的期望平均方式，更为合理的，则是利用物理随机系统的基本思想，建立集合意义上的矩演化或概率密度演化途径（李杰，2006）。

基于这样的基本认识和理念，本文作者和他的学生们系统展开了混凝土细观随机损伤模型和宏观弹塑性损伤力学模型的研究。在这些研究基础上，建立了混凝土随机损伤本构理论的基本构架。

4.3　细观随机损伤模型

大量的试验观察表明，混凝土的非线性变形来源于两种基本的物理机制：微裂缝（微缺陷）的扩展和水泥浆体的塑性滑移。在变形过程中，损伤演化与塑性滑移之间相互影响、相互耦合。任何正确的混凝土本构关系，必须对这两种基本的物理变形机制作出合理的反映。一般说来，混凝土在复杂应力作用下的破坏形态可概括为三种基本形式：拉伸破坏、剪切破坏以及高静水压力下的压碎破坏。在不考虑高静水压力导致的应变强化的前提下，混凝土材料的损伤和破坏主要源于两种不同的物理机制：受拉损伤机制和受剪损伤机制。相应地，混凝土的细观损伤也可分为受拉损伤和受剪损伤，前者受最大拉应变控制，而后者则由于剪应变引起。可以分别采用受拉损伤变量 D_t 和受剪损伤变量 D_s 来描述上述两种机制对混凝土材料宏观力学性能的影响。

为在细观上反映这两种破坏机制，我们采用在细观尺度意义上其断裂应变服从某一概率分布的微弹簧来表征细观单元，发展了两类细观随机断裂—滑移模型（李杰，张其云，2001；李杰，杨卫忠 2007），如图 7 所示。两类单元分别对应受拉损伤破坏机制和受剪损伤破坏机制。在建议模型中，在细观层次上将混凝土离散为具有一定特征高度和截面积的小柱体，并用微弹簧加以表示。拉伸（或剪切）微弹簧均假定为理想弹脆性材料，其极限应变为一随机变量。

在受拉伸破坏机制中，小柱体的破坏是由于骨料和水泥砂浆之间界面被拉开或集料及凝胶体中初始微缺陷扩展而产生，表现为细观受拉弹簧的随机断裂。而在以压应力为主的受剪破坏机制中，小柱体破坏起源于骨料和水泥砂浆之间界面或初始微缺陷因剪应力作用导致的界面拉开，在模型中表现为受剪弹簧的随机断裂。为了反映水泥砂浆内部以及砂浆、骨料界面间塑性滑移的影响，分别引入细观拉伸塑性变形元件和细观剪切塑性变形元件。基本的受拉损伤单元或受剪损伤单元分别由拉伸微弹簧和拉伸塑性变形元件或剪切微弹簧和剪切塑性变形元件串联组成。

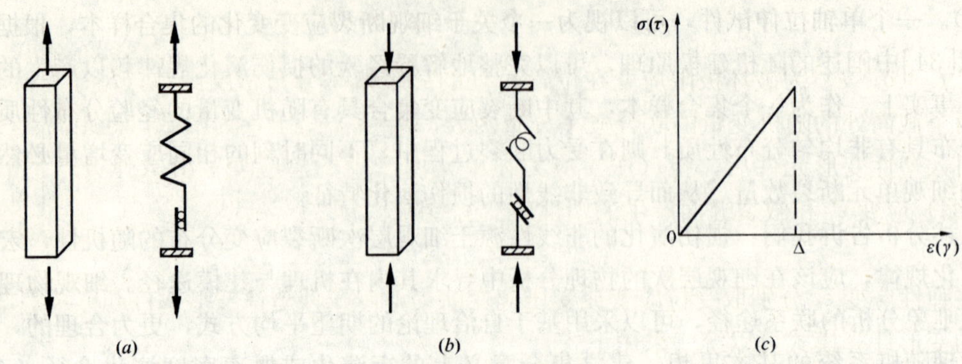

图 7　细观物理模型

(a) 拉伸单元；(b) 剪切单元；(c) 微弹簧的本构关系

由于细观受拉单元与受剪单元受力机制上的类似性，可以采用统一的方式给出其数学描述。

引用 Rabotnov 的经典损伤定义损伤变量（Krajcinovic 1996），即

$$D_i = \frac{A_D}{A} \quad (i = t, s) \tag{3}$$

式中，A_D 为因细观拉伸（或剪切）损伤单元破坏而导致混凝土退出工作的面积；A 为无损混凝土的截面积，即试件的横截面积。

假设离散后模型中细观单元的截面积均相等，式（3）可变换为

$$D_i(\varepsilon_e) = \frac{1}{M}\sum_{i=1}^{M} H(\varepsilon_e - \Delta_i) \tag{4}$$

式中，Δ_i 为第 i 个细观单元发生拉伸或剪切破坏时相应的应变，可视为服从某一分布的随机变量；H（•）为 Heaviside 函数，即

$$H(\varepsilon_e - \Delta_i) = \begin{cases} 0, \varepsilon_e \leqslant \Delta_i \\ 1, \varepsilon_e > \Delta_i \end{cases} \tag{5}$$

当模型中细观单元的数目 M 趋向于无穷大时，则在宏观横截面上的细观单元体可以看作一维连续体。若 M→+∞时式（4）的极限存在，则细观单元破坏时的应变为一连续随机场 $\Delta(x)$。不失一般性，x 可认为介于 0 和 1 之间，即，$x \in [0,1]$。相应地，式（4）可以表示为如下形式（Kandarpa & Kirkner 1996）

$$D_i(\varepsilon_e) = \int_0^1 H[\varepsilon_e - \Delta(x)]\mathrm{d}x \tag{6}$$

式中，$\Delta(x)$为在位置 x 处的随机破坏应变，考虑数学上的处理方便，可假定为一维均匀随机场。

由于 Δ_i 的随机场性质，$D_i(\varepsilon_e)$为一随机函数。若 $\Delta(x)$的一维、二维分布密度函数均存在，利用概率论中随机变量函数的均值和方差的计算方法，可得到受拉（或受剪）损伤变量的均值和方差，分别为

$$\mu_D(\varepsilon_e) = \int_0^{\infty}\int_0^1 H(\varepsilon_e - \delta) f_{\Delta}(\delta; x)\mathrm{d}x\mathrm{d}\delta = F(\varepsilon_e) \tag{7}$$

$$V_D^2(\varepsilon_e) = \left[2\int_0^1 (1-\gamma)F_\Delta(\varepsilon_e,\varepsilon_e;\gamma)\mathrm{d}\gamma\right] - F^2(\varepsilon_e) \tag{8}$$

式中，f_Δ（δ；x）为破坏极限应变在位置 x 处的一维概率分布密度函数；F_Δ（ε_e，ε_e；γ）为在两个截面处的随机变量的联合概率分布函数，$\gamma = |i-j|$，为两截口处距离。

引入塑性变形元件后，拉（压）应力在单元内产生的总应变 ε 由微弹簧产生的拉伸（或压缩）应变 ε_e 和微裂缝面产生的塑性应变 ε_p 组成，即

$$\varepsilon = \varepsilon_e + \varepsilon_p \tag{9}$$

在弹性应变 ε_e 分别取受拉应变或受压应变前提下，拉伸塑性变形元件和剪切塑性变形元件的变形计算模式可以统一采用下式

$$\varepsilon_p = \frac{\delta}{1-D}\varepsilon_e \tag{10}$$

式中，δ 为塑性变形系数

细观随机断裂—滑移模型的本质，是将材料细观物理性质的随机性作为损伤演化的依据之一，这就自然而然地将混凝土材料内秉的非线性与随机性以及两者的耦合作用纳入到一个统一的模型中来，从而实现了非线性与随机性的综合反映。

4.4 弹塑性随机损伤本构关系

根据损伤力学基本理论（Lemaitre & Rodrigue 2005），定义有效应力张量$\bar{\boldsymbol{\sigma}}$满足无损伤材料的弹塑性本构关系，即

$$\bar{\boldsymbol{\sigma}} = \boldsymbol{C}_0 : \boldsymbol{\varepsilon}^e = \boldsymbol{C}_0 : (\boldsymbol{\varepsilon} - \boldsymbol{\varepsilon}^p) \tag{11}$$

式中：标记“：”为二阶缩并积；$\boldsymbol{\varepsilon}$ 为总应变张量，一般分解为弹性应变张量 $\boldsymbol{\varepsilon}^e$ 和塑性应变张量 $\boldsymbol{\varepsilon}^p$ 两部分，即 $\boldsymbol{\varepsilon} = \boldsymbol{\varepsilon}^e + \boldsymbol{\varepsilon}^p$；$\boldsymbol{C}_0$ 为材料的初始刚度张量。

为了反映拉应力和压应力对混凝土材料的不同影响，将上述有效应力张量 $\bar{\boldsymbol{\sigma}}$ 分解为正、负分量（$\bar{\boldsymbol{\sigma}}^+$，$\bar{\boldsymbol{\sigma}}^-$）之和的形式

$$\bar{\boldsymbol{\sigma}}^+ = \boldsymbol{P}^+ : \bar{\boldsymbol{\sigma}} \tag{12a}$$

$$\bar{\boldsymbol{\sigma}}^- = \bar{\boldsymbol{\sigma}} - \bar{\boldsymbol{\sigma}}^+ = \boldsymbol{P}^- : \bar{\boldsymbol{\sigma}} \tag{12b}$$

式中：四阶对称张量 $\boldsymbol{P}^+$ 和 $\boldsymbol{P}^-$ 称为 $\bar{\boldsymbol{\sigma}}$ 的正、负投影张量，表示为 $\bar{\boldsymbol{\sigma}}$ 的特征值$\hat{\sigma}_i$ 和特征向量 $\boldsymbol{p}_i$ 的函数

$$\boldsymbol{P}^+ = \sum_i H(\hat{\sigma}_i)(\boldsymbol{P}_{ii} \otimes \boldsymbol{P}_{ii}) \tag{13a}$$

$$\boldsymbol{P}^- = \boldsymbol{I} - \boldsymbol{P}^+ \tag{13b}$$

其中，标记“$\otimes$”为张量积；H（·）为 Heaviside 函数；$\boldsymbol{P}_{ii} = \boldsymbol{p}_i \otimes \boldsymbol{p}_i$ 为二阶对称张量；$\boldsymbol{I}$ 为四阶一致性张量。

在等温绝热状态下，通常可以假定材料的弹性 Helmholtz 自由能势和塑性 Helmholtz 自由能势不耦合。在此前提下，材料的总弹塑性 Helmholtz 自由能势可表示为弹性部分 ψ^e 和塑性部分 ψ^p 之和的形式，即（李杰，吴建营，2005；Li & Wu 2006）：

$$\psi(\boldsymbol{\varepsilon}^e, \boldsymbol{q}^p, d^+, d^-) = \psi^e(\boldsymbol{\varepsilon}^e, d^+, d^-) + \psi^p(\boldsymbol{q}^p, d^+, d^-) \tag{14}$$

式中：$\boldsymbol{q}^p$ 为描述材料塑性特性的内变量；d^+，d^- 分别表示受拉损伤变量和受剪损伤变量，采用小写符号的原因将会在下文中得到解释；ψ^e 为材料的弹性 Helmholtz 自由能势

$$\psi^e(\boldsymbol{\varepsilon}^e, d^+, d^-) = \psi^{e+}(\boldsymbol{\varepsilon}^e, d^+) + \psi^{e-}(\boldsymbol{\varepsilon}^e, d^-) \tag{15}$$

$$\psi^{e+}(\boldsymbol{\varepsilon}^e, d^+) = (1-d^+)\psi_0^{e+}(\boldsymbol{\varepsilon}^e) = \frac{1}{2}(1-d^+)\bar{\boldsymbol{\sigma}}^+ : \boldsymbol{\varepsilon}^e \tag{16a}$$

$$\psi^{e-}(\boldsymbol{\varepsilon}^e, d^-) = (1-d^-)\psi_0^{e-}(\boldsymbol{\varepsilon}^e) = \frac{1}{2}(1-d^-)\bar{\boldsymbol{\sigma}}^- : \boldsymbol{\varepsilon}^e \tag{16b}$$

ψ^p 为材料的塑性 Helmholtz 自由能势

$$\psi^p(\boldsymbol{q}^p, d^+, d^-) = \psi^{p+}(\boldsymbol{q}^p, d^+) + \psi^{p-}(\boldsymbol{q}^p, d^-) \tag{17}$$

$$\psi^{p+}(\boldsymbol{q}^p, d^+) = (1-d^+)\psi_0^{p+}(\boldsymbol{q}^p) = (1-d^+)\int_0^{\varepsilon^p} \bar{\boldsymbol{\sigma}}^+ : \mathrm{d}\boldsymbol{\varepsilon}^p \tag{18a}$$

$$\psi^{p-}(\boldsymbol{q}^p, d^-) = (1-d^-)\psi_0^{p-}(\boldsymbol{q}^p) = (1-d^-)\int_0^{\varepsilon^p} \bar{\boldsymbol{\sigma}}^- : \mathrm{d}\boldsymbol{\varepsilon}^p \tag{18b}$$

材料的损伤过程和塑性流动过程都是不可逆热力学过程，由热力学第二定律可知，其能量耗散应为非负值，且必须满足热力学的不可逆条件，即 Clausius-Duheim 不等式。在等温绝热条件下，该不等式为

$$\dot{\gamma} = \boldsymbol{\sigma} : \boldsymbol{\varepsilon} - \psi \geqslant 0 \tag{19}$$

将式（14）微分并代入上式，将有：

$$\left(\boldsymbol{\sigma} - \frac{\partial \psi^e}{\partial \boldsymbol{\varepsilon}^e}\right) : \dot{\boldsymbol{\varepsilon}}^e + \left(-\frac{\partial \psi}{\partial d^+}\right)\dot{d}^+ + \left(-\frac{\partial \psi}{\partial d^-}\right)\dot{d}^- + \left(\boldsymbol{\sigma} : \dot{\boldsymbol{\varepsilon}}^p - \frac{\partial \psi^p}{\partial \boldsymbol{q}^p}\dot{\boldsymbol{q}}^p\right) \geqslant 0 \tag{20}$$

由于 $\dot{\boldsymbol{\varepsilon}}_e$ 的任意性，要满足上述不等式要求，应有

$$\boldsymbol{\sigma} = \frac{\partial \psi^e(\boldsymbol{\varepsilon}^e, d^+, d^-)}{\partial \boldsymbol{\varepsilon}^e} \tag{21}$$

将式（15）定义的材料弹性 Helmholtz 自由能势代入（21）式可以得到：

$$\boldsymbol{\sigma} = (1-d^+)\frac{\partial \psi_0^{e+}(\boldsymbol{\varepsilon}^e)}{\partial \boldsymbol{\varepsilon}^e} + (1-d^-)\frac{\partial \psi_0^{e-}(\boldsymbol{\varepsilon}^e)}{\partial \boldsymbol{\varepsilon}^e} \tag{22}$$

由式（16）、（12）可以得到如下关系式：

$$\frac{\partial \psi_0^{e+}(\boldsymbol{\varepsilon}^e)}{\partial \boldsymbol{\varepsilon}^e} = \boldsymbol{P}^+ : \bar{\boldsymbol{\sigma}} = \bar{\boldsymbol{\sigma}}^+ \tag{23a}$$

$$\frac{\partial \psi_0^{e-}(\boldsymbol{\varepsilon}^e)}{\partial \boldsymbol{\varepsilon}^e} = \boldsymbol{P}^+ : \bar{\boldsymbol{\sigma}} = \bar{\boldsymbol{\sigma}}^- \tag{23b}$$

于是，弹塑性损伤本构关系可以表示为：

$$\begin{aligned}\boldsymbol{\sigma} &= (1-d^+)\bar{\boldsymbol{\sigma}}^+ + (1-d^-)\bar{\boldsymbol{\sigma}}^- \\ &= (\boldsymbol{I} - \boldsymbol{D}) : \bar{\boldsymbol{\sigma}} = (\boldsymbol{I} - \boldsymbol{D}) : \boldsymbol{C}_0 : \boldsymbol{\varepsilon}^e\end{aligned} \tag{24}$$

式中：$\boldsymbol{D}$ 为四阶损伤张量，其表达式为：

$$\boldsymbol{D} = d^+ \boldsymbol{P}^+ + d^- \boldsymbol{P}^- \tag{25}$$

上式表明：上述弹塑性损伤本构模型实际上是一类张量损伤模型。

由于引入了两类内变量即损伤变量 d^+、d^- 和塑性变形 $\boldsymbol{\varepsilon}^p$，因此式（24）尚不能构成完整的混凝土本构关系，还必须建立内变量的演化法则。这里，首先要求从不可逆热力学的基本原理出发，基于损伤能释放率建立损伤准则（李杰，吴建营，2005）。其次，关于

塑性变形 $\boldsymbol{\varepsilon}^{\mathrm{p}}$，原则上可以采用在有效应力空间建立塑性应变演化方程：

$$\dot{\boldsymbol{\varepsilon}}^{\mathrm{p}}=\dot{\lambda}^{\mathrm{p}}\left(\frac{\bar{s}}{\|\bar{s}\|}+\alpha^{\mathrm{p}}1\right)=\dot{\lambda}^{\mathrm{p}}(1_{\bar{s}}+\alpha^{\mathrm{p}}1) \tag{26}$$

但考虑到直接运用塑性力学方法在屈服状态判断时需要进行迭代，从而使大型结构非线性分析时的计算效率严重下降，采用经验表达的方法来考虑塑性变形更为可取，如式（10）。

而对于损伤变量 d^+、d^- 的演化法则，则需要根据能量耗散原理和细观机制分析给出。由式（20）可知，损伤演化过程中的能量耗散 $\dot{\gamma}^{\mathrm{d}}$ 应该满足：

$$\dot{\gamma}^{\mathrm{d}}=Y^+\dot{d}^+ +Y^-\dot{d}^-\geqslant 0 \tag{27}$$

式中：Y^+ 和 Y^- 为与受拉损伤变量 d^+ 和受剪损伤变量 d^- 功共轭的热力学广义力，即受拉损伤能释放率和受剪损伤能释放率，分别表示为：

$$Y^+=-\frac{\partial\psi}{\partial d^+}=-\frac{\partial\psi^+}{\partial d^+}=\psi_0^{e+}(\boldsymbol{\varepsilon}^{\mathrm{e}}) \tag{28a}$$

$$Y^-=-\frac{\partial\psi}{\partial d^-}=-\frac{\partial\psi^-}{\partial d^-}=\psi_0^-(\boldsymbol{\varepsilon}^{\mathrm{e}},\boldsymbol{q}^{\mathrm{p}})=\psi_0^{e-}(\boldsymbol{\varepsilon}^{\mathrm{e}})+\psi_0^{p-}(\boldsymbol{q}^{\mathrm{p}}) \tag{28b}$$

根据上式和前述弹、塑性 Helmholtz 自由能势表达式，可以给出

$$Y^+=\frac{1}{2E_0}\left\{\frac{2(1+v_0)}{3}3\bar{J}_2^+ +\frac{1-2v_0}{3}(\bar{I}_1^+)^2-v_0\bar{I}_1^+\bar{I}_1^-\right\} \tag{29a}$$

$$Y^-=b_0[\alpha I+\sqrt{3J}]^2 \tag{29b}$$

基于上述损伤能释放率，可以建立受拉损伤准则和受剪损伤准则，分别为：

$$\bar{g}^+(Y_n^+,r_n^+)=Y_n^+-r_n^+\leqslant 0 \tag{30a}$$

$$\bar{g}^-(Y_n^-,r_n^-)=Y_n^--r_n^-\leqslant 0 \tag{30b}$$

式中：r^+、r^- 分别为受拉和受剪损伤能释放率阈值，用于控制损伤的发展。

对于一维受力状态，可由式（29）导出

$$Y_1^+=\frac{E_0}{2}(\varepsilon^{e+})^2 \tag{31a}$$

$$Y_1^-=b_0[(\alpha+\sqrt{3})E_0\varepsilon^{e-}]^2 \tag{31b}$$

对于一般的多维受力状态，不妨假设存在

$$\varepsilon_{\mathrm{eq}}^{e+}=\sqrt{\frac{2Y^+}{E_0}} \tag{32a}$$

$$\varepsilon_{\mathrm{eq}}^{e-}=\frac{1}{(\alpha+\sqrt{3})}\sqrt{\frac{Y^-}{b_0}} \tag{32b}$$

不难证明：

$$\varepsilon_{\mathrm{eq}}^{e+}=\sqrt{\frac{Y^+}{Y_1^+}}\varepsilon^{e+}=\theta^+\varepsilon^{e+} \tag{33a}$$

$$\varepsilon_{\mathrm{eq}}^{e-}=\sqrt{\frac{Y^-}{Y_1^-}}\varepsilon^{e-}=\theta^-\varepsilon^{e-} \tag{33b}$$

上述两式说明：将一维受力损伤能释放率中的弹性应变放大θ倍，即可得到多维受力状态的损伤能释放率，称ε_{eq}为能量等效应变。根据损伤一致性条件：对于初始损伤相同的两种受力状态，若损伤能释放率相同，则相应损伤相等。因此，通过引入能量等效应变，可以将一维受力状态的损伤演化法则，应用于多维受力状态。

前已指出：在唯象学的意义上，是难以从本质上反映损伤演化的内在物理机制的。要实现这一目的，必须引入多尺度分析的基本思想，从细观损伤的角度建立损伤演化的基本模型。前述细观随机断裂—滑移模型，恰恰满足了这一目的。根据损伤一致性条件和能量等效应变概念，可知在一般意义上存在如下损伤演化规律：

$$D^{+}(\varepsilon^{e+})=\int_0^1 H[\varepsilon_{eq}^{+}-\Delta^{+}(x)]\mathrm{d}x \tag{34a}$$

$$D^{-}(\varepsilon^{e-})=\int_0^1 H[\varepsilon_{eq}^{-}-\Delta^{-}(x)]\mathrm{d}x \tag{34b}$$

由于$\Delta(x)$的随机场性质，$D(\varepsilon_e)$为一随机函数。

事实上，确定性损伤本构关系与随机损伤本构关系所反映的是同一个物理规律，其间唯一的区别在于损伤变量是取确定性变量还是随机变量。当以上述两式作为损伤演化准则代入（24）式，得到的是一般意义上的弹塑性随机损伤本构关系。而当采用统计平均方式、以D的均值作为损伤演化准则代入（24）式时，我们将得到确定性弹塑性损伤本构关系。注意到式（7），有确定性损伤演化准则

$$d^{+}=E(D^{+})=\int_0^{\varepsilon_{eq}^{+}}\frac{1}{\sqrt{2\pi}\zeta_t x}\cdot\exp\left[-\frac{(\ln x-\lambda_t)^2}{2\zeta_t^2}\right]\mathrm{d}x \tag{35a}$$

$$d^{-}=E(D^{-})=\int_0^{\varepsilon_{eq}^{-}}\frac{1}{\sqrt{2\pi}\zeta_s x}\cdot\exp\left[-\frac{(\ln x-\lambda_s)^2}{2\zeta_s^2}\right]\mathrm{d}x \tag{35b}$$

式中，λ，ζ，ξ是确定破坏应变随机场$\Delta(x)$的概率分布参数，可以由单轴受力全过程实验结合随机建模准则确定。

显然，均值意义上的损伤演化准则是真实损伤演化过程的一种近似。在结构非线性分析中引用此类准则仅能给出具有某种均值意义的结构反应，而不能反映因初始随机性造成的结构响应涨落。在另一方面，如果选择唯象学意义上的损伤演化准则、并视其中物理参数为随机变量，也可以得到随机损伤本构关系，只是如此得到的本构关系，其损伤演化缺乏细观的物理机制解释，因而，也难以在本质上正确地把握或确立随机变量。

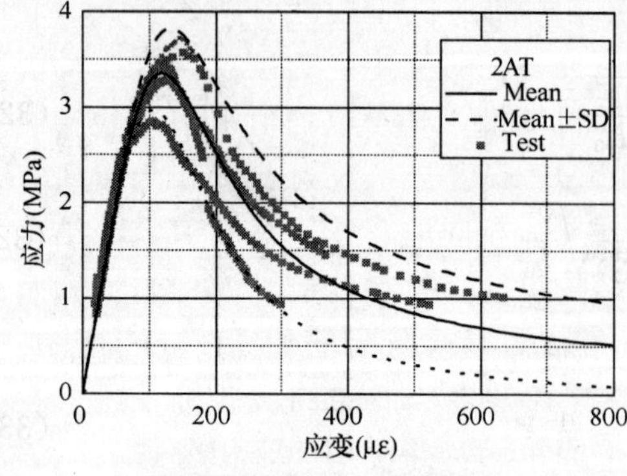

图8　单轴受拉随机损伤本构模型与实验结果的对比

值得指出：在具体工程应用中，根据需要，完全可以依据具体情况、选取确定性损伤本构关系或随机损伤本构关系应用于不同的工程对象。

图8和图9分别给出了按上述随机损伤本构模型给出的单轴受拉和单轴受压计算结果，图中同时给出了我们进行的混凝土单轴

受力应力—应变全曲线的部分试验结果。显然，主要试验点落在应力均值加、减一倍均方差的范围之内。这正是随机损伤模型的优势所在：不仅可以在均值意义上反映混凝土 $\sigma—\varepsilon$ 关系，也能在概率意义上预测其离散范围，从而，更为全面地反映混凝土材料受力行为的本质非线性与随机性。

图 10 给出利用上述随机损伤本构模型计算得出的二维峰值应力（即应力强度）与我们所进行的混凝土双轴受力试验结果（李杰，任晓丹，杨卫忠 2007）的比较，为证实可信性，图中同时列出了 Kupfer、Lee、过镇海等研究者的试验结果。显然，建议模型不仅可以给出主应力强度均值，也可以给出其方差，从而可以衡量应力强度变异性的大小。从图 10 可见，多数试验点落在均值加减一倍均方差之内，显示了建议模型的合理性。

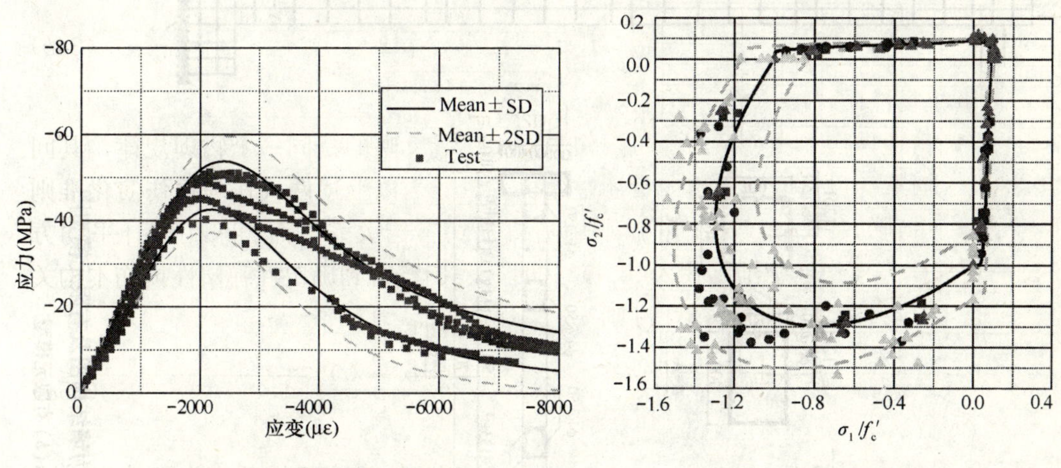

图 9　单轴受压随机损伤本构模型与实验结果的对比

图 10　二维强度比较图

4.5　混凝土结构的随机非线性分析

利用上述弹塑性随机损伤本构关系并结合概率密度演化方法（李杰，陈建兵 2003；Li & Chen 2006，2007），可以进行混凝土结构的随机非线性分析。为此，我们设计进行了混凝土双连梁短肢剪力墙结构实验研究。在研究中，为了反映真实的混凝土性质，将混凝土强度和弹性模量作为基本的随机变量，其参数均值采用试验实测平均值：轴心抗压强度 42.9MPa，弹性模量 39705MPa，变异系数为 10%。为了研究混凝土随机性与非线性对结构反应的影响，进行了 4 片相同的双连梁短肢剪力墙结构静力全过程试验。为保证 4 片墙体的一致性，在结构设计、材料配制、模型施工以及加载制度上进行了严格的控制。双连梁短肢剪力墙结构形式与配筋如图 11(*a*)所示。试验设备采用了同济大学建工系试验室的 10000kN 大型多功能结构试验机系统。在有限元分析过程中，采用四边形 8 节点二次等参平面单元，并使用 4 个积分点的减缩积分，非线性反应求解采用改进弧长法。具体计算在大型通用有限元软件 ABAQUS 的二次开发平台上完成。有限元模型见图 11(*b*)所示。

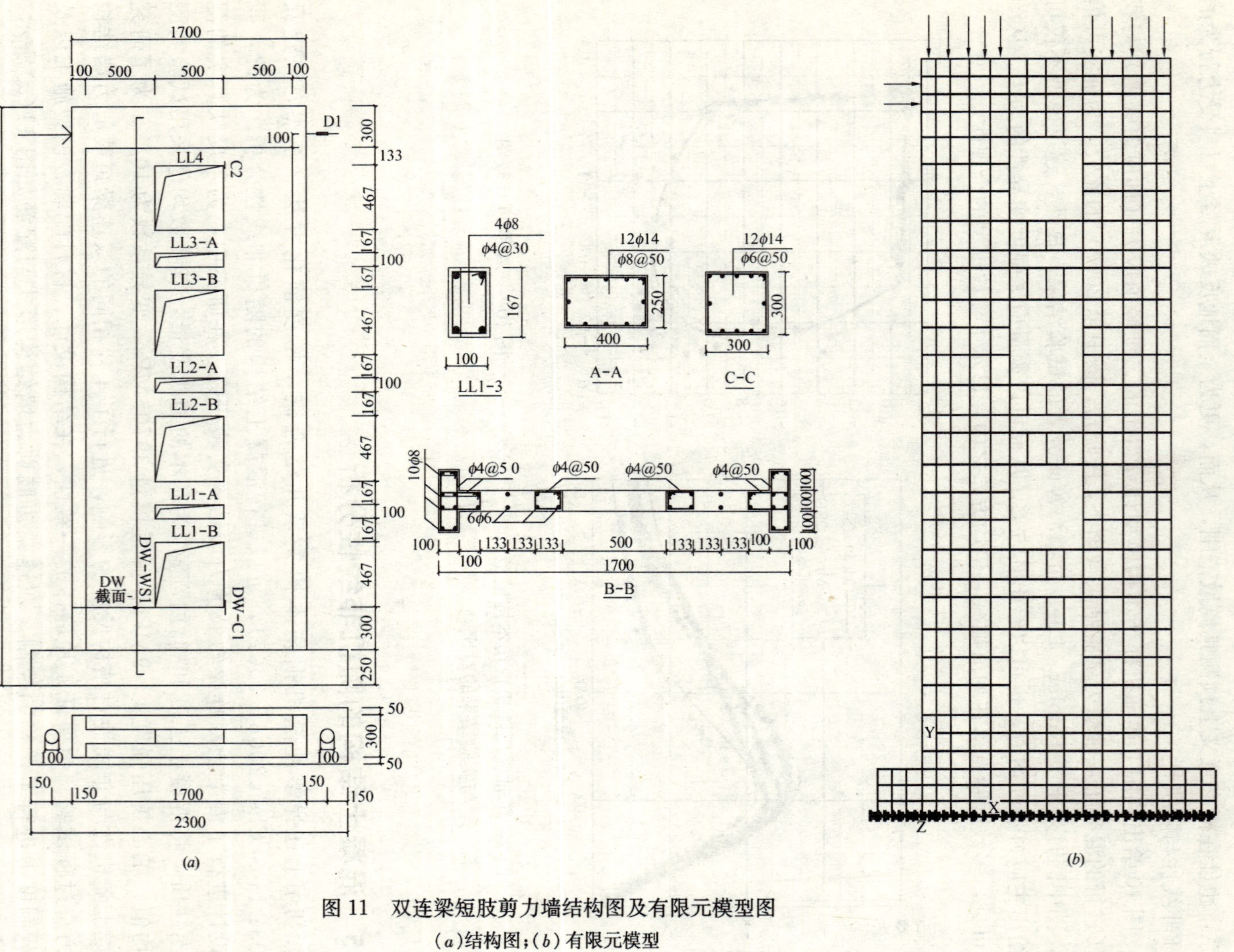

图 11 双连梁短肢剪力墙结构图及有限元模型图

(a)结构图;(b)有限元模型

结构荷载—位移曲线分析结果如图 12(*a*)所示。可见：试验的荷载—位移曲线均落在理论分析均值加减一倍标准差的范围内，在结构宏观层次反应上，计算结果与试验结果在二阶统计量上符合良好，证明了基于随机性和非线性耦合这一物理背景开展研究工作的正确性。

顶点位移在 10mm，50mm 和 130mm 的时刻、水平承载力的概率密度曲线如图 12(*b*) 所示。可见：概率密度曲线经历了一个不断变化的过程。

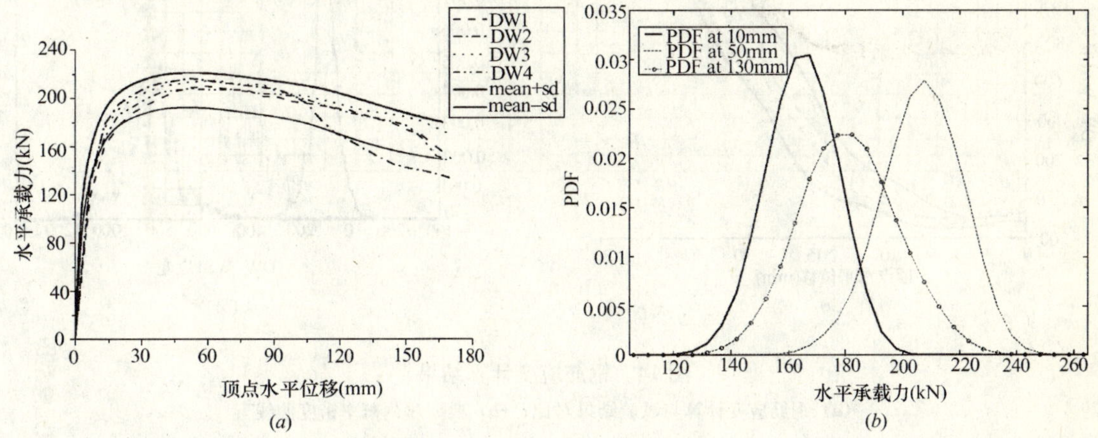

图 12　荷载—位移计算结果

(*a*) 荷载—位移计算与试验结果对比；(*b*) 典型时刻概率密度曲线

结构典型截面处的混凝土和钢筋试验应变曲线及相应理论分析结果如图 13(*a*)和图 14(*a*)所示。可见：试验应变曲线均基本落在理论分析均值加减二倍标准差范围内。在应变层次上计算结果也与试验结果在二阶统计量上吻合理想，进一步证明了我们所发展的弹塑性损伤本构模型的正确性。从图 13(*a*)和图 14(*b*)所示的典型概率密度曲线可见：在反应后期，概率密度曲线明显不再是通常见到的单峰光滑曲线，而可能是多峰曲线。说明混凝

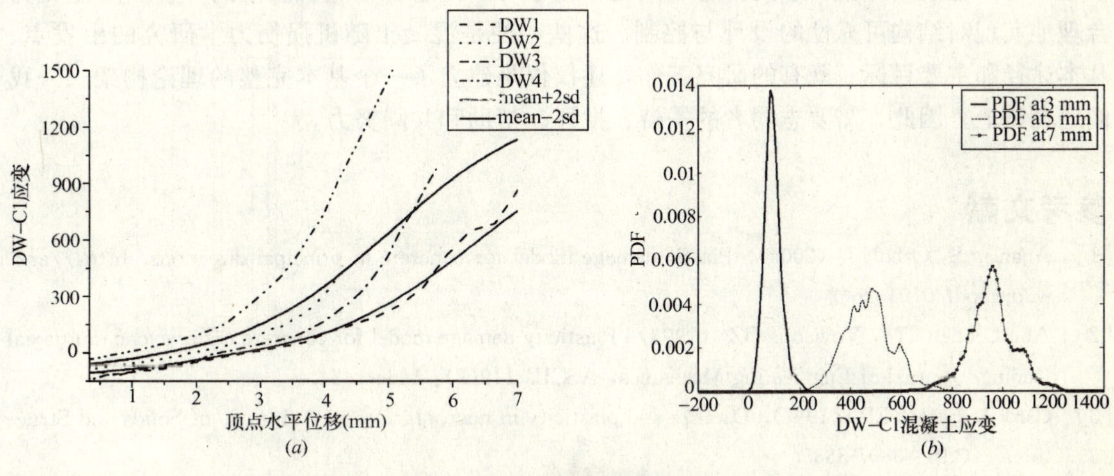

图 13　混凝土应变计算结果

(*a*) 混凝土应变计算与试验结果对比；(*b*) 典型时刻概率密度曲线

土细观的非线性演化路径十分复杂：随机性影响非线性的发展路径、非线性的发展又对随机损伤的演化起到反作用，随机性与非线性耦合，构成了丰富多彩的非线性随机演化过程。

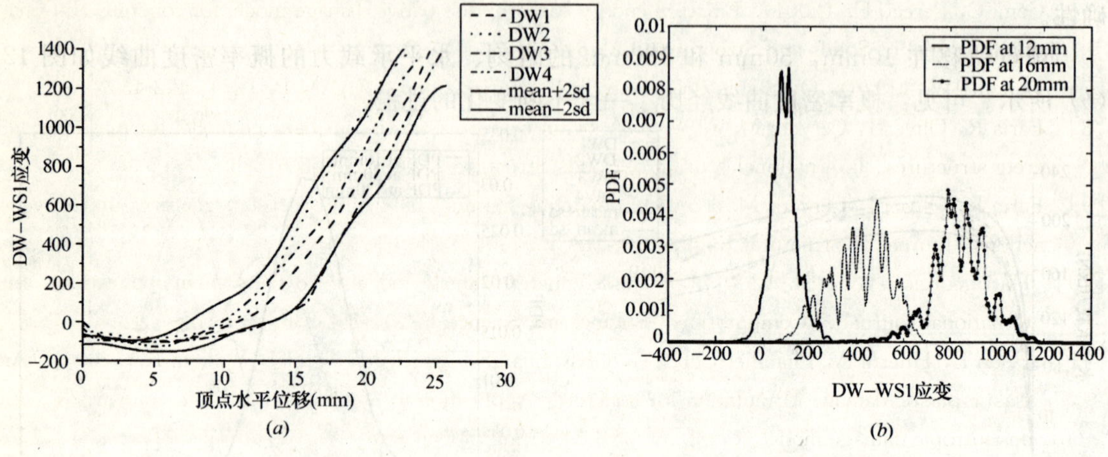

图 14　钢筋应变计算结果

(*a*) 钢筋应变计算与试验结果对比；(*b*) 典型时刻概率密度曲线

4.6　结语

损伤（以及塑性滑移）导致混凝土应力—应变本构关系的非线性；混凝土初始损伤及损伤发展过程的随机性导致损伤演化规律的非线性；利用细观随机损伤模型，可以反映基本的损伤演化规律；基于宏观连续介质损伤力学原理，可以建立应用于结构分析的混凝土随机损伤本构模型；确定性损伤本构关系与随机损伤本构关系可以互相转化；通过引入有限单元法或宏单元力学模型，可以实现从本构到结构的分析；在这三个不同分析尺度的联系途径上，应强调随机要素的传递与演化；基于对演化过程中随机涨落的分析，可望更为合理地实现对结构可靠性的设计与控制。这些，便是混凝土随机损伤力学研究的出发点、基本途径和主要目标。既有的研究工作，还仅仅是建立了一个基本完整的理论构架——我们还在路上。因此，需要志同者的扶持、批评、帮助和共同努力。

参考文献

[1] Ananiev S, Ozbolt J. (2004). Plastic-damage model for concrete in principal directions. http://arxiv.org/pdf/0704.2662.

[2] Abu-Lebdeh TM, Voyiadjis GZ. (1993). Plasticity-damage model for concrete under cyclic multiaxial loading. Journal of Engineering Mechanics, ASCE. 119(7): 1465-1484.

[3] Carol I, Bazant ZP. (1997). Damage and plasticity in microplane theory. Int. J. of Solids and Structures. 4(29): 3807-3835.

[4] Carol I, Rizzi E, Willam K. (2001). On the formulation of anisotropic elastic degradation. I: Theory based on a pseudo-logarithmic damage tensor rate; II: Generalized pseudo-Rankine model for tensile

damage. International Journal of Solids and Structures. 38(4)：491-546.

[5] Carmeliet J，Hens H.（1994）. Probabilistic nonlocal damage model for continua with random field properties. Journal of Engineering Mechanics ASCE. 120(10)：2013-2027.

[6] Comi C，Perego U.（2001）. Fracture energy based bi-dissipative damage model for concrete. International Journal of Solids and Structures. 38：6427-6454.

[7] Dougill JW.（1976）. On Stable Progressively Fracturing Solids. ZAAM P.（27）：432-437.

[8] Faria R，Oliver J，Cervera M.（1998）. A strain-based plastic viscous-damage model for massive concrete structures. International Journal of Solids Structures. 35(14)：1533-1558.

[9] Faria R，Oliver J，Cervera M.（2004）. Modeling material failure in concrete structures under cyclic actions. Journal of Structural Engineering ASCE. 130(2)：1997-2005.

[10] Hatzigeorgioiu G，et al.（2001）. A simple concrete damage model for dynamic Fem applications. International Journal of Computational Engineering Science. 2(2)：267-286.

[11] Jason L，Huerta A，Pijaudier-Cabot G，Ghavamian S，Jirasek M，Carol I，William K.（2006）. An elastic plastic damage formulation for concrete：Application to elementary tests and comparison with an isotropic damage model. Computational Modeling of Concrete. 195(52)：7077-7092.

[12] Ju JW，（1989），On energy-based coupled elastoplastic damage theories：constitutive modeling and computational aspects. International Journal of Solids Structures，25(7)：803-833.

[13] Kandarpa S，Kirkner DJ.（1996）. Stochastic damage model for brittle materiel subjected to monotonic loading. Journal of Engineering Mechanics. 126(8)：788-795.

[14] Krajcinovic D.（1996）. Damage mechanics. Second edition. Elsevier B. V.

[15] Lee J，Fenves GL.（1998）. Plastic-damage model for cyclic loading of concrete structures. ASCE. Journal of Engineering Mechanics Division. 124：892-900.

[16] Lemaitre J. Rodrigue D，(2005). Engineering Damage Mechanics：Ductile. Creep，Fatigue and Brittle Failures，Springer.

[17] Li J，Chen JB.（2006）. The probability density evolution method for dynamic response analysis of non-linear stochastic structures. International Journal for Numerical Methods in Engineering. 65：882-903.

[18] Li J，Chen JB.（2008）. The principle of preservation of probability and the generalized density evolution equation. Structural Safety. 30:65-77.

[19] Li J，Wu JY.（2006）. Energy-based CDM model for nonlinear analysis of confined concrete structures. American Concrete Institute，SP-238,209-221.

[20] Lubarda V A，Krajcinovia D，Mastilovic S.（1994）. Damage model for brittle elastic solids with unequal tensile and compressive strengths. Engineering Fracture mechanics. 49：681-697.

[21] Lubliner J，Oliver J，Oliver S and Onate E.（1989）. A plastic-damage model for concrete. International Journal of Solids Structures. 25(3)：299-326.

[22] Mazars J.（1984）. Application de la mecanique de l'endommangement au comportement non lineaire et a la rupture du beton de structure. These de Doctorate d'Etat，L. M. T.，Universite Paris，France.

[23] Mazars J.（1986）. A description of micro- and macro-scale damage of concrete structures. Engineering Fracture Mechanics. 25：729-737.

[24] Mazars J，Pijaudier-Cabot G.（1989）. Continuum damage theory：Application to concrete. ASCE Journal of Engineering Mechanics. 115(2)：345-365.

[25] Ortiz MA. (1985). Constitutive theory for inelastic behavior of concrete. Mechanics of Material. 4: 67-93.

[26] Prisco M, Mazars J. (1996). 'Crush-crack' a non-local damage model for concrete. Mechanics of Cohesive-frictional Materials. 321-347.

[27] Resende L. (1987). A damage mechanics constitutive theory for the inelastic behavior of concrete. Computer Methods in Applied Mechanics and Engineering. 60(1): 57-93.

[28] Simo JC, Ju JW. (1987). Strain- and stress-based continuum damage models-I. Formulation. International Journal of Solids Structures. 23(7): 821-840.

[29] Vallinppan S, Yazdchi M, Khalili N. (1999). Seismic analysis of arch dams: a continuum damage mechanics approach. International Journal for Numerical Methods in Engineering. 45(11): 1695-1724.

[30] Wu JY, Li J, Faria R. (2006). An energy release rate-based plastic-damage model for concrete. International Journal of Solids and Structures. 43(3-4): 583-612.

[31] Wu JY, Li J. (2007). Unified plastic-damage model for concrete and its applications to dynamic nonlinear analysis of structures. Structural Engineering and Mechanics, 23(5).

[32] Wu Jianying & Li Jie, (2008). On the mathematic and thermodynamic aspects of strain equivalence based anisotropic damage model. Mechanics of Materials. accepted and in press

[33] 江见鲸,李杰,金伟良.(2007).高等混凝土结构理论.北京:中国建筑工业出版社.

[34] 李杰.(1996).随机结构系统——分析与建模.北京:科学出版社.

[35] 李杰.(2004).混凝土随机损伤力学的初步研究.同济大学学报.32(10):1270-1277.

[36] 李杰.(2006).随机动力系统的物理逼近,中国科技论文在线,1(9),95-104.

[37] 李杰,陈建兵.(2003).随机结构非线性动力响应的概率密度演化方法.力学学报.35(6):716-722.

[38] 李杰,卢朝辉,张其云.(2003).混凝土随机损伤本构关系——单轴受压分析.同济大学学报.31(5):505-509.

[39] 李杰,吴建营.(2005).混凝土弹塑性损伤本构模型研究I:基本公式.土木工程学报.38(9):14-20.

[40] 李杰,任晓丹,杨卫忠.(2007).混凝土二维本构关系试验研究.土木工程学报.40(4):6-14.

[41] 李杰,杨卫忠,(2007).基于能量等效应变的混凝土随机损伤本构关系.已投稿.

[42] 李杰,张其云,(2000).“混凝土随机损伤本构关系研究进展”,结构工程师,第54期增刊,54-61

[43] 李杰,张其云.(2001).混凝土随机损伤本构关系研究.同济大学学报.29(10):1-8.

[44] 宋玉普.(2002).多种混凝土材料的本构关系和破坏准则.北京:中国水利水电出版社

[45] 吴建营,(2004).基于损伤能释放率的混凝土弹塑性损伤本构模型及其在结构非线性分析中的应用[D].上海:同济大学博士学位论文.指导教师:李杰.

[46] 吴建营,李杰.(2006a).反映阻尼影响的混凝土弹塑性损伤本构模型.工程力学.23(11):116-121.

[47] 吴建营,李杰.(2006b).考虑应变率效应的混凝土动力弹塑性损伤本构关系.同济大学学报.34(11):1427-1430.

[48] 杨卫忠,(2007),混凝土弹塑性随机损伤本构关系理论与试验研究[D].上海:同济大学博士学位论文.指导教师:李杰.

[49] 张其云.(2001).混凝土随机损伤本构关系研究[D].上海:同济大学博士学位论文.指导教师:李杰.

第5章　Chapter 5

抗震分析与设计的改进能力谱方法

An Extension of the Capacity Spectrum Method for Seismic Analysis and Design

W. D. Iwan

California Institute of Technology, USA

Email: wdiwan@caltech. edu

Abstract: This paper examines the use of the Capacity Spectrum Method (CSM) in the analysis and design of structural systems subjected to seismic excitation. Deficiencies of the conventional CSM approach are pointed out and a modified approach is introduced that overcomes these deficiencies while retaining the most desirable features of CSM analysis. A statistical methodology is used to define more accurate effective linear parameters and a new solution procedure is presented.

5.1 Introduction

The Response Spectrum has become an important tool in the design and analysis of structures subjected to loads such as those resulting from earthquakes. The concept of a frequency spectrum of an earthquake was first proposed by Biot in his Caltech Ph. D thesis in 1932 (*Biot*, 1932). Biot concluded that it was difficult to identify the actions of an earthquake with that of the acceleration curve studied previously, and that it was much more convenient to divide the problem and analyze separately the elastic properties of a building and the frequency distribution of the earthquake. To accomplish this, he proposed the "Energy Spectrum" for an earthquake. The Response Spectrum was first introduced by Benioff of Caltech in 1934 (Benioff, 1934), but he referred to it as a Pendular Spectrum. His idea was expressed as follows: "suppose we substitute for the engineering structures a series of undamped pendulum seismometers having frequencies ranging from the lowest fundamental frequency of engineering structures to the highest significant overtones……(then) during an earthquake each component seismometer would write a characteristic seismogram. Plotting the maximum recorded deflection of each pendulum against its frequency, we obtain a curve which may be termed the undamped pendular spectrum of the earthquake. ", This concept gradually evolved and in 1959 Housner of Caltech introduced the Response Spectrum in its present form (*Housner*, 1959).

Due to its conceptual simplicity, it was not surprising that the Response Spectrum became the dominant means to characterize earthquake excitation from an engineering perspective. Numerous researchers examined the Response Spectra from collections of measured earthquakes and derived smoothed averaged or enveloped spectra that could be readily used in the design of structures. Eventually, the Design Response Spectrum became a standard means of characterizing the seismic loads on a structure for purposes of design and analysis (*Newmark and Hall*, 1982). This was true for many codes and standards as well as for engineering practice.

In spite of its great usefulness, the Response Spectrum had one very significant limitation; it was based on the assumption of *linear structural response*. It was generally understood that most of the structures that were designed or analyzed by the Response Spectrum method would ultimately exhibit nonlinear behavior in a strong earthquake. But the method of analysis had become so prevalent that great effort was undertaken to adapt the concept to nonlinear structures. Eventually, two different approaches emerged, the Inelastic Response Spectrum and the Capacity Spectrum Method (CSM). The Inelastic Response Spectrum (*Newmark and Hall*, 1982) used the concept of the response of independent oscillators to an earthquake excitation as in the elastic Response Spectrum, but made the oscillators inelastic by using elasto-plastic or bilinear hysteretic restoring force models. The Capacity Spectrum Method, which is the subject of this paper, took a quite different approach.

In 1975 Freeman, Nicoletti, and Tyroli published a paper that set forth the concepts of the Demand Spectrum of an earthquake and the Capacity Spectrum of a structure (*Freeman, et al*, 1975). The Demand Spectrum was derived from the standard linear Response Spectrum by plotting the pseudo response acceleration as a function of the response displacement. This graph is referred to as the "Acceleration Displacement Response Spectrum" (ADRS). The Capacity Spectrum was obtained by plotting the mass-normalized force applied to the structure as a function of the deformation (displacement) of the structure. The response of the structure was then determined by the point of intersection of the Demand and Capacity Spectra curves (the *Performance Point*). The CSM is very appealing, as it seems intuitive that the "demand" on a structure by an earthquake should somehow be equal to its "capacity" if the structure survives the excitation. However, there are some subtle aspects of the CSM that are not so intuitive. For example, since a linear Response Spectrum is used to define the Demand but the structure is actually nonlinear, what assumptions should be made regarding the effective linear natural frequency or stiffness of the structure and its effective linear damping? The answer to this question will make a great deal of difference in the final performance of the structure being analyzed. This paper describes one approach to answering this question in a rigorous manner.

5.2 Conventional capacity spectrum method

The application of the conventional Capacity Spectrum Method is shown in Fig. 1 (*ATC*-40, 1996).

In the conventional CSM the area within the structural hysteresis loop is used to compute an effective linear damping (response amplitude dependent) of the inelastic structure and this linear damping is used to reduce the level of the Initial ADRS Demand to the "Reduced" ADRS Demand as shown. The effective linear stiffness or period of the inelastic system is taken to be the "secant" stiffness or period labeled T_{secant} in Fig. 1. Since the effective linear damping is generally related to the maximum response displacement, iteration is usually needed to find the final Performance Point.

The assumptions of the conventional CSM may seem logical, but there are a number of deficiencies in the conventional approach. These include: 1) inadequate theoretical or numerical basis for the effective damping, 2) the secant stiffness is known theoretically to be a poor estimate of the effective stiffness of most nonlinear systems, and 3) equivalent linear parameters for a nonlinear system are interrelated so that modifying one parameter requires modification of both.

The deficiencies of the conventional CSM can be seen from Fig. 2 which presents the results of an analysis of the Rinaldi Receiving Station ground motion record from the Northridge earthquake of 1994. The figure shows the ratio of the actual computed nonlinear response to the response predicted from the conventional CSM for elasto-plastic systems with different levels of maximum ductility. It is clear form the figure that the actual response can be quite different from that predicted using the CSM which, in this case, generally underestimates the actual response demand.

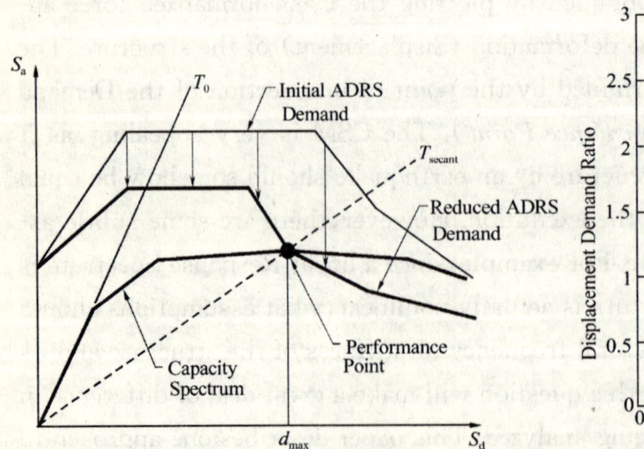

Fig. 1 Conventional Capacity Spectrum Method

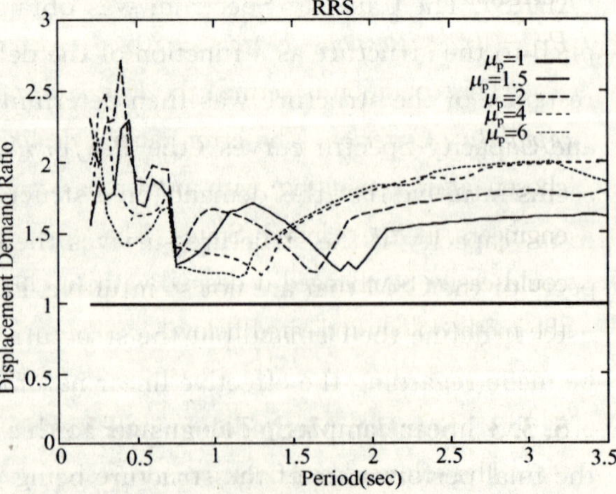

Fig. 2 Ratio of actual to conventional CSM response for Rinaldi excitation.

5.3 A new approach to specification of effective linear parameters

5.3.1 Problem Statement

In light of the above indicated deficiencies of the conventional CSM, it is desirable to find a way to improve the performance of the method without completely abandoning it. For this reason, we adopt the following problem statement:

Find a new set of effective or equivalent linear stiffness (period) and damping parameters that gives better response predictions than the conventional CSM without abandoning the general concept of the method.

Previous efforts at defining improved effective linear parameters for the response of inelastic structural systems subjected to earthquake excitation (*Iwan*, 1980) have shown that both the effective linear period and damping employed in the conventional CSM are too large. These two errors may tend to compensate for each other in some circumstances, but in general they lead to unacceptably large response errors. In order to overcome this problem, we describe herein a procedure to select effective linear parameters according to a more rigorous methodology.

5.3.2 The Linearization Optimization Criterion

The effective linear parameters will herein be selected in such a manner as to maximize the probability that the response error lies within what will be referred to as the "Range of Engineering Acceptability." Thus the parameters will be selected such that

$$Pr[-10\% < Response\ Error < +20\%] = Maximum$$

This condition is termed the *Engineering Acceptability Criterion*. The error range selected was chosen after consulting with numerous structural engineers in the United States. The error range could easily be changed if desired without changing the methodology indicated below.

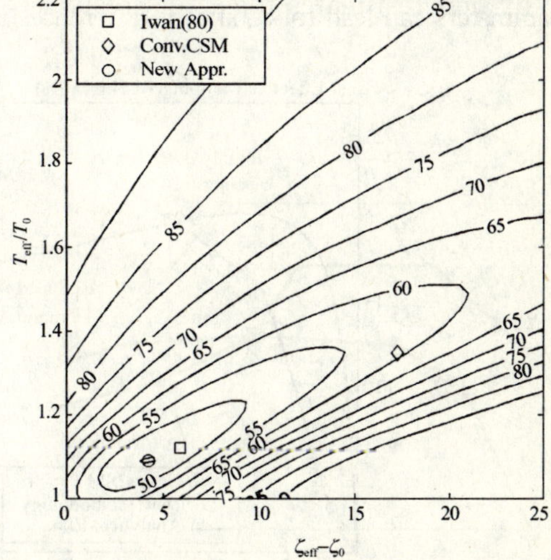

Fig. 3　Probability of exceedance of the range of engineering acceptability (BLH α=10%)

5.3.3 Solution Methodology and Results

The solution methodology is to search through the set of all possible effective linear parameters to find those parameters that maximize the Engineering Acceptability Criterion. This is a somewhat tedious numerical problem, but is otherwise straight forward. A typical result is shown in Fig. 3.

From Fig. 3 it is seen that there tends to be a valley running diagonally from the origin along which the error is minimized. This explains why the use of effective linear period and damping parameters that are both too large does not result in as great an error as might be anticipated. But the figure also shows that considerable improvement can be made in the CSM approach if a different set of effective linear parameters is employed.

An extensive study of the optimal effective linear parameters for a variety of inelastic structural systems was conducted and analytical expressions were derived to express the effective linear period and damping as a function of ductility. Both far-field and near-field earthquake accelerograms were considered. A general set of analytical expressions that were derived are indicated below.

For $\mu<4.0$:

$$T_{eff}/T_o-1=0.126(\mu-1)^2-0.0224(\mu-1)^3$$
$$\zeta_{eff}-\zeta_o=5.07(\mu-1)^2-1.08(\mu-1)^3$$

For $4.0\leqslant\mu\leqslant6.5$

$$T_{eff}/T_o-1=0.171+0.119(\mu-1)$$
$$\zeta_{eff}-\zeta_o=11.7+1.58(\mu-1)$$

Analytical expressions for effective period and damping for $\mu>6.5$ were also derived by considering the theoretical asymptotic behavior of these parameters.

An example of the results of this study is shown in Fig. 4. As can be seen from the figure, there is considerable difference between the optimal effective linear period and damping parameters and those used in the conventional CSM. Generally, both the optimal period and damping parameters are lower than their CSM counterparts except for large values of ductility where the optimal effective damping is greater than that employed in the conventional CSM. The differences in effective linear parameters can lead to significant differences in response.

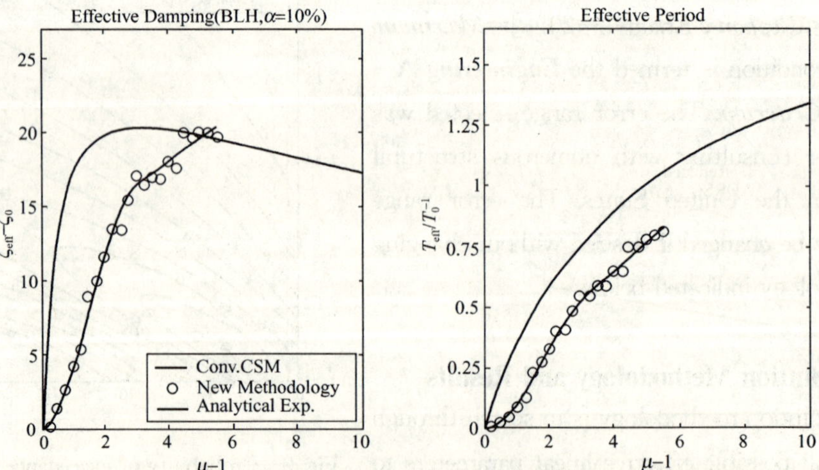

Fig. 4 Optimal effective linear parameters

Using the newly defined effective linear parameters, it is possible to compare the probability of achieving the Engineering Acceptability Criterion for both the new and conventional effective linear parameters. This result is shown in Fig. 5.

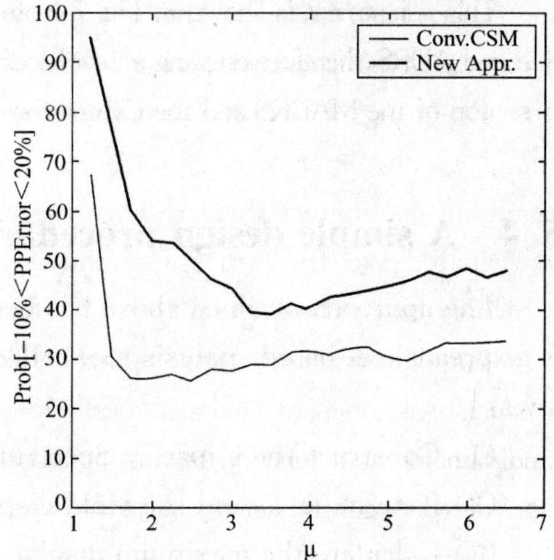

Fig. 5　Performance Point error for conventional and optimal effective linear parameters.

5.3.4 Application of the Optimal Parameters

Once a new set of optimal effective linear parameters is determined, it might appear to be straightforward to apply these within the CSM framework. However, there is a problem that must first be overcome. By not employing the secant period (stiffness), the Performance Point will not automatically lie on the Capacity Curve. This can lead to misinterpretation of the results and a more complicated solution procedure. However, this problem can easily be resolved by making a slight modification to the conventional ADRS. This modification leads to what is referred to as the Modified Acceleration Displacement Response Spectum or *MADRS*.

Fig. 6 shows the Capacity Spectrum for a structure characterized by a bilinear hysteretic restoring force and the associated Initial ADRS and Reduced ADRS using an appropriate set of optimal parameters. The intersection of the optimal effective period line and the Reduced ADRS represents the solution for the system response displacement (SD). However, it is noted that this point does not naturally fall on the Capacity Spectrum as was the case in the conventional CSM. Since the system forces as characterized by the PSA must agree with the Capacity Spectrum, the Performance Point must be adjusted downward until it falls on the Capacity Spectrum.

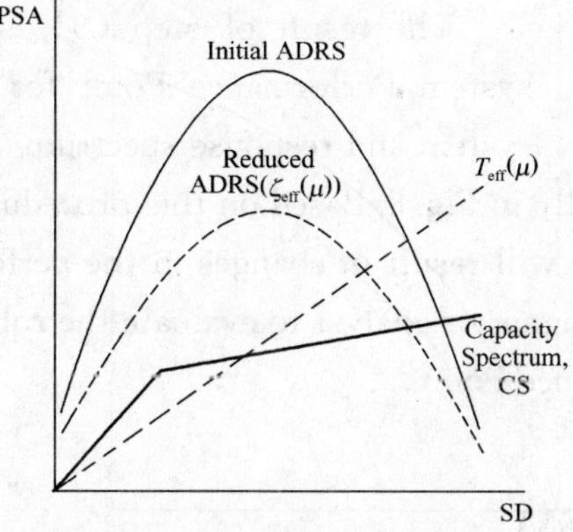

Fig. 6　Unmodified CSM with optimal parameters.

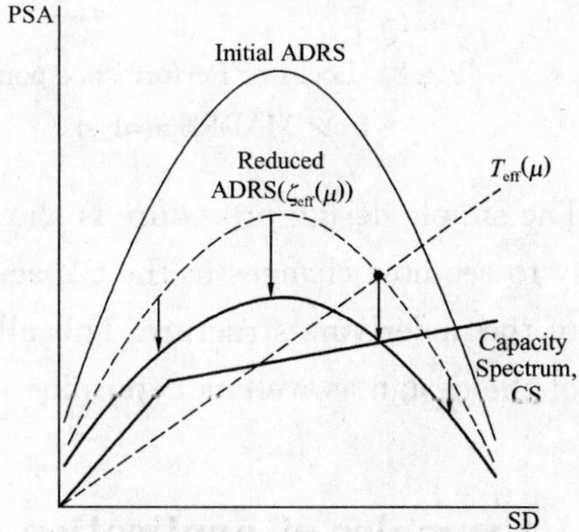

Fig. 7　Procedure for constructing the Modified ADRS

This adjustment is shown in Fig. 7 Note that the adjustment can be made for every point on the Reduced ADRS, therefore creating a new Modified ADRS or MADRS. By making this adjustment, the intersection of the MADRS and the Capacity spectrum becomes the Performance Point of the system.

5.4 A simple design procedure

The approach outlined above has been incorporated into a simple design procedure that is a spreadsheet-based analysis tool called AutoCSM. The procedure may be outlined as follows:

(1) Construct the Capacity Spectrum from a pushover analysis of the structure.

(2) Determine T_0 and the yield displacement from the Capacity Spectrum.

(3) Calculate the maximum displacements corresponding to a set of discrete values of ductility.

(4) Construct the secant stiffness for each ductility selected in step (3).

(5) Construct a family of site ADRS for damping coefficients consistent with the selected ductilities.

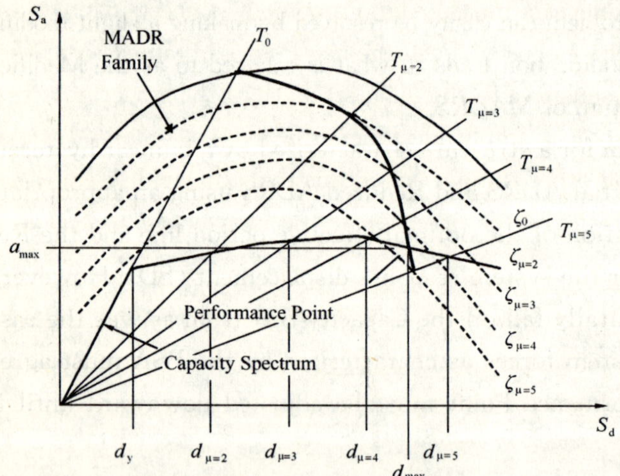

Fig. 8 Locus of Performance points from MADRS analysis

(6) Compute the MARDS from the ADRS for the selected damping-ductility pairs.

(7) Find the intersections of the MADRS and secant stiffness lines.

(8) Connect the points to obtain a Locus of Performance Points.

(9) Locate the intersection of the Locus of Performance Points and the Capacity Spectrum.

The result of step (9) is the system Performance Point for this system and response spectrum.

The simple design procedure is shown graphically in Fig. 8. Based on this procedure, it is easy to see how changes in the Capacity spectrum will result in changes in the performance of the underlying structure. This allows the designer or analyst to evaluate the robustness of the design as well as determine its Performance Point.

5.5 Examples of application of autoCSM

Two examples of the application of AutoCSM are given. One is for a specific earthquake response spectrum and the other is for a popular design spectrum. The examples indicate the flexibility of the procedure and the insights that can be obtained from its application.

The first example employs the maximum velocity direction ground motion of the 1940 El Centro earthquake and a bilinear hysteretic structure. The results are shown in Fig. 9 The Locus of Performance Points for the conventional CSM method is shown as a thick dashed line in the figure while the corresponding locus for the MADRS method is shown as a thick solid line.

It is quite evident that the two methodologies give very different performance results for the same structure. For the level of yield strength considered in this example, the conventional CSM method predicts a response displacement of nearly 20 cm while the MADRS approach predicts a displacement of less than 7 cm. This observation alone might mean the difference between acceptable and unacceptable performance of the subject structure. However, other important observations can also be made from the figure. Note that for the MADRS the Locus of Performance Points is nearly vertical. This implies that the maximum response displacement of the system is relatively insensitive to the yield strength of the system so that a change in yield strength due to a miscalculation or model uncertainty would not be too critical. On the other hand, the Locus of Performance Points obtained form the conventional CSM analysis is extremely sensitive to the yield strength of the structural system. There is a sudden increase in response amplitude when the mass normalized yield strength drops below about 0. 1g. Among other things, this means that the yield strength of the system must be known with a high degree of precision if the conventional CSM approach were used in design or analysis.

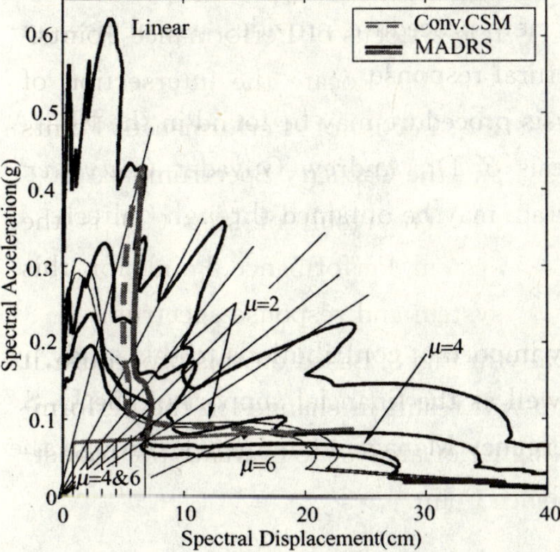

Fig. 9　Example of earthquake-specific MADRS and Locus of Performance Points. El Centro 1940 ground motion, maximum velocitydirection

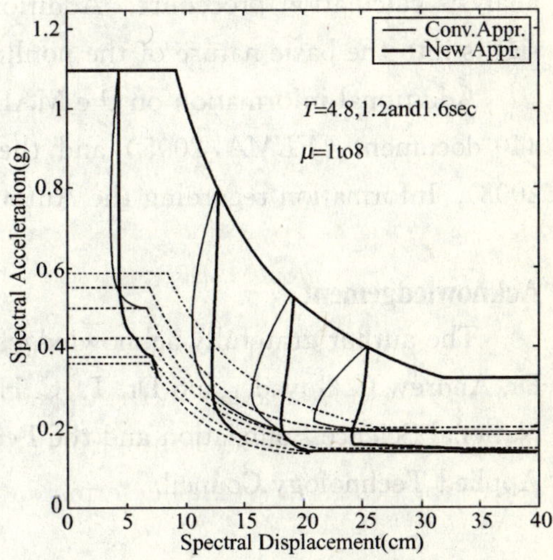

Fig. 10　Locus of Performance Points for 1997 UBC Design Spectrum

Fig. 10 shows the results of the conventional CSM and MADRS analysis when applied to a generic response spectrum such as might be given in a building code. In this case, the Design Response Spectrum is from the 1997 Uniform Building Code. Again, there are notable differences between the results of the two analysis methodologies, especially for longer period structures. For short period structures, the Locus of Performance Points for both methods is nearly vertical indicating that the response displacement is fairly insensitive to the response ductility. However, at longer periods there is considerable divergence between the two methods. The Locus of Performance Points for the MADRS method is still nearly vertical for lower levels of ductility while the conventional CSM predicts an initial reduction in displacement response with increased ductility. As the ductility increases further, both approaches predict a sharp transition of the Locus of Performance Points to a nearly horizontal line. This transition corresponds to rapidly increasing response displacements likely leading to structural failure. Although the two approaches predict a similar transition, it is noted that the MADRS approach indicates that the unstable horizontal locus occurs at a lower value of yield strength than that predicted by the conventional CSM. This difference can be important in assessing the safety of an existing structural system.

5.6 Conclusions

Using the proposed statistical approach for determining equivalent linear parameters, it is possible to obtain significantly improved response predictions while still employing the conceptual framework of the conventional Capacity Spectrum Method. Application of the CSM using the new equivalent linear parameters can be incorporated into a simple design/analysis calculation procedure. Additionally, the new procedure provides useful new insights into the basic nature of the nonlinear structural response.

Additional information on the MADRS analysis procedure may be found in the FEMA 440 document (*FEMA*, 2005) and the PhD thesis of Dr. Andrew Guyader (*Guyader*, 2003). Information regarding the AutoCSM program may be obtained through Caltech.

Acknowledgement

The author gratefully acknowledges the many important contributions to this study by Dr. Andrew C. Guyader and Dr. T.-C. Huang, as well as the financial support of the U. S. National Science Foundation and the Federal Emergency Management Agency through the Applied Technology Council.

References

[1] ATC-40 (1996). "eismic evaluation and retrofit of concrete buildings" Applied Technology Council, Redwood City, California

[2] Biot, M. A. (1932). "Transient oscillations in elastic systems" *Ph. D Thesis, California Institute of Technology*

[3] Benioff, H. (1934). "The physical evaluation of seismic destructiveness" *Bull. Seism. Soc. Am*. 24, 4, 398-403.

[4] FEMA 440. (2005) "Improvement of Nonlinear Static Seismic Analysis Procedures" *Federal emergency Management Agency*, Washington, D. C.

[5] Freeman, S. A. , Nicoletti, J. P. and Tyrell, J. V. (1975). "Evaluations of Existing Buildings for Seismic Risk - A Case Study of Puget Sound Naval Shipyard, Bremerton, Washington", *Proceedings of U. S. National Conference on Earthquake Engineering, Berkeley, U. S. A.*

[6] Guyader, A. C. (2003). "A statistical approach to equivalent linearization with application to performance-based engineering" *PhD Thesis, California Institute of Technology*

[7] Housner, G. W. (1959). "Behavior of structures during earthquakes" *J. of the Eng. Mech. Div, Proc. of the Amer. Soc. of Civil Eng.* EM04, October, 1959.

[8] Iwan, W. D. (1980) "Estimating Inelastic Response Spectra from Elastic Spectra" *International Journal of Earthquake Engineering and Structural Dynamics*, 8, *pp.* 375-388, 1980.

[9] Newmark, N. M. and Hall, W. J. (1982) "Earthquake Spectra and Design" *Earthquake Engineering Research Institute Engineering monographs on earthquake criteria, structural design, and strong motion records*, Berkeley, CA.

第 6 章　Chapter 6

输电塔线体系抗震抗风（雨）研究*

李宏男[1]　白海峰[2]　任月明[3]

（1. 大连理工大学土木水利学院，辽宁 大连 116023；
2. 中铁九局集团有限公司技术中心，辽宁 沈阳 110013；
3. 东北电力设计院，吉林 长春 130021）

提　要：根据输电塔线体系地震作用、风（雨）激励和结构特征，对输电塔线体系地震反应、风（雨）致结构动力响应和疲劳可靠度等问题进行了基础理论研究，建立了地震反应分析的整体简化动力模型、风（雨）激励响应体系分析模型和疲劳可靠性分析方法；提出了雨荷载的计算方法以及与风湍流共同作用于输电塔线体系的荷载组合原则。理论分析和试验验证表明：导线对输电塔的动力响应贡献程度会随杆塔档距的增加而增大，且随着场地条件的变软亦有增加的趋势；输电塔线体系的运动耦合性对结构动力特性与响应的附加效应不容忽视；降雨对输电塔线体系的响应影响明显；风雨共同作用，结构局部动态受压失稳是造成结构体系连续倒塔破坏的成因；疲劳可靠度指标与峰值应力假定和风荷载的相干程度有关；结构失效概率随着雨荷载的参与、风暴发生频次和持续时间的增长而增加。

关键词：输电塔线体系；地震作用；风（雨）荷载作用；动力特性与响应；疲劳可靠性

Fundamental Theoretical Study on Antiseismic and Wind-Proof of Transmission Tower-Line System

H. N. Li[1], H. F. Bai[2], Y. M. Ren[3]

(1. School of Civil and Hydraulic Engineering
Dalian university of technology, Dalian 116023, China;
2. Technology centre of China Railway No. 9 Group Co., Ltd. Shenyang 110013, China;
3. Northeast elective power design institute, Chang chun Jilin 130021, China)

Abstract: According to earthquake action, wind (rain) excitation and struictural characteristics,

---

* 基金项目：国家自然科学基金重点项目（编号：50638010）、教育部博士点基金项目（编号：20060141027）、教育部长江学者创新团队发展计划（编号：IRT0518）和高等学校学科创新引智计划资助项目（B08014）。

作者简介：李宏男（1958.7—），男，博士、长江学者特聘教授，博士生导师。E-mail: hnli@dlut.edu.cn
白海峰（1965.10—），男，博士，高级工程师。主要从事结构动力学方面的研究。

the problems related with earthquake response, wind-rain induced response and fatigue reliability of TLS are conducted fundamentally theoretical studies, so that the simple dynamic model for earthquake response analysis of whole structre, the analytical model of wind-rain excited structural response and analysis method of structural fatigue reliability are built, and the calculation method of rain forcing and its combination principles of acting on TTLS together with wind are presented. Theoretical and experimental analyses indicate that: as the increasing of neighbourhood tower distance and the softening subground, the wire contribution of dynamic response to transmission tower is inclined to enlarge; the accessional effects which the motion coupled properties of TLS influence on structural dynamic characteristics should not be neglected; it is obvious that rainfall occurred simultaneously together with wind have distinct effect on responses of TLS; the reasons which cause the structural consecutive collapse is wind and rain combination excitation as well as dynamic-press-unstability of local structures; the fatigue reliability indices are related to stress peak distribution assumption and coherence degree of wind loading, and structural failure probability can be increased with rain load participation, occurrence times and duration of storms.

Key words: transmission tower-line system; earthquake action; wind (rain) loading action; dynamic characteristics and responses; fatigue reliability

6.1 引言

输电塔线体系作为高压电能输送的载体，是重要的生命线工程。地震作用和风荷载是输电塔-线结构体系的主要设计荷载，在地震和风荷载作用下结构体系的破坏或功能失效状况表明，一方面存在对荷载作用机理和结构的动力响应特性认识不足的问题；另一方面说明现行设计理论还存在严重缺陷，对结构防灾控制措施的研究有待加强。因此，从提高设计水平和抵抗自然灾害能力的角度出发，结合国内外相关理论研究和工程建设背景，对输电塔线体系的抗灾减灾性能进行系统的基础性理论研究具有科学意义和实用价值。

输电塔线体系的抗震设计理论研究以往大多局限于静力荷载、断线冲击荷载、等效静荷载等作用工况［1～3］，而对动荷载作用则涉及甚少。这种理论体系的形成有其客观原因：对中、小档距的输电塔线体系而言，导线质量与杆塔质量相差悬殊，其动力响应对输电塔的贡献不明显，可忽略导线的影响，因此，我国现行《110～500kV 架空送电线路设计技术规程》[4]中并没有给出考虑导线动力响应贡献的输电塔设计分析方法；然而，对于大跨越输电塔线体系而言，无论是杆塔档距跨度，还是导线的单位质量和回路数设置，使得结构体系成为空间耦联体系，建立包括导线动力响应贡献的分析理论显得尤为重要。针对上述问题，作者［5～7］提出了大跨越输电塔体系的合理抗震计算简图，并分析了复合地震动作用下的结构地震反应[8, 9]，给出了反应谱分析法；在考虑了土-塔相互作用[10, 11]的基础上，阐述了导线对输电塔动力响应贡献的重要性。

除少数大跨越外，输电塔线体系结构抗风设计通常采用准静态设计方法[4, 12, 13]，即在设计基准期内，根据输电线路等级选取服从某种概率分布的最大平均风速作为荷载的设计标准值，采用风振系数或阵风荷载因子考虑风湍流的放大作用，以等效荷载的形式进行结构设计。准静态设计对于呼称高度较低，输电回路单一，湍流效应不显著的输电塔线体系具有简便实用，安全储备偏于保守的特点而得以沿用。然而，随着电网输电向超高压或特高压化发展，输电塔线体系趋于大型化，环境荷载效应发生了质的改变［14，15］。主要表现为：(1) 输电塔高度的增加，纵向刚度的降低，使其动力特性表现为高耸结构的特征；(2) 大跨越体系的导（地）线、绝缘子所占结构质量比增加，且与输电塔发生耦合振动，结构响应复杂；(3) 最新研究表明[16]，降雨具有荷载属性，风雨共同作用造成的结构体系破坏事件更为频繁。准静态设计解决此类结构分析显然存在局限性。因此，综合考虑风荷载均匀流、湍流以及降雨对输电塔线体系的作用机理、动力响应特征，对结构设计理论的深入研究尤为必要。

此外，由于大型输电塔在输电网络中所占的比重日益增大，结构的疲劳使用寿命预测和疲劳可靠度分析开始成为结构设计的内容之一［14，15］。以往对钢结构的疲劳多侧重于损伤研究，而很少做疲劳可靠度评价，对高耸格构式塔结构，尤为如此。Rojiani 和 Wen［17］考虑风荷载参数和强度变化等不确定性因素，提出风荷载作用下钢结构可靠性问题；Ditlievesen［18］用断裂力学理论分析动力可靠度的首次穿越问题；Deoliya 等［19］开始分析微波通讯塔风振疲劳可靠性问题，并初步建立了塔状结构疲劳可靠度的分析方法；徐建波等［20］通过风洞试验，考虑非 Gaussian 宽带效应，对桅杆结构的疲劳损伤特性进行分析。可见，高耸格构式输电塔架的疲劳研究处于探索阶段，理论与应用研究尚需深入。

本文在现有研究成果的基础上，侧重输电塔线体系、地震作用和风雨荷载激励的概念，对输电塔线体系抗震抗风的基础性理论问题进行研究。研究的目的在于解决不同档距导线对输电塔体系抗震性能的贡献，确定输电塔体系振动时导线的影响界限，并提出考虑导线影响的输电塔体系简化抗震分析方法；在建立风雨共同作用荷载模型的基础上，分析输电塔线体系的动力响应特征和结构失稳破坏的成因；从疲劳累积损伤理论出发，结合风（雨）荷载作用特征，通过不同荷载类型的疲劳损伤统计，建立疲劳可靠性分析评价方法。

6.2 输电塔体系抗震研究

6.2.1 抗震计算模型与运动方程

文献［5～7］分析表明：对于具有任意层导线的输电塔线体系，一跨导线的侧向和纵向振动模型可简化成如图 1 所示的形式。在地震地面运动作用下，由导线及杆塔组成的复合结构体系振动是弹性—重力耦联振动。经过推演，可分别得到考虑导线与杆塔相互作用结构体系侧向与纵向振动方程（暂不考虑阻尼）

$$[M]_{(x)}\{\ddot{x}\}_{(x)}+[K]_{(x)}\{x\}_{(x)}=-[M]_{(x)}\{E\}\ddot{x}_{g}(t) \tag{2-1}$$

和

$$[M]_{(y)}\{\ddot{x}\}_{(y)}+[K]_{(y)}\{x\}_{(y)}=-[M]_{(y)}\{E\}\ddot{y}_{g}(t) \tag{2-2}$$

式中，$\{x\}_{(x)}$ 和 $\{x\}_{(y)}$ 分别为结构体系侧向和纵向振动位移向量；$\{E\}$ 为单位向量；$\ddot{x}_g$ 和 $\ddot{y}_g$

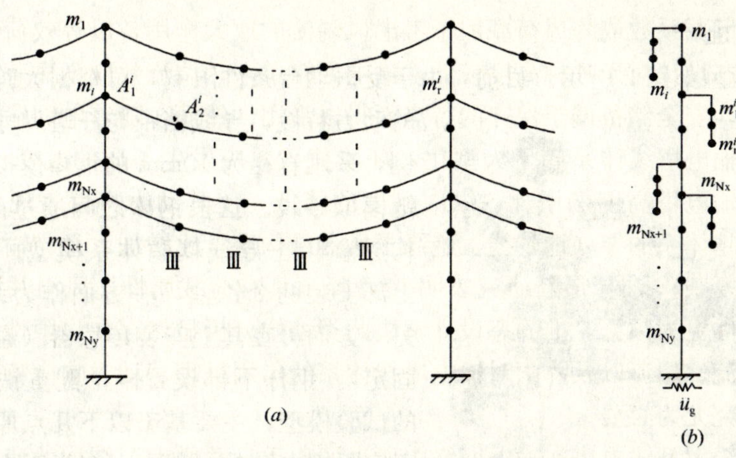

图1　输电塔-导线耦联体系抗震计算模型

分别为侧向和纵向地面加速度；$[M]_{(x)}$ 和 $[M]_{(y)}$ 分别为结构体系侧向和纵向质量矩阵；$[K]_{(x)}$ 和 $[K]_{(y)}$ 分别为结构体系侧向和纵向刚度矩阵。上述向量和矩阵的表达式可参见文献［21］和文献［22］。

6.2.2　结构体系状态方程

为计算输电塔线体系地震动反应，采用 Matlab 语言编制了基于 Rayleigh 阻尼形式的时程分析求解计算程序。为表达方便起见，可省略式（1）和式（2）中的方向脚标，将地面运动表示成 $\ddot{u}_g$ 并引入阻尼矩阵，可将结构体系的振动方程式（1）和式（2）写成如下统一状态方程形式，

$$[M]\{\ddot{x}\}+[C]\{\dot{x}\}+[K]\{x\}=-[M]\{E\}\ddot{u}_g \tag{2-3}$$

式中，$[C]$ 为结构体系阻尼矩阵。变换公式（3），可将其转化为加速度向量表达式

$$\{\ddot{x}\}=-[M]^{-1}[C]\{\dot{x}\}-[M]^{-1}[K]\{x\}-[M]^{-1}[M]\{E\}\ddot{u}_g \tag{2-4}$$

令 $\{z\}=\begin{Bmatrix}x\\ \dot{x}\end{Bmatrix}$，则 $\{\dot{z}\}=\begin{Bmatrix}\dot{x}\\ \ddot{x}\end{Bmatrix}$；又令 $\{\dot{z}\}=\begin{Bmatrix}\dot{x}\\ \ddot{x}\end{Bmatrix}=[A_1]\{z\}+[B_1]\{v\}=[A_1]\begin{Bmatrix}x\\ \dot{x}\end{Bmatrix}+[B_1]\{v\}$，并补充 $\{\dot{x}\}=\{\dot{x}\}$，可将公式（4）表示为

$$\begin{Bmatrix}\dot{x}\\ \ddot{x}\end{Bmatrix}=\begin{bmatrix}0 & E\\ -[M]^{-1}[K] & -[M]^{-1}[C]\end{bmatrix}\begin{Bmatrix}x\\ \dot{x}\end{Bmatrix}+\begin{Bmatrix}0\\ -\{E\}\end{Bmatrix}\ddot{u}_g \tag{2-5}$$

此时，

$$[A_1]=\begin{bmatrix}0 & E\\ -[M]^{-1}[K] & -[M]^{-1}[C]\end{bmatrix};\ [B_1]=\begin{Bmatrix}0\\ -\{E\}\end{Bmatrix};\{v\}=\{\ddot{u}_g\}$$

则状态方程可表示为

$$\begin{cases}\{\dot{z}\}=[A_1]\{z\}+[B_1]\ddot{u}_g\\ \{z\}=[A_2]\{z\}+[B_2]\{v\}\end{cases} \tag{2-6}$$

其中，$[A_2]=[E]$，$[B_2]=0$。

6.2.3　试验验证[23]

试验实体模型如图 2 所示。目前，由于受振动台条件限制，可根据大跨越输电塔体系的动力特性，导线档距与杆塔尺寸比例和质量比例，采用直径为 10mm 的钢棒模拟输电塔，用钢链模拟导线。三根钢棒之间通过两层钢链连接，构成输电塔-导线耦联体系模型。中间一根钢棒上部安装两个铁盒，一方面作为结构体系配重，另一方面可在其上安装传感器（通过螺栓与铁盒固定）。钢棒下部设置两个配重铁块。选择这样的试验模型，主要基于以下几点原因：第一，由于实际输电塔的横截面大多为正方形，且杆塔四周杆件的型号、横截面积等均相同，因此，其沿横截面两个方向的刚度也相同。选择圆截面钢棒代替杆塔，可以较好地模拟杆塔在地震中的动力行为；第二，试验模型中的配重块重量和布置可以在较大程度上保证模型与实际杆塔的质量分布相同；第三，用钢链模拟导线，钢链与杆塔之间采用铰接，能够较好地实现导线的动力性能；第四，试验模型采用的三座杆塔通过两跨（每跨两层）导线相联，可以较好地模拟杆塔与导线间的耦联作用。图 3 为试验模型简图（图中尺寸单位：mm）。

图 2　试验布置图

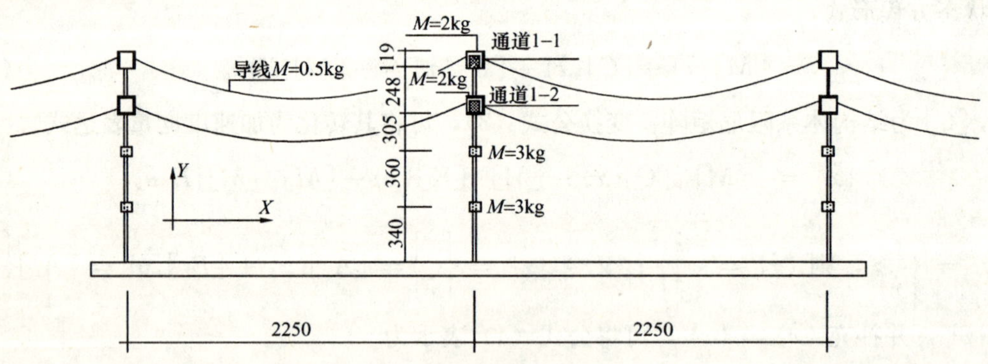

图 3　试验模型布置图

试验选用的地震动记录　　**表 1**

编　号	场地条件	地震名称	发震时间	震级	记录地点	峰值(gal)
①	软土	唐山地震	1976	7.1	天津医院	104.18
②	中等	Imperial Valley	1951	5.6	El Centro	27.56
③	硬土	唐山地震	1976	6.3	迁安县	118.91

为比较不同场地条件下导线对输电塔地震反应的贡献，本文选用了表 1 所示的具有代表性的 El Centro 波、天津医院波和迁安波作为地震动输入。试验中地震动记录的峰值均调整

到 0.1g，侧向和纵向振动时，分别给出了输电塔线体系最大加速度地震反应理论值和试验值，其结果对比与误差分析如表 2 所示。理论计算与试验时程曲线的比较参见文献 [7]。

考虑导线影响时的理论与试验结果比较　　表 2

记　录	宁河波						El Centro 波						迁安波					
振动方向	侧向振动			纵向振动			侧向振动			纵向振动			侧向振动			纵向振动		
结果分析	理论	试验	误差%	理论	试验	误差%	理论	试验	误差%	理论	试验	误差%	理论	试验	误差%	理论	试验	误差%
1-1	0.15	0.14	4.42	0.16	0.18	7.62	0.13	0.14	4.89	0.13	0.14	8.06	0.13	0.13	4.38	0.13	0.13	4.38
1-2	0.22	0.21	3.22	0.16	0.18	8.55	0.18	0.18	1.14	0.23	0.21	6.50	0.12	0.11	10.9	0.13	0.12	10.6

试验分析表明：输电塔线耦联体系的理论分析模型是合理的；考虑导线对动力响应的贡献，纵向振动响应在结果分析 1-1 和 1-2 通道最大误差分别为 8.06%和 10.6%；而侧向振动响应在结果分析 1-1 和 1-2 通道最大误差分别为 4.89%和 10.9%。

6.2.4 理论计算分析

为确定纵向振动时导线对输电塔体系动力响应贡献的影响程度，选择如图 4 所示的五种不同类型的实际输电塔进行理论分析，假定输电塔的结构形式均为空间桁架体系，塔身横截面为正方形，输电塔主要参数如表 3 所示。为分析简便，计算时将杆塔均等效地简化为 7 个质点，每档距内的导线等效地简化为 4 个质点。

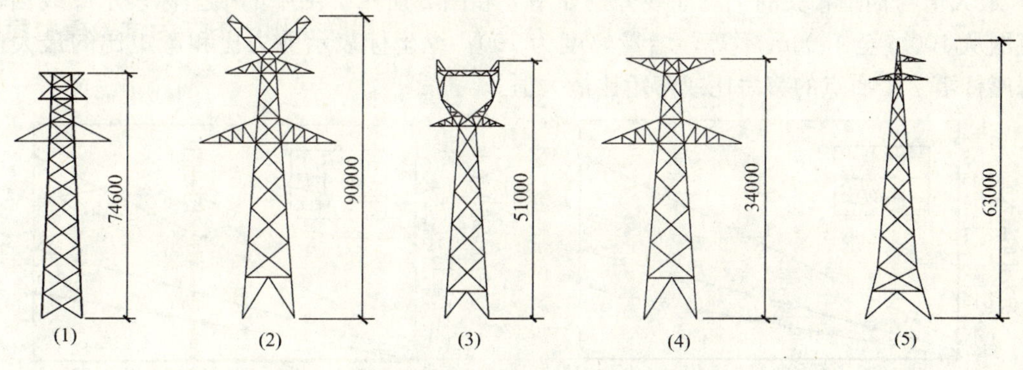

图 4　计算输电塔实例

输电塔主要参数　　表 3

	塔　高	塔　重	电缆型号	避雷线型号	电缆计算质量
塔 1	74.6m	33427kg	LGJ-500/45	GJ-70	1688kg/km
塔 2	90m	40430kg	LGJ-500/45	GJ-70	1688kg/km
塔 3	51m	15067kg	LGJ-400/50	GJ-70	1511kg/km
塔 4	34m	11390kg	LGJ-400/50	GJ-70	1511kg/km
塔 5	63m	16434kg	LGJ-400/50	GJ-70	1511kg/km

为考察地基对输电塔线体系地震反应的影响，选择硬土、中硬土和软弱土三类不同场地，每类场地采用 3 条不同的地震动记录，并将地震加速度峰值均调整为 0.2g 进行理论计算。所选用的地震动记录如表 4 所示。

计算选用的地震记录 **表 4**

场地类型	编号	地震名称	发震时间	震级	记录地点	峰值加速度 gal
软弱场地	①	San Fernando	1971. 2. 9	6. 6	PortHueneme	25. 91
	②	San Fernando	1971. 2. 9	6. 6	Univ. Avenue	56. 36
	③	唐山余震	1976. 11. 15	7. 1	天津医院	104. 18
中硬场地	④	Imperial Valley	1951. 1. 23	5. 6	El-Centro	30. 35
	⑤	Kern County	1952. 7. 21	7. 7	Taft	152. 7
	⑥	Imperial Valley	1940. 5. 18	6. 7	El-Centro	341. 7
坚硬场地	⑦	Landers	1992. 6. 28	7. 5	Baker Fire	105. 58
	⑧	Landers	1992. 6. 28	7. 5	Fort Irwin	119. 85
	⑨	唐山余震	1976. 8. 31	6. 3	迁安	118. 91

为比较导线对输电塔动力反应的贡献程度，分别采用不挂导线工况（对应的剪力和弯矩分别表示为 $V1$ 和 $M1$）和本文建议的挂导线工况（对应的剪力和弯矩分别表示为 $V2$ 和 $M2$）进行分析；同时，为比较分析不同导线档距对输电塔体系地震反应内力的影响，杆塔的档距布置从小到大进行变化。计算结果取输电塔在各类场地上三条地震动记录输出的均值，并通过最大剪力比和最大弯矩比反映导线对输电塔动力反应的影响程度。每类场地上三种地震动记录输出的输电塔 3 组典型侧向和纵向剪力比（$V2/V1$）和弯矩比（$M2/M1$）最大值与档距相关的平均曲线分别如图 5 和图 6 所示。图中的横坐标表示导线档距，其范围从 100m 至 1000m 不等，增量幅度为 100；纵坐标表示剪力比和弯矩比的最大值，即每座杆塔 7 个质点的剪力比或弯矩比最大值。

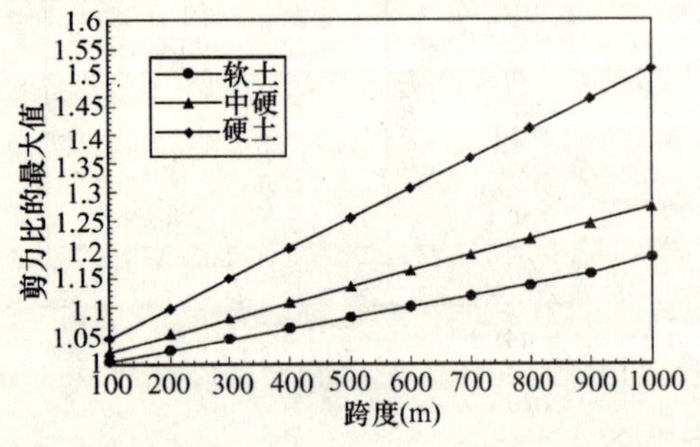

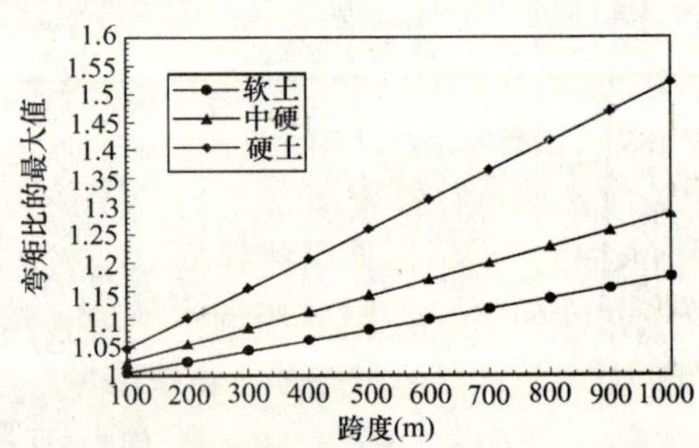

图 5 输电塔 3 侧向地震反应的平均曲线

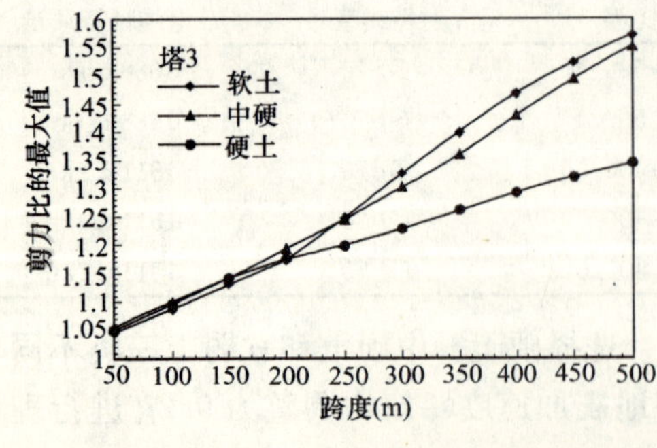

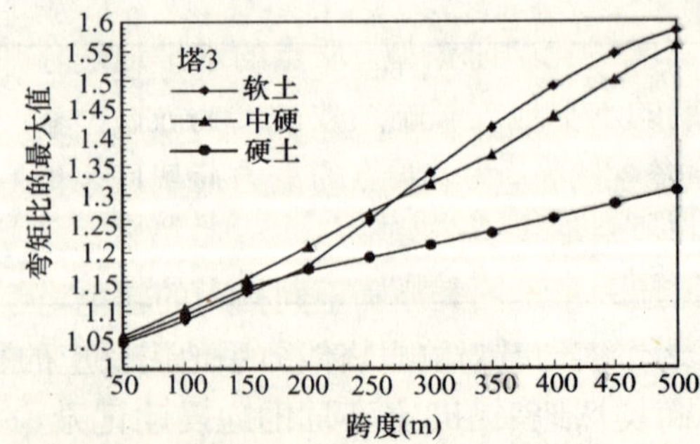

图 6 输电塔 3 纵向地震反应的平均曲线

由图 5 和图 6 可以看出，导线对输电塔体系地震反应内力的影响程度随档距的增加而增大；当导线的档距达到表 5 和表 6 中给出的数值时，导线对输电塔体系地震反应内力的影响在 5%以内。因此，可以得出结论：当导线的跨度超过表 5 和表 6 中的各数值时，进行输电塔体系的动力响应分析应考虑导线的影响。具体计算可按下节提出的方法进行。这里可将表中第二行所给出的导线的档距定义为为界限档距，记作 l_0。同时，表中还给出了导线布置为界限档距 5 种类型杆塔的分析误差均值和方差。由表中数据可以看出：计算结果的方差很小，这说明在三种场地情况下，数据的离散程度很小。

侧向塔考虑导线影响的界限档距值　　表 5

场地类别		软弱	中硬	坚硬
跨度界限值 l_0（m）		300	200	150
平均误差	剪力比	4.93%	3.57%	4.34%
	弯矩比	4.77%	3.53%	3.65%
方差	剪力比	0.001034	0.000837	0.000554
	弯矩比	0.000995	0.001443	0.001185

纵向塔考虑导线影响的界限档距值　　表 6

场地类别		软　弱	中　硬	坚　硬
界限跨度 l_0（m）		50	50	50
平均误差	剪力比	4.7%	5.8%	4.8%
	弯矩比	4.5%	5.9%	4.6%
方差	剪力比	0.001136	0.004930	0.000932
	弯矩比	0.001294	0.005428	0.000868

6.2.5　简化抗震计算方法

现行《电力设施抗震设计规范》[24]（以下简称《规范》）规定：大跨越输电塔及杆塔高度为 50m 以上的自立式铁塔的水平地震作用宜采用振型分解反应谱法计算，且杆塔动力特性的计算可不计入导线和避雷线重量。采用振型分解反应谱法计算地震作用的公式为

$$F_{ji} = \zeta\alpha_j\gamma_jX_{ji}G_i(i = 1,2,\cdots n;j = 1,2,\cdots m) \tag{2-7}$$

$$\gamma_j = \sum_{i=1}^{n} X_{ji}G_i / \sum_{i=1}^{n} X_{ji}^2G_i \tag{2-8}$$

式中，F_{ji} 为第 j 阶振型 i 质点的水平地震作用标准值；ζ 为结构系数；α_j 为第 j 阶振型自振周期的水平地震影响系数；γ_j 为 j 振型参与系数；X_{ji} 为第 j 阶振型 i 质点在 X 方向的水平相对位移；G_i 为 i 质点的重力荷载代表值，包括全部恒荷载、固定设备重力荷载和质点上的其他附加重力荷载。

上述大量不同类型和不同档距的输电塔线耦联体系的地震反应分析表明：按现行《规范》设计的大跨越输电塔体系的档距超过界限档距时，在强震作用下偏于不安全。因此，可根据计算结果提出建议简化抗震计算方法，即在式（2-8）的 G_i 上增加一个附加质量 Δm 考虑导线的影响，Δm 可由下式计算确定

$$\Delta m = f(l_x)\cdot l_x\cdot q \tag{2-9}$$

式中，Δm 为考虑导线影响后杆塔的附加质量(kg)；l_x 为导线的档距(m)；q 为导线单位长度质量(kg/km)；$f(l_x)$为附加质量系数，当进行输电塔体系侧向抗震计算时，可表示为

$$f(l_x)=\begin{cases}0.17+\dfrac{3l_x}{200l_0} & \text{软土}\\ 0.21+\dfrac{l_x}{100l_0} & \text{中硬（当 } f(l_x)>0.7\text{ 时,取 } f(l_x)=0.7\text{）}\\ 0.35+\dfrac{l_x}{20l_0} & \text{硬土}\end{cases} \tag{2-10}$$

当进行输电塔体系纵向抗震计算时，可表示为

$$f(l_x)=0.5+\frac{l_x}{200l_0}\text{（当 } f(l_x)>1.0\text{ 时,取 } f(l_x)=1.0\text{）} \tag{2-11}$$

采用简化抗震计算方法对不同档距工况下各种类型输电塔进行地震反应分析，结果与整体模型分析吻合得很好。作为例子，输电塔 3 在侧向和纵向的计算结果比较分别如图 7 和图 8 所示。可以看出：就软土和硬土场地而言，按简化方法计算得到的结果与采用整体模型计算的结果十分接近，最大相对计算误差仅在 3%左右；而对于中硬场地，结果偏差稍大，最大相对误差在 6%左右，这是因为输电塔 3 的自振周期与场地的特征周期很接近，存在共振现象。可见，简化计算方法具有足够的工程应用精度，能够满足工程应用。

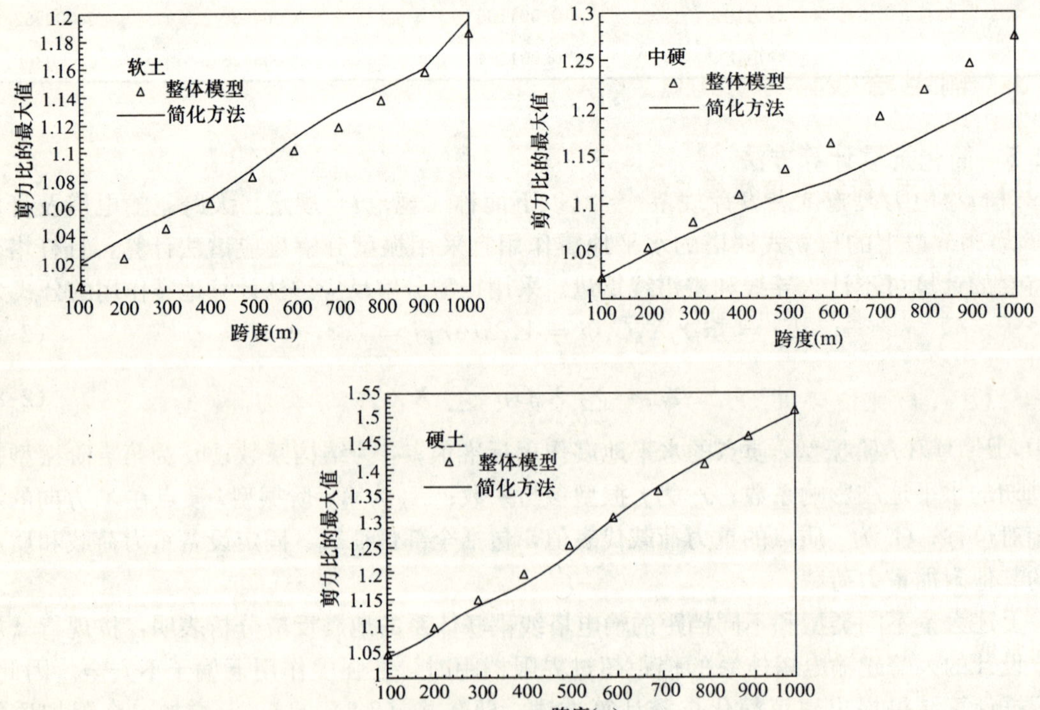

图 7　简化方法与整体模型法在输电塔侧向计算的对比图

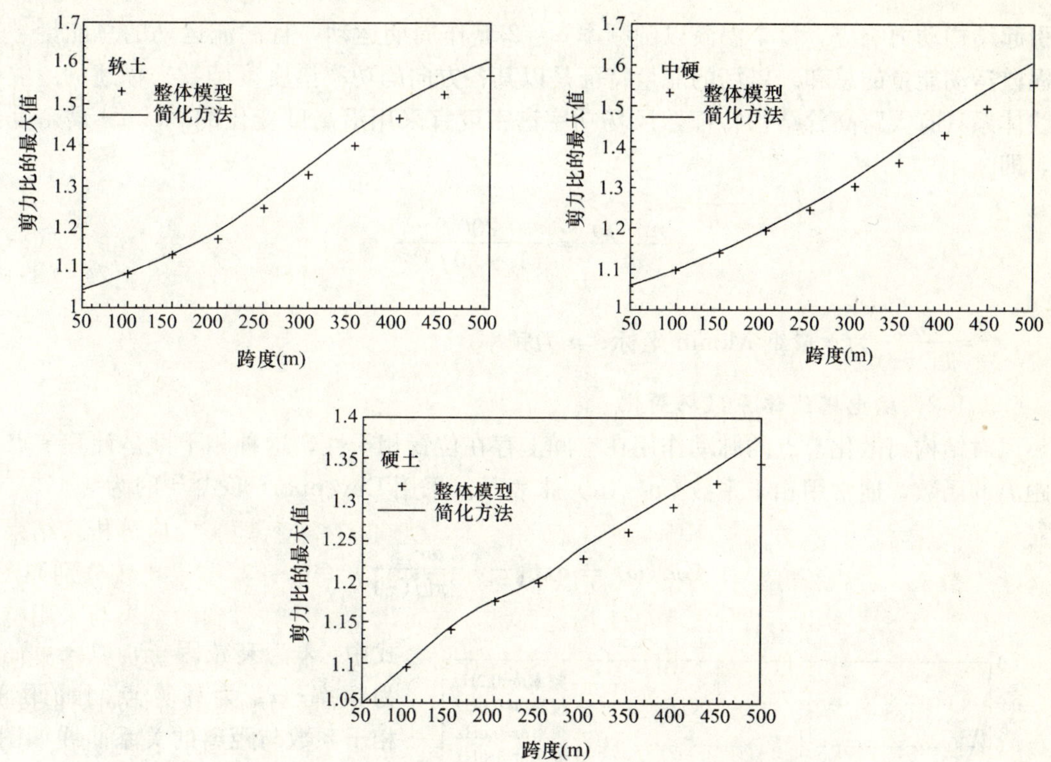

图 8　简化方法与整体模型法在输电塔纵向计算的对比图

6.3　输电塔线体系抗风(雨)研究

6.3.1　风(雨)荷载数值建模与组合

6.3.1.1　风特性

如所周知，工程应用上，一般将自然风运动假定为具有零均值的平稳随机过程，风速可表示为平均风速与脉动风速之和，即

$$U(z,t)=\overline{U}(z)+u(z,t) \tag{3-1}$$

式中，$U(z,\ t)$ 来流风速；$\overline{U}(z)$平均风速；z 风速计算高度；t 为时间。

平均风速 $\overline{U}(z)$ 是描述风场运动沿计算高度 z 变化规律的重要指标，其对数律风剖面描述为

$$\overline{U}(z)=\frac{1}{k}u_*\ln(z/z_0) \tag{3-2a}$$

式中，$k=0.4$，称为 Karman 常数；z_0 为地面粗糙长度；u_* 为摩阻速度，用参考高度 z_r（通常取 $z_r=10\text{m}$）处的平均风速 $\overline{U}(z_r)$表示为

$$u_*=\frac{k\overline{U}(z_r)}{\ln(z_r/z_0)} \tag{3-2b}$$

风脉动对结构的作用是流-固耦合作用，即流经结构的各种尺度的涡漩叠加和相互作

用引起结构动力响应。每个涡漩以圆频率 $\omega=2n\pi$ 作周期运动，且湍流运动的总能量是每个涡漩运动能量的总和。风场的能量特征是以其各方向的功率谱密度函数来描述的。输电塔线体系具有大跨高耸结构特征，风场功率谱密度宜采用沿高度变化的 Kaimal 谱密度函数，即

$$\frac{nS(f)}{u_*^2}=\frac{200f}{(1+50f)^{5/3}} \tag{3-3}$$

其中，$f=\dfrac{nz}{\overline{U}(z)}$ 为无量纲 Monin 坐标；n 为频率。

6.3.1.2 输电塔线体系风场简化

风对结构离散化节点的脉动作用在空间上存在位置相干性，这种相干性是计算节点空间距离的函数，通常用相干系数 $Coh(\omega)$ 来表示。采用 Davenport 形式[25]时为

$$Coh(\omega)=\exp\left(-\lambda\frac{\omega r_{jm}}{2\pi\overline{U}(z)}\right) \tag{3-4}$$

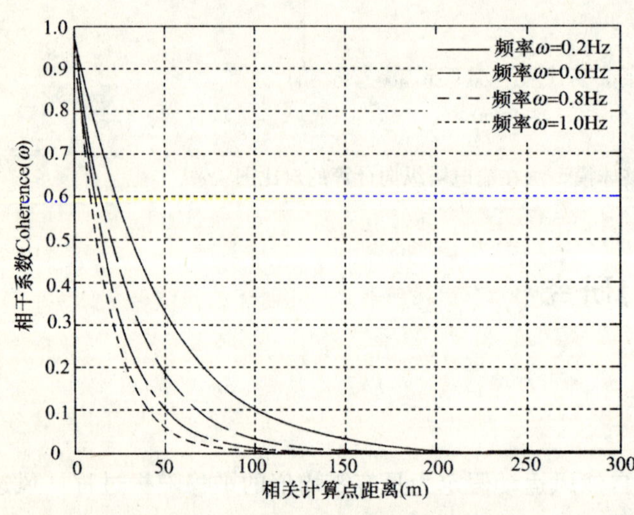

图 9 计算点距离与相干系数关系曲线

式中，λ 为衰减因子，取 $\lambda=8$；ω 为频率；r_{jm} 为计算点间的距离。相干系数与距离的关系曲线如图 9 所示。可见，结构自振频率大于 0.2Hz，结构间距大于 100m 时，风场脉动相干性低于 10%，且随频率和距离的增加明显衰减。基于以上规律，输电塔间距通常大于 100m，可只考虑单塔的塔架垂向的相干性。对输电导线可分两种情况考虑，小垂度、地面平坦时，可不计输电线路在传输方向上的高差变化，只考虑其侧向相干性；地面陡峭或相邻杆塔不等高，导线在传输方向上的高差变化较大时，应同时考虑侧向和塔架垂向的相干性。对于输电塔与导线的连结处而言，虽然距离较短，但由于导线受绝缘子、金具等的约束，振动响应不显著，可以忽略风场相干性影响。考察输电塔-线体系的结构特征，这样的简化是可以接受的。

6.3.1.3 风场数值模拟

(1) 谐波叠加法

谐波叠加法（也称谱解法）是由 Shinozuka 等[26, 27]提出的平稳随机过程数值模拟方法。设 $u_i^0(t)$ （$i=1, 2, \cdots, m$）是 m 个具有零均值的一维多变量高斯平稳随机过程（上标 0 表示目标样本函数），其互谱密度矩阵为

$$S^0(\omega)=\begin{bmatrix} S_{11}^0(\omega) & & & \\ S_{21}^0(\omega) & S_{22}^0(\omega) & & Sym \\ \vdots & \vdots & & \ddots \\ S_{m1}^0(\omega) & S_{m2}^0(\omega) & \cdots & S_{mm}^0(\omega) \end{bmatrix} \tag{3-5}$$

其中，$S_{ij}^0(\omega)(i=1,2,\cdots,m;j=1,2,\cdots,m)$ 为互相关函数 $R_{ij}^0(\tau)(i\neq j)$ 或自相关函数 $R_{ii}^0(\tau)$ 的Fourier变换。由互相关函数的不对称性可知，互谱密度矩阵 $S^0(\omega)$ 通常具有复数形式。对 $S^0(\omega)$ 进行 Cholesky 分解

$$S^0(\omega)=H(\omega)H^{*T}(\omega) \tag{3-6}$$

其中，$H(\omega)$ 是下三角矩阵，$H^{*T}(\omega)$ 是复共轭转置矩阵。$H(\omega)$ 的表达式如下

$$H(\omega)=\begin{bmatrix} H_{11}(\omega) & 0 & \cdots & 0 \\ H_{21}(\omega) & H_{22}(\omega) & \cdots & 0 \\ \vdots & \vdots & \ddots & \vdots \\ H_{m1}(\omega) & H_{m2}(\omega) & \cdots & H_{mm}(\omega) \end{bmatrix} \tag{3-7}$$

对于对角线元素，由自相关函数的性质有

$$H_{ii}(\omega)=H_{ii}(-\omega)\quad i=1,2,\cdots,m$$

对于非对角线元素可表示为

$$H_{il}(\omega)=|H_{il}^*(-\omega)|e^{i\theta_{il}(\omega)}$$

其中：$\theta_{il}(\omega)=\tan^{-1}\left\{\dfrac{\mathrm{Im}[H_{il}(\omega)]}{\mathrm{Re}[H_{il}(\omega)]}\right\}$，Im［·］和 Re［·］分别表示 $H(\omega)$ 的虚部和实部。

根据 Shinozuka 理论［26］，当 $N\rightarrow\infty$ 时，随机风场的样本时间序列可由下式来模拟，

$$u_i(t)=\sum_{l=1}^{i}\sum_{k=1}^{N}|H_{il}(\omega_k)|\sqrt{2\Delta\omega_k}\times\cos[\omega_k t-\theta_{il}(\omega_k)+\varphi_{lk}]i=1,2,\cdots,m \tag{3-8}$$

式中，N 为频率等分数，即频率域内的数据采样数目；$\theta_{il}(\omega)$ 为结构上两个不同荷载作用点之间的相位角；φ_{lk} 为介于 0～2π 之间均匀分布的独立相位角。ω_u 为上限截止频率。

综上所述，只要已知目标谱密度函数 $S^0(\omega)$，恰当地选择 N，ω_u 和 Δt，便可模拟风随机样本时间序列。

（2）互谱密度的计算

由于湍流的不均匀性，不同高度计算点的风速和相位均不相同，出现阵风首先作用于结构的较高点，经过时差 τ 后，作用于较低点的现象。即使在同一水平高度上两点，也会因相位差的不同，使脉动风速时程存在相关性，由互谱密度函数[28]表示为

$$S_{ij}(\omega)=\sqrt{S_{ii}(\omega)S_{jj}(\omega)}Coh_{ij}(\omega)\exp[\phi(\omega)](i=1,2,\cdots,m) \tag{3-9}$$

式中，$Coh_{ij}(\omega)$ 为相干函数；$\phi(\omega)$ 为两个离散点之间的相位差。文献［29］对大量的风速相位在 $\phi-f$ 坐标中进行统计分析后发现，当 f 从 0 到 0.1 变化时相位角近似线性增加；f 在 0.1 到 0.125 区间时相位角近似线性减小；f 大于 0.125 时相位角就互不相关，并在 $-\pi$ 到 π 之间随机分布。另外，对于特别高耸的输电塔线体系应注意的是，随着高度的增加

湍流的均匀性不断增加，在高度约 200m 以上，对 f 小于 0.125 时的相位角基本上为零，f 大于 0.125 时仍是在 $-\pi$ 到 π 之间随机分布。其建议的经验公式为

$$\phi(\omega)=\begin{cases}0.25\pi f & f\leqslant 0.1\\ -10\pi f+1.25 & 0.1<f\leqslant 0.125\\ random & f>0.125\end{cases} \tag{3-10}$$

式中，f 为 Monin 坐标。

对输电塔，可仅考虑风场的垂向相关性，相干函数为

$$Coh_{ij}(\omega)=\exp\left(-\frac{2nC_z|z_j-z_i|}{\overline{U}(z_i)+\overline{U}(z_j)}\right) \tag{3-11}$$

对输电导线可分为两种情况，当输电线路位于平坦地段，导线在跨度内落差和垂度较小时，可认为导线位于同一水平高度，沿输电线路方向各离散点具有相同的平均风速和风速谱，即

$$S_{11}(\omega)=S_{22}(\omega)=\cdots=S_{mm}(\omega)=S(\omega) \tag{3-12}$$

此时

$$S_{ij}(\omega)=\sqrt{S_{ii}(\omega)S_{jj}(\omega)}Coh_{ij}(\omega)\exp[i\phi(\omega)]=S(\omega)Coh_{ij}(\omega)\exp[i\phi(\omega)] \tag{3-13}$$

当离散点等间距布置时，间距 $r_{ji}=|y_j-y_i|=\Delta|j-i|$ 为常数，代入式（16）整理得

$$Coh_{ij}(\omega)=\left[\exp\left(-\frac{nC_z\Delta}{\overline{U}(z)}\right)\right]^{|j-i|}=(\cos\alpha)^{|j-i|} \tag{3-14}$$

式中，$\cos\alpha=\exp\left(-\frac{nC_y\Delta}{\overline{U}(z)}\right)$。

当输电线路位于陡坡地段或杆塔不等高时，导线落差梯度较大，或由于导线跨距较大，导线垂度效应明显时，各离散点的平均风速和风速谱均不相同，应同时考虑侧向和垂向相关性。此时其相干系数为

$$Coh_{ij}(\omega)=\exp\left(-\frac{2n\sqrt{C_y^2(y_j-y_i)^2+C_z^2(z_j-z_i)^2}}{\overline{U}(z_i)+\overline{U}(z_j)}\right) \tag{3-15}$$

式中，C_y，C_z 为指数衰减系数，Simiu 建议［27］，$C_y=16$；$C_z=10$。

6.3.1.4 作用于结构的风荷载

在已知风向、风速时程和结构类型的情况下，作用于结构的风荷载可表示为

$$F_D(t)=\frac{1}{2}\rho AC_D(\alpha)[U(t)]^2 \tag{3-16}$$

式中，ρ 为空气密度；A 为有效迎风面积；$C_D(\alpha)$为攻角 α 对应的阻力系数；$U(t)$为风速时程。

6.3.1.5 雨荷载模型的建立

(1) 雨的分类

降雨的大小是以降雨量来衡量的。一般地，按 24h 降雨量（指降雨在地面上液态水的厚度）可把降雨分为 7 个等级，如表 1 所示［30］。

降雨强度等级的划分　表 7

等　级	微量	小雨	中雨	大雨	暴雨	大暴雨	特大暴雨
降雨量（$mm \cdot 24h^{-1}$）	<0.1	0.1-10	10-25	25-50	50-100	100-200	>200

（2）降水粒子的谱分布

雨滴在空气中降落时，其形状由大小决定。半径小于 140μm 的雨滴可认为是球形，随着尺度的增加，逐渐变成椭球体和平底椭球体，当半径大于 6mm 时，球体会破碎。通常利用雨滴的等效直径（同体积球所具有的直径）D_0 来描述雨滴的大小，其分布函数可写成 $n(D_0)$。大量观测表明，雨滴谱一般服从负指数分布，使用广泛的是马歇尔－帕尔默指数分布，简称 M-P 分布或 M-P 谱［30］，即

$$n(D)=n_0\exp(-\Lambda D) \tag{3-17}$$

式中，$n_0=8\times10^3$ 个 $\cdot\, m^{-3}\cdot mm^{-1}$；$\Lambda$ 为斜率因子，$\Lambda=4.1I^{-0.21}$，这里 I 是雨强（mm/h）；D 为雨滴直径（mm）。

（3）雨滴降落的末速度

在重力作用下，水滴的下落速度不断增加，与此同时，空气阻力也随之增加，重力和阻力达到平衡时，水滴匀速下降，此时的下降速度成为水滴的末速度 V_m。文献［16］将各种公式与实测资料比较发现，当雨滴直径 $D<1.0$mm 时，式（3-7）与实测数据吻合较好；当 $1.0\text{mm}<D<3.0$mm 时，式（3-8）较接近；当 $D>3.0$mm 时，式（3-9）在应用时较好，即

$$V_m=10^6\left(\frac{0.787}{D^2}+\frac{503}{\sqrt{D}}\right)^{-1},D<1.0\text{mm}; \tag{3-18a}$$

$$V_m=(17.2-0.844D)\sqrt{0.1D},1.0\text{mm}<D<3.0\text{mm}; \tag{3-18b}$$

$$V_m=\frac{D}{0.113+0.0845D},3\text{mm}<D<6.0\text{mm}; \tag{3-18c}$$

（4）雨滴冲击力的计算

雨滴速度在与结构碰撞的极短时间内变为零，这是一个动量变化过程，雨滴与结构之间的相互作用过程遵循牛顿第二定律。设雨滴碰撞前的末速度为 V_s，质量为 m，雨滴在与结构碰撞的极短时间 τ 内速度为零。

由动量定理知

$$\int_0^{\tau}f(t)\mathrm{d}t+\int_{V_s}^{0}m\mathrm{d}v=0 \tag{3-19}$$

式中，$f(t)$ 为单个雨滴撞击力矢量；v 为雨滴速度矢量。雨滴在时间 τ 内对结构的撞击力 $F(\tau)$ 为

$$F(\tau)=\frac{1}{\tau}\int_0^{\tau}f(t)\mathrm{d}t=\frac{mV_s}{\tau} \tag{3-20}$$

降雨对输电塔线体系作用以中雨为荷载基准，这是由于通常小于中雨的降雨对结构的附加作用不超过 5%［16］。假设雨滴下落时为一球体，$m=\rho\pi d^3/6$，单个雨滴冲击力可表示为

$$F(\tau)=\frac{mV_s}{\tau}=\frac{1}{6\tau}\rho\pi d^3V_s \tag{3-21}$$

式中，$F(\tau)$ 为单个雨滴冲击力；ρ 为雨滴的密度；τ 为作用时间；d 为雨滴直径；V_s为雨滴作用于结构的速度。要将式（4）表示的雨滴荷载转化为结构分析的均布荷载，可表示为

$$F_d=F(\tau)/A\cdot b\cdot\alpha \tag{3-22}$$

式中，A 为作用面积，$A=\pi d^2/4$；b 为结构迎雨面的宽度；τ 为作用时间，取 $\tau=d/2V_s$；α 为降雨密度系数，雨滴直径在［d_1，d_2］范围内，单位体积雨滴个数为 $n=\int_{d_1}^{d_2}n(r)\mathrm{d}r$，则降雨密度系数 $\alpha=\pi d^3n/6$。

在实际应用中，降雨对结构的作用力可按竖直方向与顺风向两个方向考虑，在竖直方向，V_s 取雨滴在无风情况下的下落末速度 V_m；在顺风方向，V_s 可取顺风向风速。

6.3.1.6 风（雨）荷载的组合原则

为便于结构分析，风（雨）荷载按以下原则组合：

（1）风荷载为主要荷载，降雨只作为风的附加荷载，即只考虑风和雨的共同激励作用，不单独将降雨作为独立的荷载形式，这是由风雨激励的成因决定的；

（2）根据降雨的作用机理，假定降雨对结构的作用是均匀的平均力，不考虑可能存在的脉动分量。结构总响应为两者的叠加。

6.3.2 结构动力响应方程

考虑结构的几何非线性变形，在恒载、平均风荷载、降雨和脉动风荷载作用下输电塔线体系的平衡方程的一般形式为

$$[M]\{\ddot{x}_{t+\Delta t}\}+[C]\{\ddot{x}_{t+\Delta t}\}+\{F^r_{t+\Delta t}\}=\{F^d_{t+\Delta t}\}+\{F^s\} \tag{3-23}$$

式中，$[M]$ 为质量矩阵；$[C]$ 为阻尼矩阵；$\{\dot{x}_{t+\Delta t}\}$，$\{\ddot{x}_{t+\Delta t}\}$ 分别为速度和加速度向量；$\{F^r_{t+\Delta t}\}$ 为恢复力向量；$\{F^d_{t+\Delta t}\}$ 为脉动风与气动阻尼力向量；$\{F^s\}$ 是恒载、平均风和降雨作用力向量；下标 $t+\Delta t$ 表示时间增量。

动力响应分析过程中，假定恒载、平均风、脉动风和降雨作用下的结构非线性刚度与恒载、平均风和降雨作用时等效，因此有

$$[M]\{\ddot{x}_{t+\Delta t}\}+[C]\{\dot{x}_{t+\Delta t}\}+[K]\{x_{t+\Delta t}\}=\{F^d_{t+\Delta t}\} \tag{3-24}$$

式中，$\{x_{t+\Delta t}\}$ 为响应的位移向量；$[K]$ 为静力等效刚度矩阵。

一般地，假定结构阻尼为粘性阻尼［31］，输电塔的阻尼比取 0.1%，导线和绝缘子的阻尼比取 0.4%。每次分析的持续时间为 800s，其中最后 600s 为计算有效时间。响应量的值可由 3 组模拟结果的总体平均得到。

6.3.3 输电塔线体系风（雨）作用动力分析

6.3.3.1 实例与分析模型

输电线路为 220kV 双回路线路，发生暴风雨灾害的地段位于零星松树覆盖的丘陵地貌区，连续倒塔三塔基（编号为 26#～28#）。输电塔均为 ZGU3 型格构式塔，呼称高

33m，总高度 49m，设计档距为 500m，实际档距布置为 380m＋260m＋260m＋530m。输电塔为角钢栓接结构，导线采用两根 LGJ-300/25 型钢芯铝绞线，GJ-50 型地线；绝缘子采用合成绝缘子串，高度 2.15m，与输电塔横担连接为垂直悬挂。灾害段输电塔线体破坏状况与结构布置分别如图 10 和图 11 所示，导线与地线的规格与力学性能指标如表 8 所示。

图 10　大连地区输电塔体系风雨致侧倾倒塌图

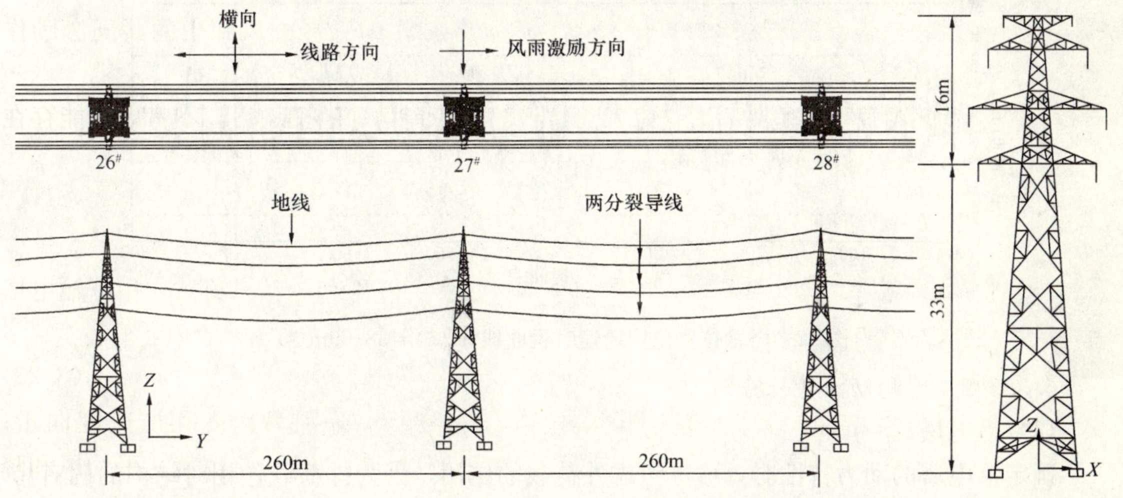

图 11　输电塔线体系结构与风雨激励图

导（地线）规格及力学性能指标　　表 8

名　称	导　线	地　线	备　注
类型	LGJ-300/25	GJ-50	
直径（m）	23.67E-3	9.0E-3	
横截面积（m^2）	333.3E-6	49.5E-6	
弹性模量（GPa）	65.0	181.42	其中：铝股 306.21E-6，钢股 27.1E-6
单位质量（kg/m）	1.058	0.424	
计算拉断力（kN）	83.41	60.56	
最大工作张力（kN）	31.7	16.2	
平均运行张力上限（kN）	19.8	9.6	

根据灾害段输电塔线体系结构构成，采用“三塔两跨”式结构体系模型。输电塔采用空间梁单元，单元的形状与规格与实际杆件相同，主节点刚性连接，次节点铰接；导

（地）线采用索单元，绝缘子串受初始张力作用，采用预应力杆单元，为分析方便，可将其模拟成截面为625mm²的方钢绝缘棒。绝缘子串与输电塔和导线的连接为两端铰接。导线在自重作用下的抛物线形垂度可按文献［32］给出的公式计算，即

$$s=\frac{Tr_0}{\mu}\left[\cosh\frac{\mu L}{2Tr_0}-1\right] \tag{3-25}$$

式中，Tr_0 为跨中张力，可取导线的最大工作张力计算；μ 为单位质量；L 为跨度；本文取 $s\approx3.0$m，且地线与导线一致。

参考高度为10m处的平均风速分别按设计标准和灾害地段实测风速取30m/s和35.5m/s，沿输电塔高度方向的平均风速按指数律剖面计算，剖面指数按荷载规范[33]取0.16。脉动风速的模拟采用Karman，考虑湍流的空间相关性，湍流密度取0.145，风速时程按文献［34］的谱叠加法模拟，每组风速时程的持续时间为800s，时间间隔为0.2s。图12为塔顶模拟风速时程曲线，其中，计算平均风速为38.59m/s。空气密度取0.120 kg/m³，输电塔杆件和电缆线的阻力系数均取 $C_D=1.0$。

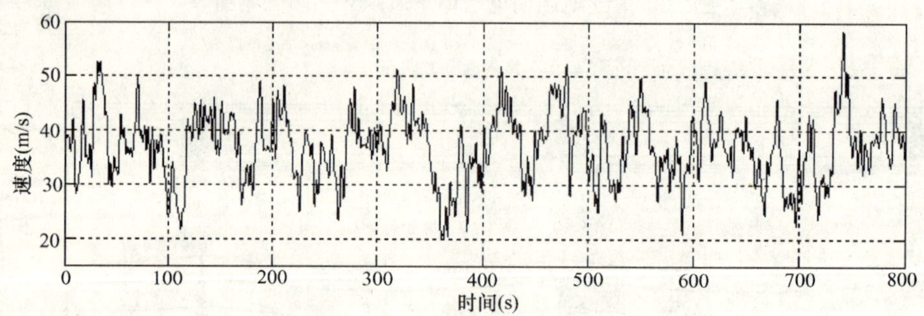

图12　塔顶位置模拟风速时程曲线($\overline{U}(z)=38.59$m/s)

6.3.3.2　结构动力特性分析

（1）输电塔

确定输电塔的动力特性时，可不考虑外荷载的作用，因为外荷载作用对结构的固有特性影响很小，频率接近于独立的量值。不挂线输电塔的各方向前三阶固有频率如表9。可以看出，除扭转方向外，相互垂直的两平面内的各阶频率比较接近，这是由于输电塔主结构为对称正方形，刚度分布均匀的缘故，微小差别则是线路垂直方向的横担质量分布所致。此外，随着模态阶数的增高，各方向的频率均有增加，且同阶各方向的频率相互接近，这是由于在高频模态的情况下，各方向振动的耦合性增强，模态阶数越高，这种特征表现得越明显。

输电塔固有模态频率（Hz）　　表9

	平面内	平面外	扭转方向
一阶模态	1.36	1.42	3.23
二阶模态	3.65	3.72	7.01
三阶模态	6.58	6.65	9.89

（2）输电电缆

在不考虑输电塔的影响，即将输电塔的刚度修正为远大于输电线路的刚度，此时分别

在恒载和恒载＋静力作用（平均风和降雨）两种工况下分析导（地）线的动力特性，其固有模态频率如表 10，括号中的数字表示恒载＋静力作用的值。可以看出，地线的各阶频率均略高于导线，这是由于导线单位长度质量大于地线单位长度质量，具有质量阻尼效应；在恒载和恒载＋静力作用这两种情况下，后者的各阶频率均略高于前者，这说明静力作用使得导（地）线的恒载初始预拉应力得到显著增强，从而使导（地）线的张力刚化效应得到体现，表现为频率的微小提高。分析还表明，输电线路的布置高度位置对其固有频率没有明显影响。

输电线路固有模态频率（Hz）　　**表 10**

输电线路类型	地线		导线	
	平面内	平面外	平面内	平面外
一阶模态	0.252（0.284）	0.254（0.287）	0.243（0.272）	0.246（0.277）
二阶模态	0.519（0.556）	0.521（0.563）	0.485（0.539）	0.497（0.546）
三阶模态	1.286（1.351）	1.295（1.367）	1.141（1.224）	1.149（1.230）

注：括号内的数值为恒载＋平均风和降雨作用时的频率。

（3）输电塔线体系

输电塔线体系的面外一阶和二阶振动模态如图 13 和图 14 所示，输电塔和导（地）线

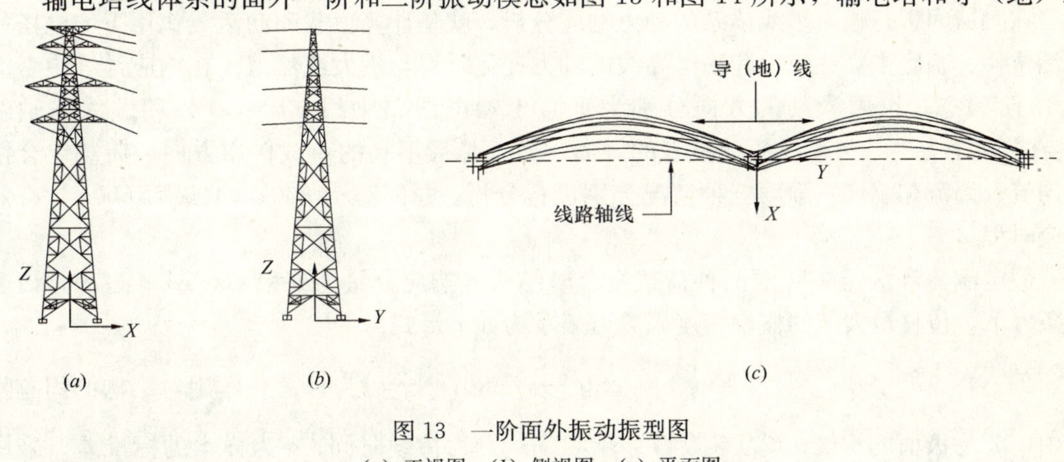

图 13　一阶面外振动振型图

(a) 正视图；(b) 侧视图；(c) 平面图

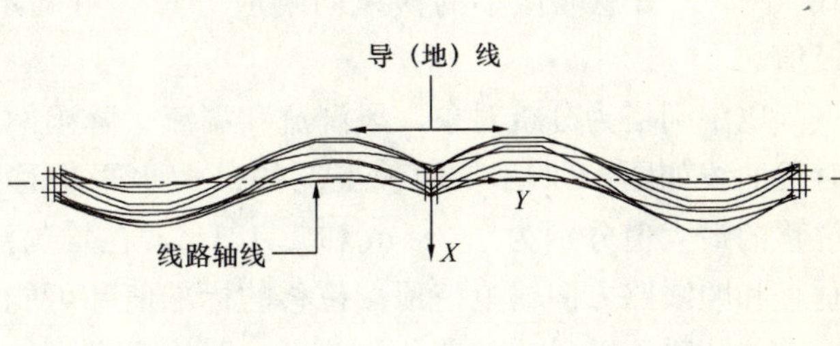

Z
X
(a)
Z
Y
(b)

图 14　二阶面外振动振型图

(a) 正视图；(b) 侧视图；(c) 平面图

的频率如表11。分析表明，输电塔线体系的振动特征值由输电线路到输电塔表现出由高到低排列的特征，这是由于输电塔与导（地）线的刚度差异形成的，导（地）线因其刚度低而具有低频特性。相比之下，输电塔则因刚度大而具有高频特性，这也是输电塔与导线的基频不能同时在某阶模态中完全得到反映的原因。由表4可以看出，输电塔线体系中输电塔与导线的频率均比两者的独立分析值略有减小，这与文献［35］的试验研究相吻合。对导线而言，导线与输电塔的耦合性有减弱导线振动的作用；对输电塔而言，导线、绝缘子等对输电塔的振动耦合性则表现为质量阻尼效应。

输电塔线体系固有模态频率（Hz） **表11**

输电线路类型	输电塔		导线	
	平面内	平面外	平面内	平面外
一阶模态	1.268	1.271	0.212	0.215
二阶模态	3.372	3.487	0.473	0.475
三阶模态	6.232	6.340	1.123	1.126

6.3.3.3 *动力响应分析*

(1) 时域分析

在时域内进行输电塔线体系的动力响应分析，就是计算恒载和动活载作用下输电塔的顶端位移、基底主立柱应力和输电导线的轴力变化时程与最大或然值。在响应计算中考虑前10阶模态，风雨激励的方向分别为垂直于输电线路轴线（$\alpha=90°$）和顺线路轴线（$\alpha=0°$）两个方向，与基频振动模态的方向一致，是最不利的荷载作用方向。荷载组合按均匀流、湍流和湍流＋降雨三种情况分别进行分析。总时程为800s，取最后600s为有效响应时程。

假定湍流和湍流＋降雨两种荷载组合时结构各响应分量为平稳Gaussian过程。该假定条件下，位移最大值的概率密度函数可表示为如下形式

$$F(\eta_e)=\exp\left[-\nu T\exp\left(-\frac{\eta_e^2}{2\sigma^2}\right)\right] \tag{3-26}$$

其中，η_e 为极值的均值；ν 为有效零穿越频率；T 为持续时间；σ 为样本的标准差。采用式（3-26）并忽略微小的横风向响应分量，可确定输电塔顶端位移脉动分量的标准差和极值。

以中间塔为分析对象，风湍流和湍流＋降雨两种作用情况时塔顶侧向水平位移分量的时程曲线如图15所示，两种工况的最大水平位移分别发生在时刻 $t_1=528$s 和 $t_2=566$s，位移的最大值分别为1.288m和1.481m。在均匀流、湍流和湍流＋降雨三种作用条件下垂直和顺线路方向输电塔顶端位移的标准值和可能最大值如表12。数据分析表明，位移的标准差几乎不随风速的变化而变化，这与风湍流的Gaussian平稳假定和定值湍流强度有关；均匀流和具有相同平均风速的湍流作用时可能位移极值有明显的差异，后者较前者的极值约增加近30%；风湍流与降雨共同作用，位移增加的趋势明显，随着降雨强度等级的提高，位移增幅可达到3%～8%；当输电塔线体系参考高度的平均风速达到35.5m/s灾害风速，降雨强度为大暴雨时，输电塔塔线体系发生强度或稳定性破坏，这与实际灾害状

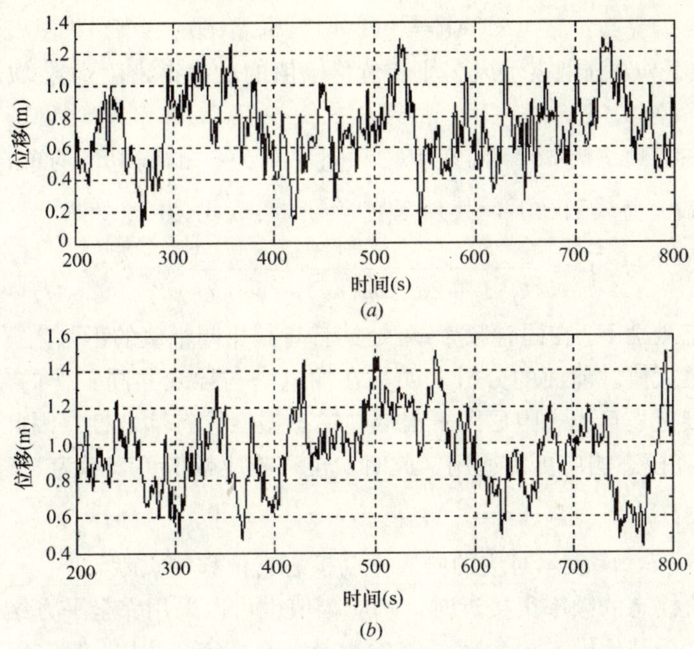

图 15　中间塔顶端侧向水平位移时程曲线

(*a*) 风湍流作用；(*b*) 风湍流与降雨共同作用

况相一致，同时也说明在考虑降雨的条件下，结构的抗力水平偏低；输电导线的位移极值随工况荷载的增加而略有增加，其增幅远小于输电塔的顶端位移，这一方面是由于导（地）线的工作张力储备充分，工作完全处于弹性范围内，另一方面是由于导（地）线的刚度均具有张力强化效应。需要说明的是输电塔主立柱应力和导（地）线的轴力在上述三种荷载组合作用下同样具有相同或相似的变化规律，这将在后续的强度与稳定分析中作进一步的阐述。

输电塔顶端与导线位移统计值（中间塔）　　**表 12**

荷载工况	输电塔				导　线	
	标准差(m)		可能位移极值(m)		可能位移极值(m)	
	YOZ 平面	XOZ 平面	YOZ 平面	XOZ 平面	X 方向	Z 方向
均匀流	0	0	0.89(1.13)	0.74(1.08)	4.18	1.37
湍　流	0.109(0.104)	0.105(0.107)	1.26(1.34)	1.19(1.37)	4.69	1.42
湍流＋降雨(暴雨)	0.113(0.106)	0.108(0.104)	1.35(1.43)	1.31(1.51)	5.12	1.53
湍流＋降雨(大暴雨)	0.102(—)	0.106(—)	1.47(—)	1.40(—)	—	—

注：表内括号中的数值为灾害风速对应的模拟工况值；“—”表示模拟工况不收敛，即结构破坏。

（2）频域分析

频域分析可采用两种方法，一种是在已进行时域分析的响应量数值描述的前提下，对响应量进行快速 Fourier 变换，得到响应量的谱密度函数；另一种是在已知传递函数和荷载激励的功率谱密度函数的情况下，直接求响应量的功率谱密度函数，此时，平稳随机激励的谱密度函数与位移响应之间的传递关系可表示为

$$S_x(\omega)=|H(\bar{\omega})|^2 S_{\text{fwind}}(\omega) \tag{3-27}$$

式中，$S_{\text{fwind}}(\omega)$和$S_x(\omega)$分别表示风湍流和位移幅值的功率谱密度函数；$H(\omega)$为频率响应函数。

对于具有模态圆频率为ω_j，刚度为k_j和阻尼比为ξ_j的输电塔的单自由度随机振动，在圆频率为$\bar{\omega}$的湍流荷载激励下，其频率响应函数可表示为

$$H(\bar{\omega})=\frac{1}{k_j[1+2i\xi_j(\bar{\omega}/\omega_j)-(\bar{\omega}/\omega_j)^2]},\xi_j>0 \tag{3-28}$$

假定广义模态激励F_{m_j}在固有频率ω_j处具有与其相同幅值的正弦激振力对应，便可确定该模态响应的最大值。前面的分析表明，在输电塔与导线共同工作的条件下，前10阶固有频率足够地接近，且模态阻尼因子很低，模态振型具有耦合性。因此，采用平方和开方法（SRSS）计算模态响应更为适用，此时，位移响应幅值可表示为

$$x=[x_1^2+x_3^2+x_7^2+x_9^2]^{1/2} \tag{3-29}$$

式中，$x_i(i=1,3,7,9)$表示对应的侧向振动模态对位移的贡献。

此外，当各阶模态的频率很接近时，响应幅值也可以采用完全平方结合法（CQC）计算，通过交叉项考虑其他模态振型对位移的贡献。在CQC法中，假定所有模态阻尼比均相同，即$\xi_j=0.2\%$，并与质量成比例，模态m和n之间的互相关系数ρ_{nm}可表示为[32]

$$\rho_{\text{nm}}=\rho_{\text{mn}}=\frac{8\xi^2(1+r)r^{3/2}}{(1-r^2)^2+4\xi^2(1+r)^2},0\leqslant\rho_{\text{nm}}\leqslant 1 \tag{3-30}$$

式中，$r=\omega_n/\omega_m$，且$\omega_m>\omega_n$。

设计风速条件下，湍流和湍流＋降雨两种荷载作用的输电塔线体系中间塔顶端侧向水平位移的功率谱密度图如图16所示。由于仅考虑湍流的脉动性，因此两种荷载作用的功

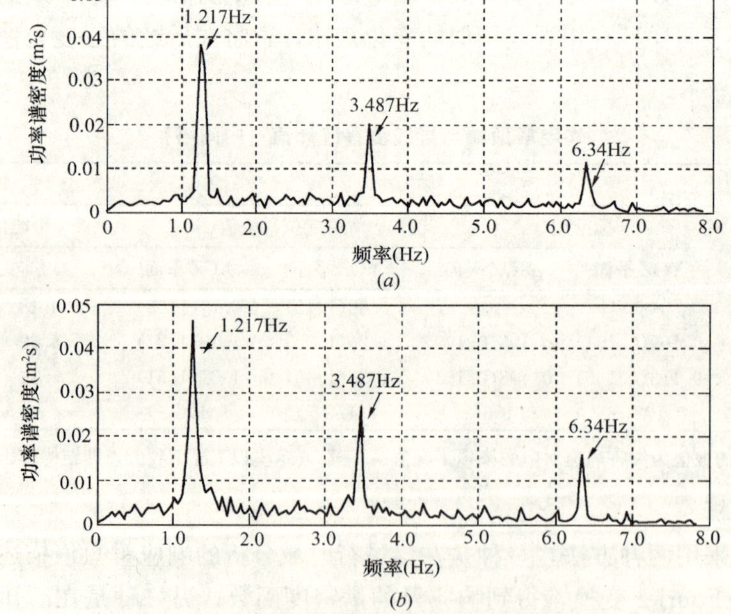

图16 中间塔顶端侧向水平位移功率谱密度

（a）风湍流作用；（b）湍流＋暴雨

率谱密度相似，不同之处在于后者较前者具有更高的能量分布，峰值略有提高。采用SRSS法和CQC法得到的顶端最大位移幅值如表13。可见，两种方法的计算结果十分接近，且与时程分析的统计值相吻合，说明两种方法对输电塔线体系的响应量分析具有很好的适用性。

基于频域分析的输电塔顶端侧向位移最大值（中间塔）　表 13

荷载工况	设计风速		灾害风速	
	SRSS法	CQC法	SRSS法	CQC法
湍流	1.271	1.286	1.426	1.430
湍流+降雨（暴雨）	1.479	1.485	1.527	1.533

6.3.3.4　强度与稳定性分析

强度分析是衡量输电塔线体系在风雨激励作用下是否出现强度破坏的重要依据。强度分析包括输电塔主杆件和输电导线两部分，分别代表两种不同的破坏成因和方式。输电塔主杆件的破坏属于结构性破坏，是结构体系破坏的直接原因；导线拉断属于局部性破坏，会改变塔线体系的动力平衡，导致结构体系破坏，是结构体系完全瘫痪的间接原因。然而，从电网运行安全的角度出发，无论哪种破坏形式出现，均表明灾害发生，是结构设计的控制性指标。这里，分别按设计风速和实际风速条件下，并考虑降雨的荷载作用进行分析判断。

输电塔主杆件应力分析主要考虑基底主杆件中的弯曲应力与拉压应力的组合，这是因为主杆件为拉弯或压弯杆件。根据主杆件的受力特征，其应力控制条件可表示为

$$\frac{N_{\max}}{A_S}+\frac{M_{\max}}{W_X}\leqslant[\sigma] \tag{3-31}$$

式中，$N_{\max}$和$M_{\max}$分别为基底处主杆件的轴力与弯矩的极大值，应该注意的是在动力响应分析中两者对应的时刻应保持一致；A_S和W_X分别为主杆件的横截面积和截面抗弯模量；$[\sigma]$为杆件设计应力允许值，可按杆塔设计规范采用[36]。不同荷载组合条件下主立杆件的强度分析如表14。

输电塔基底主立柱杆件强度分析　表 14

荷载工况	设计风速			灾害风速			允许应力(MPa)
	最大轴力(N)	最大弯矩(N−m)	正应力(MPa)	最大轴力(N)	最大弯矩(N−m)	正应力(MPa)	
均匀流	120274.7	778.9	59.34	143515.1	927.8	70.77	315
湍流	153963.6	987.8	75.78	179481.8	1146.5	88.23	315
湍流+降雨（暴雨）	158546.7	1023.7	78.2	185676.7	1176.6	91.1	315
湍流+降雨（大暴雨）	164852.2	1145.1	82.9	—	—	—	315

注：$A_S=2.737\times10^{-3}\mathrm{m}^2$；$W_X=5.058\times10^{-5}\mathrm{m}^3$。

由表14的数据可以看出，均匀流作用和湍流作用时主立柱杆件的轴力与弯矩最大值具有明显的差异，两者间的相对差接近30%，湍流作用与湍流+降雨作用的差别相对较

小，在3%—6%之间；主立柱杆件的受力特征主要表现为轴力，弯矩所占的比重相对较小，说明输电塔杆件更接近于桁架杆件；从控制应力与截面应力的比较来看，杆件的工作应力远小于截面控制应力，因此可以推断，输电塔在灾害风速和降雨作用下的破坏并非为杆件的强度破坏，而应该是局部杆件的动态受压失稳所致。

输电导（地）线的强度分析可按送电线路设计规程[4]对运行张力工作率上限值的控制得到保证，而不具体分析输电线的截面正应力，即

$$T_{max} \leqslant [T_{up}], \text{或} T_{max}/R_n \leqslant [\gamma] \tag{3-32}$$

式中，T_{max}为导(地)线的最大运行张力；$[T_{up}]$为规范允许的运行张力上限；R_n为名义张力强度；$[\gamma]$为规范允许的工作率。不同工况条件下导(地)线的强度分析如表15。

导（地）线的张力变化范围 表15

输电线路类型	工作张力 T（kN）		名义张力强度 Rn（kN）	工作率T/Rn（%）
	最小值	最大值		
导线	14.38	19.75	83.41	17.2～23.7
地线	10.07	13.18	60.56	16.6～21.8

由表15的数据可以看出，不论是设计荷载还是灾害荷载作用，输电线路的运行张力最大值均不超过名义张力强度的25%，满足送电线路设计规程[4]的要求。这说明输电塔线体系连续倒塌的直接原因并非导（地）线的拉断所造成的，这与该线路破坏后只有输电线路脱落，而未出现断裂的实际状况相吻合，从而排出了输电线路的拉断导致输电塔线倒塌的可能性。

6.4 输电塔线体系风（雨）致疲劳损伤研究

6.4.1 结构疲劳损伤准则

在随机荷载作用下，结构疲劳损伤采用疲劳累积损伤理论，即Palmgren-Miner[37]线性疲劳损伤准则。该准则认为：确定某种材料在常应力交变荷载作用下的疲劳特性，通常以试验杆件出现可见开裂为特征。杆件在不同的应力水平下，疲劳统计特征可表示为S-N曲线形式，即

$$NS^m = k \tag{4-1}$$

式中，S为应力幅值S_a；N为疲劳寿命（或疲劳循环数）；m和k为经验常数。应用表明，疲劳寿命N对于大多数材料服从对数正态分布或Weibull分布。

假定随机脉动荷载引起的疲劳过程可以模拟成一个序列，则每个连续变幅值应力过程对材料产生总累积疲劳损伤D可表示为

$$D = \frac{T\eta}{k}E[s^m] = \frac{T\eta}{k}\int_{-\infty}^{+\infty} s^m f_s(s)\mathrm{d}s \tag{4-2}$$

式中，T为结构的总的疲劳寿命；η为峰值期望率；$f_s(s)$为应力幅值或应力峰值的概率密

度函数。应力响应分布函数及其期望值可按文献[38]给出的类型和方法计算。

6.4.2 结构疲劳可靠性分析

6.4.2.1　基本假定

为分析方便，提出如下假定：

（1）假定风荷载的脉动分量为沿高度和宽度均匀变化的随机过程；

（2）假定塔架由理想化的三维空间杆件组装而成；

（3）疲劳损伤以初始裂纹的扩展为基础，同 Palmgren-Miner 准则的规定相一致；

（4）输电塔的随机振动为线性振动。

6.4.2.2　疲劳失效概率 P_F 的确定

（1）完全分布法

确定结构疲劳设计标准，应首先确定设计应力范围 S 和疲劳使用寿命 N_D（疲劳循环数），并保证疲劳破坏发生的概率最小。如果已知 N 的概率分布函数，则 N_D 可由响应量的接受设计可靠性概率 $P\ [N>N_D]\ =1-P_f$ 确定。根据 Weibull 概率准则[37, 39]，疲劳寿命 N 分布函数可由 Weibull 分布描述

$$P[N > x] = F_N(x) = 1 - \exp[-\left(\frac{x}{u_N}\right)^{\alpha_N}] \tag{4-3}$$

式中，u_N 和 α_N 分别为尺度和形状参数。此时疲劳损伤的可接受概率 p_f 对应的设计寿命为

$$N_D = u_N\,[\ln\frac{1}{(1-p_f)}]^{1/\alpha_N} \tag{4-4}$$

如果近似地取 $\ln(1/\,1-p_f)\cong p_f$，则式（4-4）可以写成

$$N_D \cong u_N\ (p_f)^{1/\alpha_N} \tag{4-5}$$

此时，则 N 的均值可表示为

$$E[N] = u_N\Gamma\left(1+\frac{1}{\alpha_N}\right) = N_D\Gamma\left(1+\frac{1}{\alpha_N}\right)(p_f)^{-1/\alpha_N} \tag{4-6}$$

式中，$\Gamma(\cdot)$ 为 Gamma 函数，根据式（4-5），结构或杆件的设计寿命 M_D 的均值应满足

$$M_D = \frac{E[M]}{\gamma} \tag{4-7}$$

式中，γ 为疲劳安全因子；$E[M]$ 为结构的疲劳寿命期望值；M_D 为结构的设计使用寿命。

根据式（4-6）和式（4-7），结构或杆件在规定设计使用期内疲劳失效概率可由以下公式计算

$$P_f = [\gamma/\,\Gamma\left(1+\frac{1}{\alpha_N}\right)]^{-\alpha_N} \tag{4-8}$$

（2）一次二阶矩法

在结构使用期内，对具体的平均风速 U 随机环境荷载造成的疲劳损伤期望值可表示为

$$E[D] = \int_U \frac{T_U\eta_U}{k}E[\overline{S}_U^m]f_U(U)\mathrm{d}U \tag{4-9}$$

式（4-9）可以写成求和的形式

$$E[D]=\frac{1}{k}\sum T_{\mathrm{U}}\eta_{\mathrm{U}}E[\overline{S}_{\mathrm{U}}^{\mathrm{m}}]n_{\mathrm{U}} \tag{4-10}$$

式中，$E[D]$ 为年损伤期望；T_{U} 为速度为 U 的风暴持续时间；n_{U} 为 $\mathrm{d}U$ 内的风暴数；k 为特定风速度范围内的风暴总数。在式（4-10）中，令 $\overline{S}=BS$，其中，S 为应力范围的最佳估计，B 为应力分析中不确定性因素的量化随机变量，则式（4-10）可以写成

$$E[D]=\frac{B^{\mathrm{m}}}{k}T_{\mathrm{U}}\eta_{\mathrm{U}}E[S^{\mathrm{m}}]n_{\mathrm{U}} \tag{4-11}$$

假定结构的总损伤等于或大于 Δ（随机变量），结构发生破坏，令 $\psi=[\sum T_{\mathrm{U}}\eta_{\mathrm{U}}E\ [S^{\mathrm{m}}]\ n_{\mathrm{U}}]^{-1}$，则可以得到结构的总寿命 M

$$M=\frac{k\psi\Delta}{B^{\mathrm{m}}}=kB^{-\mathrm{m}}\psi\Delta \tag{4-12}$$

定义疲劳极限状态函数 Z 为变量 M 和 M_{D} 的函数，假定两变量均为对数正态分布的随机变量，则

$$Z=M-M_{\mathrm{D}} \tag{4-13}$$

在式（4-13）中，$Z\leqslant0$ 表示结构失效，则其失效概率可表示为

$$P_{\mathrm{f}}=P(Z\leqslant0) \tag{4-14}$$

利用式（4-12）和式（4-14），极限状态也可以写成

$$l_{\mathrm{n}}\left(\frac{\psi k\Delta}{B^{\mathrm{m}}M_{\mathrm{D}}}\right)=0 \tag{4-15}$$

根据一次二阶矩理论，结构的可靠性指标 β 可表示为

$$\beta=\frac{\mu_Z}{\sigma_Z}=\frac{l_{\mathrm{n}}\psi+\mu_{\ln\Delta}+\mu_{\mathrm{lnk}}-m\mu_{\mathrm{lnB}}-\ln M_{\mathrm{D}}}{\sqrt{(\sigma_{\mathrm{lnk}}^2+m^2\sigma_{\mathrm{lnB}}^2+\sigma_{\ln\Delta}^2)}} \tag{4-16}$$

可见，对于假定的 M_{D} 值，可采用完全分布法和一次二阶矩法计算结构的可靠性指标和失效概率。

6.4.2.3 结构体系疲劳可靠度计算方法

格构式输电塔是由单根杆件拼装而成的，进行结构可靠度分析时可将输电塔视为结构体系，而将单根杆件或塔架的节段单元作为结构体系可靠度分析的单元或子结构，则结构体系的可靠度可由单元或子结构可靠度指标按系统可靠度的计算规则计算[40]。该方法将结构体系分为两类，即串联体系和并联体系。串联体系认为：如果组成结构的杆件或子结构的任何一部分失效，则结构体系失效；而并联体系则认为：只有组成结构的杆件或子结构全部失效时，结构体系失效。事实上，结构体系往往是串联与并联体系的结合体。

对串联体系而言，

$$R_{\mathrm{sys}}=\prod_j R_j \tag{4-17}$$

对并联体系而言

$$R_{\mathrm{sys}}=1-\prod_j\ (1-R_j) \tag{4-18}$$

式中，R_{sys} 为结构体系的可靠度；R_j 为第 j 根组成杆件或第 j 个子结构的可靠度。

格构式输电塔的结构失效可理解为：任何层（即节段单元）的失效都将导致输电塔结构失效；每层中如果少于三根立柱的有效支撑，则所有层的有效支撑失效。基于上述理解，则输电塔的可靠性分析包括以下几个步骤：

（1）计算第 j 层立柱的失效概率 P_{fj}^{c}。对于所考虑的环境荷载形式，在具体层内的所有柱式杆件失效概率相同，因此，层内立柱节点（即杆件的上下端节点）的失效概率可表示为

$$P_{fj}^{c}=1-\prod_{i=1}^{2}(1-P_{fij}^{I}) \tag{4-19}$$

式中，P_{fij}^{I} 为第 i 根立柱与第 j 层相交节点的失效概率。

（2）计算第 j 层 P_{fj}^{L} 的失效概率。由于每层由 4 根主立柱杆件组成，层的失效是 1 根以上的立柱失效的结果。此时，层的失效概率由双正态分布描述为[41]

$$P_{fj}^{L}=\begin{bmatrix}4\\4\end{bmatrix}[P_{fj}^{C}]^{4}+\begin{bmatrix}4\\3\end{bmatrix}[1-P_{fj}^{C}][P_{fj}^{C}]^{3}+\begin{bmatrix}4\\2\end{bmatrix}[1-P_{fj}^{C}][P_{fj}^{C}]^{2} \tag{4-20}$$

（3）确定结构整体失效概率 P_{f}^{sys}。

具有 N 层的输电塔，结构整体失效概率等于任何一层失效的概率，此时结构相当于串联体系，其失效概率为

$$P_{f}^{sys}=1-\prod_{j=1}^{N}(1-P_{fj}^{L}) \tag{4-21}$$

因此，输电塔的可靠性概率可以写成如下形式

$$R_{sys}=1-P_{f}^{sys} \tag{4-22}$$

6.4.3 可靠度评价与因素分析

6.4.3.1 输电塔结构与模拟环境的描述

现以某实际输电塔结构设计为例，在模拟环境荷载作用下，采用上述方法进行疲劳可靠分析，并考察相关因素对结构疲劳可靠性指标的影响。输电塔为 SZT1（4）型双回路直线塔，结构如图 17 所示（单位：cm），输电塔线体系的输电等级为 500kV 线路，分析跨度为 550m，其结构杆件、导线与地线的构成如表 16 所示。

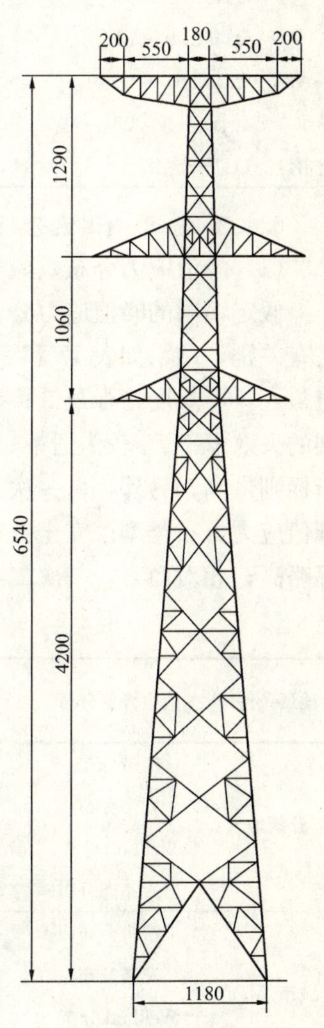

图 17　输电塔

假定平均风速沿高度变化服从指数型剖面，指数系数按我国现行荷载规范[33]对于开阔地、城市郊区和城市中心分别取 0.12、0.22 和 0.30。通常情况下，与结构疲劳可靠性相关的风速范围在 15m/s 和 45m/s 之间，风荷载的确定可根据《110—500kV 架空送电线路设计技术规程》[4]给出的方法计算。

输电塔架与线路的结构组成 表 16

节段	主要杆件	次要杆件
1	Q345L90×7,Q345L90×8	L63×5,L56×4,L70×5,L40×3,L40×4,Q345L140×10
2	Q345L75×6,Q345L100×8	L56×4,L50×4,L40×3,L40×4,Q345L140×10
3	Q345L75×6,Q345L90×7	L63×5,L56×4,L45×4,L40×3, Q345L140×10
4	Q345L90×7	L90×8,L90×7,L80×6,L75×6,L70×5,L63×5,L56×4,L45×4,L40×3,Q345-8×80
5	Q345L125×8	L90×8,L90×7,L80×6,L75×6,L70×5,L45×4,L40×3,Q345L140×10,Q345-8×575,Q345-8×630
6	Q345L125×10	L70×5, L40×3,Q345L140×10
7	Q345L140×12	L100×8,L90×7,L80×6,L75×6,L40×3,Q345-8×655
8～13	Q345L160×12(14,16)	L80×6,L75×6,L63×5,L56×4,L50×4,Q345L180×12(14)
导线	4×LGJ-400/35	
避雷线	LGJ-185/43	

6.4.3.2 影响因素分析

(1) 峰值应力分布效应

假定不同的峰值应力分布，得到结构的年疲劳损伤、疲劳失效概率和疲劳可靠度指标略有差别，结果如表 17 和表 18 所示。比较分析可知，窄带假定得到的疲劳失效概率 P_f 值最大，可靠度指标最低；Weibull 分布与窄带分布假定得到的 P_f 值相接近；宽带分布得到的失效概率 P_f 较为适中；对数正态分布得到的 P_f 值略低于宽带分布得到的值，可靠度指标则最高；另外，由一次二阶矩法和完全分布法确定的 P_f 值不同，其变化程度取决于峰值应力分布类型。完全分布法与一次二阶矩法相比，一般得出的 P_f 值较低，可靠性指标略高；相对而言，一次二阶矩法偏于保守。

地形条件对疲劳可靠度指标的影响 表 17

地形条件	应力峰值分布	年疲劳损伤	疲劳失效概率		疲劳可靠度指标	
			一次二阶矩法	完全分布法	一次二阶矩法	完全分布法
开阔地	窄带分布	5.361×10^{-4}	3.953×10^{-5}	2.212×10^{-5}	3.72	3.87
	宽带分布	4.078×10^{-4}	2.796×10^{-4}	2.033×10^{-4}	3.81	3.89
	Weilbull 分布	5.206×10^{-4}	3.529×10^{-4}	1.782×10^{-4}	3.75	3.93
	对数正态分布	2.948×10^{-4}	4.789×10^{-5}	4.377×10^{-5}	4.24	4.27
城市郊区	窄带分布	6.495×10^{-4}	4.561×10^{-4}	2.808×10^{-4}	3.68	3.81
	宽带分布	5.187×10^{-4}	3.497×10^{-4}	2.123×10^{-4}	3.75	3.88
	Weilbull 分布	5.111×10^{-4}	4.103×10^{-4}	2.2315×10^{-4}	3.71	3.86
	对数正态分布	4.987×10^{-4}	2.583×10^{-4}	1.633×10^{-4}	3.84	3.95
城市中心	窄带分布	1.603×10^{-3}	1.449×10^{-3}	8.217×10^{-4}	3.35	3.52
	宽带分布	1.352×10^{-3}	8.420×10^{-4}	5.283×10^{-4}	3.51	3.64
	Weilbull 分布	1.556×10^{-3}	1.409×10^{-3}	7.823×10^{-4}	3.36	3.53
	对数正态分布	7.312×10^{-4}	4.749×10^{-4}	2.411×10^{-4}	3.67	3.85

降雨条件对疲劳可靠度指标的影响　**表 18**

荷载条件	峰值应力分布	年疲劳损伤	疲劳失效概率		疲劳可靠度指标	
			一次二阶矩法	完全分布法	一次二阶矩法	完全分布法
风荷载单独作用	窄带分布	5.119×10^{-4}	3.996×10^{-4}	2.583×10^{-4}	3.73	3.84
	宽带分布	3.984×10^{-4}	2.317×10^{-4}	1.141×10^{-4}	3.86	4.01
	Weilbull 分布	4.654×10^{-4}	3.679×10^{-4}	2.399×10^{-4}	3.74	3.85
	对数正态分布	3.131×10^{-4}	6.635×10^{-5}	3.522×10^{-5}	4.17	4.32
风雨荷载共同作用	窄带分布	2.782×10^{-3}	2.635×10^{-3}	2.023×10^{-3}	3.17	3.25
	宽带分布	1.581×10^{-3}	1.420×10^{-3}	1.012×10^{-3}	3.36	3.46
	Weilbull 分布	3.215×10^{-3}	3.069×10^{-3}	2.154×10^{-3}	3.12	3.23
	对数正态分布	8.677×10^{-4}	8.139×10^{-4}	4.741×10^{-4}	3.52	3.67

（2）地形条件影响

对表 17 中的数据分析还表明，地形条件对输电塔的疲劳累积影响不容忽视。就城市中心而言，疲劳失效概率 P_f 值高于城郊地貌，同样城郊地貌也高于开阔地地形条件。这说明；随着地面粗糙度系数的增加，所在地形对近地风速有减弱的趋势，对输电塔线体系的作用相对减弱，从而可靠性指标得到增强。从表 17 可以看出，仅按可靠性指标评价，该塔架在开阔地条件下满足一级安全等级对可靠性指标的要求（$\beta\geqslant3.7$），而在城市中心区域只能满足二级安全等级的可靠性指标的要求（$\beta\geqslant3.2$）。可见，地形条件对疲劳损伤可靠性具有显著影响。

（3）降雨条件的影响

由表 18 中的数据分析可知，考虑降雨的荷载条件对输电塔的疲劳有显著影响，在其他条件保持不变的情况下，单独按风荷载设计的输电塔，其疲劳可靠性指标满足安全等级为一级的要求（$\beta\geqslant3.7$），而考虑风雨共同作用的情况，则输电塔的疲劳可靠性指标满足的安全等级有所降低。这在某种程度上也揭示了输电塔线体系在风雨同时作用的暴风雨中偶有倒塌的现象。

6.5　结论

通过对输电塔线体系抗震、抗风（雨）激励的动力响应和结构疲劳可靠度等方面的基础理论研究，在荷载特性和结构体系特征方面得出如下结论：

（1）在地震作用下，输电塔线体系整体简化动力分析模型在兼顾体系剪切刚度和弯曲刚度的同时，可以在很大程度上提高计算效率，简便可行；对常用输电塔线体系理论分析和试验验证表明，导线对输电塔的动力响应贡献程度会随杆塔档距的增加而增大，且随着场地条件的变软亦有增加的趋势，其附加效应不容忽视；所提出的结构体系简化计算方法具有较高的工程应用精度，可供设计规范修订参考。

（2）在风雨荷载作用下，输电塔线体系的动力特性、动态响应规律和结构构成的强度变化规律主要表现为：（i）输电塔线体系的固有频率与独立进行输电塔和导线分析的固有频率相比，输电塔和导线的固有频率均略有减小，这是由于塔线运动耦合的结果；同时输电塔线体系在稳态荷载作用下的固有频率较恒载单独作用时略有增加，表现出输电线张力刚化效应。(ii) 在均匀流、湍流和湍流+降雨三种不同荷载条件下，输电塔线体系的结构响应量呈明显增加的趋势，表明结构的荷载响应湍流有别于均匀流，降雨作用对响应量具有明显的激励贡献。(iii) 结构体系的动力响应主要来源于低阶模态贡献，且基频模态的贡献占绝对比重，采用前几阶模态响应量的组合计算具有足够的工程实用精度。(iv) 强度与稳定分析表明，通常情况下，输电塔和输电线路的强度储备均满足设计强度要求。结构体系的连续倒塌破坏是风雨共同激励，导致输电塔局部杆件动态受压失稳造成的。为防止异常气候条件下环境荷载的倒塌灾害，应有效地增强关键部位主立柱杆件的受压稳定性。

(3) 在风雨荷载作用下，以结构疲劳损伤和设计使用寿命为控制条件的输电塔架疲劳可靠性分析表明：(i) 应力峰值分布假定对疲劳可靠性指标有明显影响。窄带分布和Weibull分布得到的失效概率相接近且较大，对数正态分布得到的失效概率较小，宽带分布得到的失效概率适中；可靠性指标则与失效概率的变化规律相反。(ii) 一次二阶矩法得到的失效概率通常大于完全分布法的值，前者作为设计依据偏于保守。(iii) 地形条件对输电塔的失效概率的影响也呈现出规律性的变化，随平均风剖面指数系数的增加，杆塔失效概率增大，疲劳可靠性指标降低，且与地形的表面粗糙度的变化相关。(iv) 在输电塔疲劳分析中考虑降雨参与作用时，疲劳失效概率大于风荷载单独作用的情况，其可靠性指标降低一个安全等级（仅就可靠度而言）。

参考文献

[1] ASCE. Committee on Electrical Transmission Structures: Loadings for Electrical Transmission Structures, ASCE Journal of Structural Division. 1982, 108(5).

[2] ASCE: Guideline for Electrical Transmission Line Structural Loading, ASCE Manuals and Reports on Engineering Practice, 1991No. 74, New York.

[3] Moze J. D. et al. Longitudinal Load Analysis of Transmission Line Systems, IEEE Transactions on Power Apparatus and Systems, Vol. PAS—96, NO. 5, 1977.

[4] 国家电力公司华东电力设计院. 110－500架空送电线路设计技术规程[S]. 北京：中国电力出版社，1999.

[5] 李宏男，王前信. 大跨越输电塔体系的动力特性[J]. 土木工程学报，1997，30(5)：28～36.

[6] Li Hong-Nan, Wang Qianxin, M. P. Singh. Seismic Response Analysis Method for Coupled System of Transmission Lines and Towers, Part I: Out of Plane. Earthquake Engineering and Engineering Vibration, 1996, 16(Sup): 67-75.

[7] Li Hong-Nan, Wang Qianxin, M. P. Singh. Seismic Response Analysis Method for Coupled System of Transmission Lines and Towers, Part II: In Plane. Earthquake Engineering and Engineering Vibration, 1996, 16(4): 37-45.

[8] 李宏男，王前信. 水平与摇摆地震动作用下大跨越输电体系反应分析[J]. 工程力学，1991，8(4)：68-69.

[9] Li Hongnan, Wang Suyan, Wang Qianxin. Response of Transmission Tower System to Horizontal

and Rocking Earthquake Excitations. Earthquake Engineering and Engineering Vibration，1997，17(4)：34-43.

[10] 李宏男，肖诗云. 纵向地震作用下输电塔相互作用体系分析[J]. 岩土工程报，1998，20(6)：102-104.

[11] Li Hong-Nan Xiao Shiyun. Model of transmission tower-pile-soil dynamic interaction under earthquake：in-plane. ASME PVP，2002，445(2)：143-147.

[12] AS 3995-1994. Australian standard for design of steel lattice towers and masts[S]. Standards Australia，North Sydney，Australia，1994.

[13] ASCE. Minimum design loads for buildings and other structures [S]. ASCE 7-98. Reston，VA，2000.

[14] 李宏男，白海峰. 高压输电塔-线体系抗灾研究的现状与发展趋势[J]. 土木工程学报，2007,40(2)：39～46.

[15] Li H. N.，Bai H. F. High-voltage transmission tower-line system subjected to disaster loads [J]. Progress in Nature science，2006，16 (9)：899-911.

[16] 李宏男，任月明，白海峰. 输电塔体系风雨激励的动力分析模型[J]. 中国电机工程学报，2007，27(30)：43-48.

[17] Rojiani K. and Wen Y. K. Reliability of steel buildings under winds. J. Struct. Div.，ASCE，1981，107(1)：203-221.

[18] Ditlevsen O. Random fatigue crack growth-a first-passage problem. Eng. Fracture Mech.，1986，23：467-477.

[19] Deoliya R. and Datta T. K. Fatigue Reliability Analysis of Microwave Antenna Towers due to Wind. J. Struct. Eng. 2001，127(10)：221-1230.

[20] 徐建波，邓洪洲，王肇民. 考虑非高斯和宽带修正的桅杆风振疲劳分析[J]. 同济大学学报. 2004，32(7)：889-893.

[21] 李宏男，石文龙，贾连光. 考虑导线影响的输电塔侧向简化抗震计算方法[J]. 振动工程学报，2003，16(2)：233-237.

[22] 李宏男，石文龙，贾连光. 导线对输电塔体系纵向振动的影响界限及简化抗震计算方法[J]. 振动与冲击，2004，23(2)：1-7.

[23] Li Hong-Nan，Shi Wenlong，Wang Guoxin and Jia Lianguang. Simplified models and experimental verification for coupled transmission tower-line system to seismic excitations，Journal of Sound and Vibration，2005(to appear).

[24] 中华人民共和国电力工业部.《电力设施抗震设计规范》(GB 50260-96). 北京：中国计划出版社，1996.

[25] A. G. Davenport. The dependence of wind load upon meteorological parameters. In：Processings of the International Research Seminar on Wind Effects on Buildings and structures (Toronto，1968). Toronto：University of Toronto press；1968. p. 19～82.

[26] Shinozuka M.，Jan C. M. Digital simulation of random processes and its applications. Sound Vib. 1972. 25(1)：111～128.

[27] Shinozuka M.，Yun C. B.，et al. Stochastic methods in wind engineering. Wind Eng. Ind. Aerodyn. 1990. 36：829～843.

[28] Simiu E，Scanlan R H. Wind effects on structures. Third edition [M]. New York：John Wiley and Sons；1996.

[29] 王之宏. 风荷载的模拟研究[J]. 建筑结构学报，1994，15(1)：44～52.

[30] 盛裴轩，毛节泰，李建国等. 大气物理学[M]. 北京：北京大学出版社，2003.

[31] Yasui H.，Marukawa H.，Momomura Y. et al. Analytical study on wind-induced vibration of power transmission towers. Wind Eng. Ind. Aerodyn. 1999，83(2)：431-441.

[32] Battista R. C.，Rodrigues R. S.，et al. Dynamic behavior and stability of transmission line towers under wind forces. Wind Eng. Ind. & Aerodyn.，2003，91:1051-1067.

[33] 中国建筑科学研究院. 建筑结构荷载规范 GB500009-2001[S]. 北京:中国建筑工业出版社，2002.

[34] 白海峰，李宏男. 大跨越输电塔线体系随机脉动风场模拟研究[J]. 工程力学，2007，24(7)：146-151.

[35] 邓洪洲，朱松晔，陈晓明等. 大跨越输电塔线体系气弹模型风洞试验[J]. 同济大学学报，2000，31(2):132-137.

[36] 国家电力公司华东电力设计院. 架空送电线路杆塔结构设计技术规定(DL/T5154-2002)[S]. 北京：中国电力出版社，1999.

[37] ASCE Committee on Fatigue and Fracture Reliability. Fatigue reliability：development of criteria for design. J. Struct. Div.，ASCE，1982，108(1)：71-88.

[38] 白海峰，李宏男. 输电线路杆塔疲劳可靠性研究[J]. 中国电机工程学报，2008，28(6)，25-31.

[39] Wirsching P. H. and Light M. C. Fatigue under wide band random stresses. J. Struct. Div.，ASCE，1980,106 (7)：1593-1607.

[40] Deoliya R. and Datta T. K. Reliability of microwave towers against extreme winds. Struct. Eng. Mech.，1998，6(5):555-570.

[41] Rao S. S. Reliability based design，McGraw-Hill，New York.，1992.

第 7 章 Chapter 7

现代钢结构抗火设计方法与研究进展*

李国强

（同济大学，上海 200092）

提 要：本文论述了火灾对钢结构的危害和钢结构抗火设计的目的与意义，介绍了传统的钢结构抗火设计方法及其存在的问题，指出基于火灾高温结构承载极限状态的现代钢结构抗火设计方法是今后工程应用的发展趋势。本文还介绍了我国为推进现代钢结构抗火设计方法，在火灾升温特性、火灾下钢构件的升温、高温钢材性能、高温钢构件极限承载力和火灾下结构所承受的荷载等重要方面取得的研究进展。

关键词：钢结构；抗火设计；传统方法；现代方法；研究进展

Development and Research on Contemporary Approach for Fire-Resistant Design of Steel Structures

G. Q. Li

(Tongji University, shanghai, 200092, China)

Abstract: In this paper, the disastrous effect of fire on steel structures and the purpose of fire-resistant design for steel structures are introduced. The conventional approach for fire-resistant design of steel structures and its disadvantage are discussed. The trend to develop and apply the contemporary approach based on load-bearing limit-state at elevated temperatures for fire-resistant design of steel structures is pointed out. The research achievement on the important aspects of the contemporary approach for fire-resistant design of steel structures are presented, including temperature elevation of smoke in fire, temperature elevation of steel components in fire, mechanical properties of steel at elevated temperatures, load-bearing capacity of steel components at elevated temperatures and demanding loads for

---

* 基金项目：国家自然科学基金资助（项目批准号：50738005），国家自然科学基金创新研究群体资助（项目批准号：50621062）。

作者简介：李国强（1963-），湖南人，教授，博士生导师，主要从事多高层钢结构及钢结构抗火研究。

E-mail: gqli@mail.tongji.edu.cn。

building structures to support in fire condition.

Key words: steel structure; fire-resistant design; conventional approach; contemporary approach; research development

7.1　引言

7.1.1　火灾对钢结构的危害

钢材虽为非燃烧材料，但钢不耐火，温度为 500℃时，钢材的屈服强度将降至室温下强度的一半，温度达到 600℃时，钢材将丧失大部分强度和刚度。因此，钢结构建筑一旦发生火灾，结构很容易遭到破坏甚至倒塌。例如，1967 年美国蒙哥马利市的一个饭店发生火灾，钢结构屋顶被烧塌；1993 年我国福建泉州一座钢结构冷库发生火灾，造成 $3600m^2$ 的库房倒塌；1996 年江苏省昆山市的一轻钢结构厂房发生火灾，$4320m^2$ 的厂房烧塌；1998 年北京某家具城发生火灾，造成该建筑（钢结构）整体倒塌；2003 年上海某钢结构厂房发生火灾，造成整体结构倒塌（图 1，图 2）；2001 年“9.11 事件”中，美国纽约世界贸易中心（WTC）两座 110 层 411m 高的钢结构大楼因飞机撞击而引发倒塌（图 3），造成 2830 人死亡；2005 年西班牙马德里一栋 32 层的钢结构建筑因火灾倒塌（图 4）；2006 年比利时布鲁塞尔国际机场发生火灾，造成钢结构飞机维修库倒塌（图 5），损失达数十亿欧元；2007 年连接美国旧金山和奥克兰两大城市的高速公路发生油罐车火灾事故，造成钢结构高架高速公路坍塌（图 6）。表 1 是近年我国一些钢结构建筑在发生火灾时倒塌的实例。

图 1　火灾中倒塌的钢结构厂房

图 2　火灾中倒塌钢结构厂房的构件情况

图 3　美国世贸中心火灾

图4 马德里一栋32层的钢结构建筑火灾

图5 比利时布鲁塞尔国际机场发生火灾

图6 高速公路发生油罐车火灾

近年我国一些钢结构建筑发生火灾倒塌的实例 表1

日期	火灾地点	结构型式	倒塌面积（mm^2）	经济损失（万元）
1997.10.29	内蒙古伊盟岱沟选煤厂	钢厂房，栈桥	487	378
1997.08.02	江西南昌市洪客隆商场	钢屋顶	5513	1748
1998.05.05	北京丰台玉泉营家具城	钢结构	23000	2088
1998.07.05	广东东莞常平镇美力玩具厂	轻钢仓库	1500	571
1999.01.09	北京丰台华大灯具批发市场	钢屋架	6391	490
1999.10.04	吉林珲春金发木业有限公司	轻钢仓库	1456	1736
1999.11.05	新建克拉玛依市第四净化水厂	钢网架	1890	120
2000.05.05	山西长冶市五一桥装潢材料市场	钢屋架	403	109
2000.10.29	广东中山市三乡镇慈航镇玩具厂	轻钢屋顶	3200	273
2001.09.24	辽宁辽阳市纺织有限公司	钢梁屋顶	8253	121
2001.01.16	山东威高集团医用高分子制品有限公司	钢厂房	9000	676
2001.08.13	广东中山市中塑塑料加工厂	钢屋架	2150	115
2002.02.27	山东德州市百货大楼	钢屋架	5800	685
2002.02.05	四川三台三角生活用纸制造公司	钢屋架	890	114
2003.04	山东某公司食品厂	钢厂房	6000	3000

火灾即使不引起钢结构建筑整体倒塌，也有可能造成结构严重破坏。例如，1970年美国50层的纽约第一贸易办公大楼发生火灾，楼盖钢梁被烧扭曲10cm左右；1990年英国一幢多层钢结构建筑在施工阶段发生火灾，造成钢梁、钢柱和楼盖钢桁架的严重破坏；2001年台北东方科学园区一幢高层钢结构建筑发生火灾，造成钢结构严重破坏，包括梁柱连接断裂（图7）、梁局部屈曲变形（图8）、梁整体挠曲变形（图9）、以及楼板大挠曲变形（图10）。

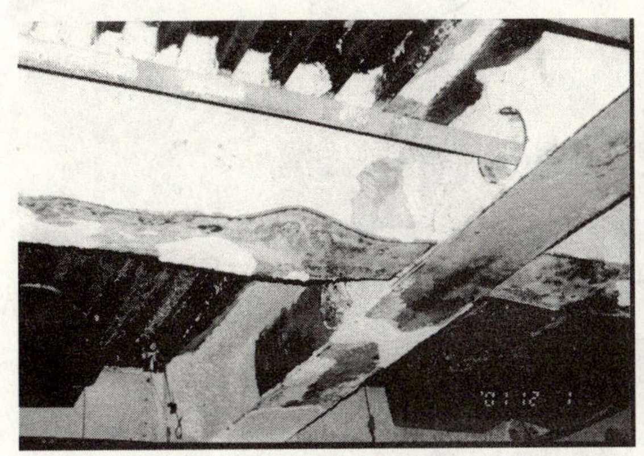

图7 火灾引起柱连接断裂

图8 火灾引起梁局部屈曲

图9 火灾引起梁整体挠曲变形

图10 火灾引起楼板大挠曲变形

7.1.2 火灾下钢结构为什么会破坏

钢材虽为非燃烧材料，却不耐火，在火灾高温下，钢结构的强度和刚度将迅速降低（图11），而火灾下环境温度迅速升高（表2），钢结构内部温度随之升高，其承载能力将下降，当结构承载力下降到不足以承受其上的荷载作用时，结果即发生破坏。图12是一钢梁在火灾下破坏的示例。

ISO 834火灾标准升温曲线温度时间关系 **表2**

时间(min)	0	5	10	15	30	60	90	120	180	240	360
温度升高 T_g-T_0(℃)	0	556	659	718	821	925	986	1029	1090	1133	1193

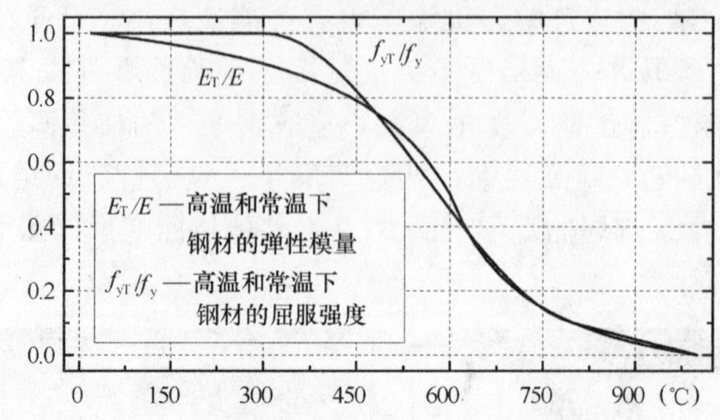

图 11　f_{yT}/f_y、E_T/E 随温度的变化

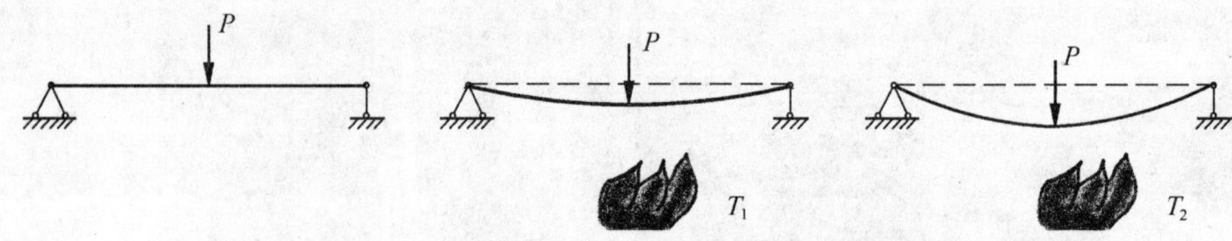

图 12　钢梁在火灾下的破坏（$T_2>T_1$）

7.2　钢结构抗火设计要求

7.2.1　结构耐火极限要求

长期以来，建筑防火被认为是建筑师设计时需考虑的问题，结构工程师设计时考虑结构防火（对结构更适合称抗火）问题的不多。确实，建筑的防火分隔、避难层的设置、安全疏散出口的布置等为建筑防火设计问题，目的在于减轻火灾损失，减少人员伤亡。然而，作为防火分隔的防火墙靠结构支承，如果火灾中支承结构破坏，防火墙也起不了防火分隔作用；还有避难层下的结构如果达不到耐火时间要求而破坏，造成的人员伤亡将更为严重；此外，建筑结构构件（如梁、楼板、楼梯等）在火灾中如果破坏，会影响人员的疏散和消防人员进入建筑内灭火，而结构在火灾中倒塌，也会造成人员伤亡。另外，对于功能重要的建筑，如机场，车站等，如火灾下结构破坏，将使建筑功能长时间失效，而造成巨大经济损失。因此各国建筑防火设计规范都有建筑结构构件耐火时间（或耐火极限）的规定。表 3 是我国规定的各类建筑结构构件的耐火极限[1,2]。

建筑结构构件的燃烧性能和耐火极限　　**表 3**

耐火等级 / 耐火极限（h） / 构件名称	单、多层建筑				高层建筑	
	一级	二级	三级	四级	一级	二级
承重墙	3.00	2.50	2.00	0.50	2.00	2.00
柱、柱间支撑	3.00	2.50	2.00	0.50	3.00	2.50

续表

<table>
<tr><th rowspan="2">耐火等级 / 耐火极限（h） / 构件名称</th><th colspan="6">单、多层建筑</th><th colspan="2">高层建筑</th></tr>
<tr><th>一级</th><th>二级</th><th colspan="2">三级</th><th colspan="2">四级</th><th>一级</th><th>二级</th></tr>
<tr><td>梁、桁架</td><td>2.00</td><td>1.50</td><td colspan="2">1.00</td><td colspan="2">0.50</td><td>2.00</td><td>1.50</td></tr>
<tr><td rowspan="2">楼板
楼面支撑</td><td rowspan="2">1.50</td><td rowspan="2">1.00</td><td>厂、库房</td><td>民用</td><td>厂、库房</td><td>民用</td><td rowspan="6">1.50</td><td rowspan="6">1.00</td></tr>
<tr><td>0.75</td><td>0.50</td><td>0.50</td><td>不要求</td></tr>
<tr><td rowspan="2">屋顶承重构件
屋面支撑、系杆</td><td rowspan="2">1.50</td><td rowspan="2">0.50</td><td>厂、库房</td><td>民用</td><td colspan="2" rowspan="2">不要求</td></tr>
<tr><td>0.50</td><td>不要求</td></tr>
<tr><td rowspan="2">疏散楼梯</td><td rowspan="2">1.50</td><td rowspan="2">1.00</td><td>厂、库房</td><td>民用</td><td colspan="2" rowspan="2">不要求</td></tr>
<tr><td>0.75</td><td>0.50</td></tr>
</table>

7.2.2　钢结构抗火设计的目的

各类钢结构抗火设计，一般期望达到如下目的：

(1) 减轻结构在火灾中的破坏，避免结构在火灾中局部倒塌造成灭火及人员疏散困难；

(2) 避免结构在火灾中整体倒塌造成人员伤亡；

(3) 减少火灾后结构的修复费用，缩短灾后结构功能恢复周期，减少间接经济损失。

7.3　传统钢结构抗火设计方法

7.3.1　设计要求

传统钢结构抗火设计，针对结构各构件进行，要求结构构件满足下列要求：

结构构件具有的耐火极限（能力）≥规定的耐火极限要求　　(1)

式（1）中，规定的耐火极限见表 3，而结构构件具有的耐火时间能力则由试验确定。

7.3.2　足尺构件试验

中国国家标准《建筑构件耐火试验方法》[3]规定：

构件的耐火时间（能力）为，构件在试验中从受火到失去稳定性的时间。

试验条件需符合下列规定：

(1) 试件：应与实际使用的尺寸相同（足尺试件）；

(2) 荷载大小：应符合下列情况之一

①试件工作荷载；

②试件设计荷载；

③试件按现行设计方法确定的试验荷载。

(3) 加载形式：梁——四点加载

　　　　　　柱——三点加载

(4) 试件边界约束：应能反映实际情况；

(5) 受火条件：梁——三面受火

　　　　　　柱——四面受火

(6) 试件升温：ISO 834 火灾标准升温（见表 2）。

一般情况下，无防火保护的钢构件满足不了耐火极限要求，而需采取防火保护措施，因此传统钢结构抗火设计方法是通过构件试验，定量地确定防火保护措施，使得钢结构的耐火时间大于或等于规定的耐火极限。

目前钢结构主要防火保护措施为涂覆防火涂料（图 13）与包覆防火板（图 14）。

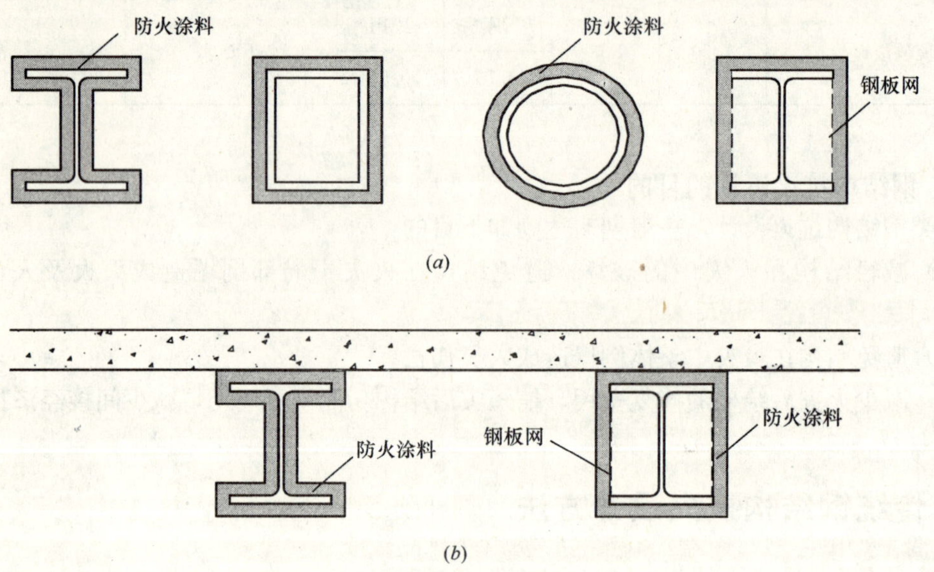

图 13　采用防火涂料钢结构防火保护构造

(a) 柱；(b) 梁

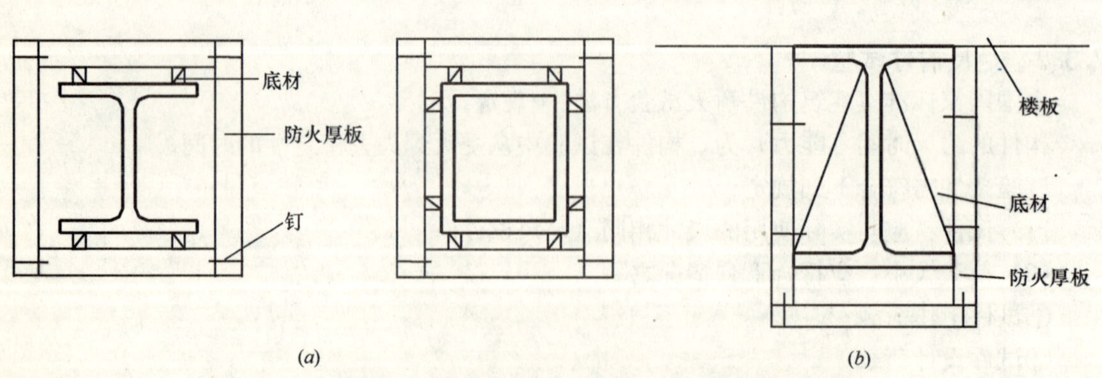

图 14　采用防火板钢结构防火保护构造

(a) 柱；(b) 梁

7.3.3　标准构件试验

由于足尺构件试验成本大，且有时难以实现，故传统钢结构抗火设计一般仅做标准梁或标准柱试验[4]，施加固定的标准荷载，然后测定其在标准火灾升温条件下的耐火时间（图 15）。将由此获得的施加在标准构件上一定厚度的防火保护措施与相应构件耐火时间能力的关系，作为用于实际钢结构抗火设计，满足规定耐火极限要求的依据。

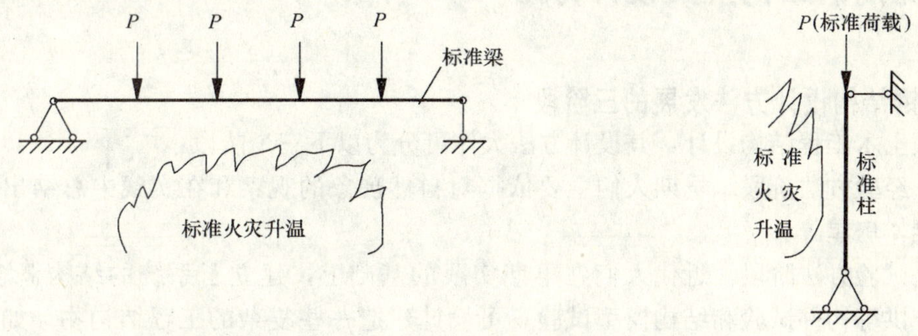

图 15　标准构件耐火试验

7.4　传统钢结构抗火设计方法的不足

（1）实际上，将构件从结构中孤立出来，施加一定的荷载，然后按一定的升温曲线加温，来测定构件耐火时间的方法，存在很多问题。首先，构件在结构中的受力，很难通过试验模拟，实际构件受力各不相同，试验难以概全，而受力的大小对构件耐火时间的影响较大，一般构件受力越大，构件的耐火时间越小（图 16）；其次，构件在结构中的端部约束在试验中难以模拟，而端部约束也是影响构件耐火时间的重要因素；再次，构件受火在结构中会产生温度应力，而这一影响在构件试验中也难以准确反映。

（2）标准构件测定的耐火时间，不能准确代表足尺构件的耐火时间。首先，构件截面的尺寸和形状，对构件受火升温影响很大（见图 17）；其次，构件大小的不同，会引起构件受力与其承载力比值（称为荷载比）的不同，而荷载比对构件的耐火时间能力的影响很大。

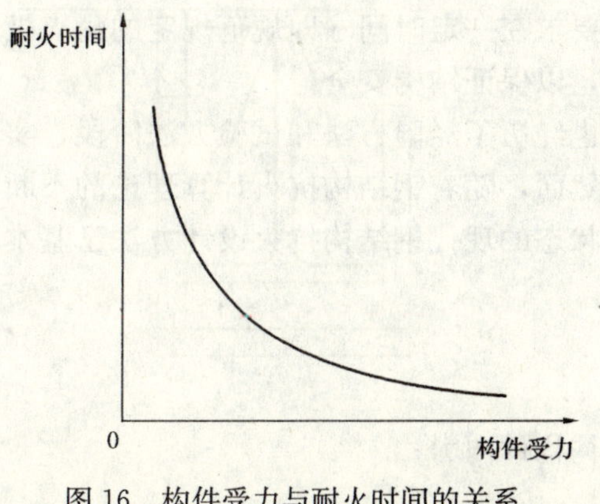

图 16　构件受力与耐火时间的关系

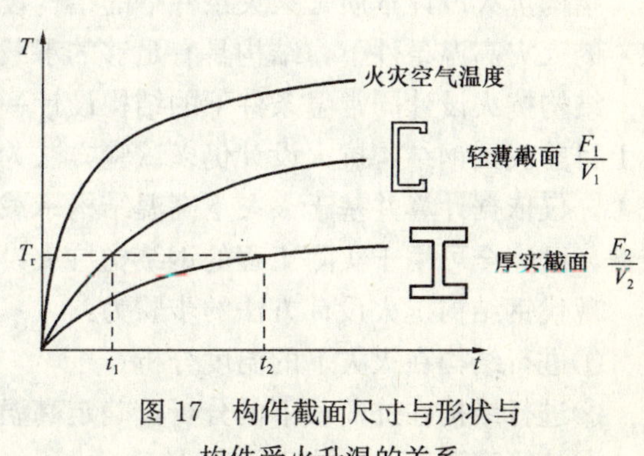

图 17　构件截面尺寸与形状与构件受火升温的关系

(3) 结构的耐火时间基于ISO 834升温曲线确定，而现有的研究表明真实火灾与火荷载密度、通风条件、建筑形式、室内空间尺寸大小等因素密切相关，ISO 834曲线并不能反映火灾的真实情况，而火灾升温曲线对结构耐火时间有影响。

7.5 现代钢结构抗火设计方法

7.5.1 钢结构设计方法发展的三阶段

回顾土木工程结构设计，其设计方法大体可分为以下三个阶段：

(1) 经验方法阶段。早期人们主要依据对自然现象的观察和在实践中总结出的经验，建造桥梁，房屋等。

(2) 试验方法阶段。近代人们在早期实践的基础上，建立了系统的结构概念，结合梁、柱、拱等构件试验和结构模型试验，可设计建造一些复杂的工程结构物，如1890年在英国爱丁堡建成的世界上第一个钢结构大桥（Forth Bridge），可认为是采用试验方法建造的。

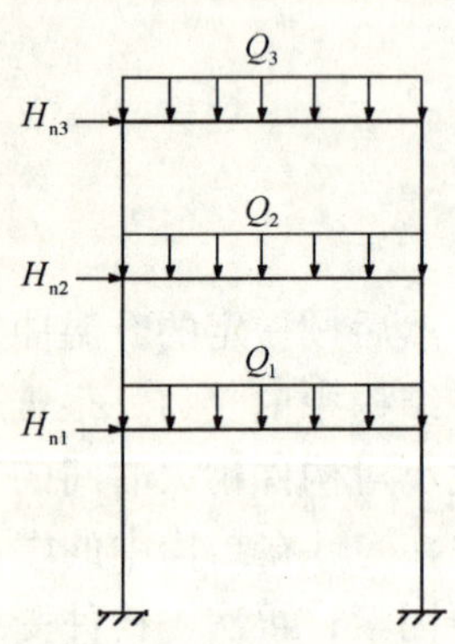

图18　结构所受荷载

(3) 计算方法阶段。现代人们建立了完善的工程结构计算理论，有限元理论更是解决了复杂工程结构（如大坝、大垮桥梁、高层建筑等）的分析计算问题，这些结构分析和承载力计算理论经试验验证后，被各种设计规范采纳，各种结构仅依据计算即可完成设计。这种依据计算进行结构设计的方法，可称为现代结构设计方法，其用于结构设计的步骤为：

①进行结构在外荷载下的分析（图18），确定结构构件内力；

②计算结构构件承载力；

③判别构件承载力是否大于构件内力。如是，则结构安全，如否，则结构不安全。

7.5.2 钢结构抗火设计方法的发展

结构抗火设计本质是火灾条件下的结构设计，要求在一定时间（即规范规定的耐火极限）的火灾高温条件下，结构具有足够的承载能力，以保证结构安全。

结构抗火设计同常温条件下的结构设计一样，也经历了经验方法和试验方法阶段，实际上目前我国钢结构抗火设计仍以试验方法为主。然而，随着钢结构抗火计算理论的不断完善，仅依据计算并基于火灾下高温结构承载极限状态的现代钢结构抗火设计方法已基本成熟，已完全可用于实际工程的钢结构抗火设计。

现代钢结构抗火设计方法的步骤为：

①进行结构在火灾下的温度分析；

②进行结构在外荷载下的分析，确定高温下结构构件内力；

③计算高温下结构构件承载力；

④判断构件承载力是否大于构件内力。如果，则结构安全，如否，则结构不安全。

图 19 是现代钢结构抗火设计步骤。

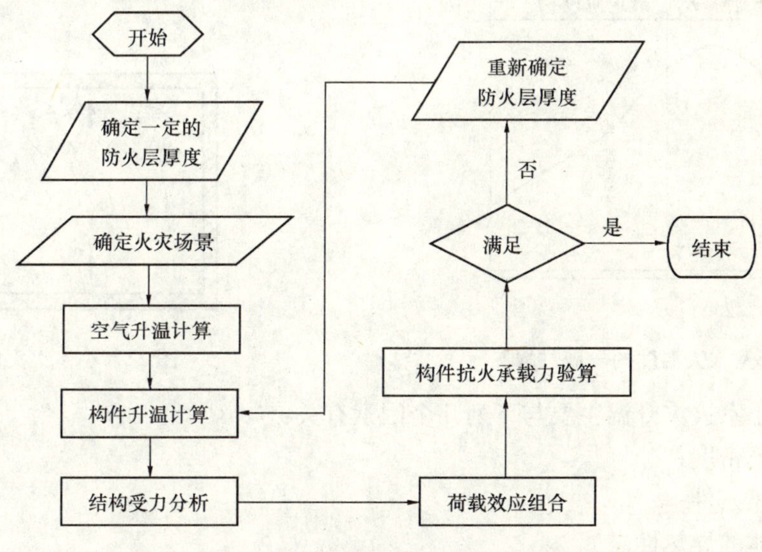

图 19　现代钢结构抗火设计步骤

7.6　现代钢结构抗火设计研究进展

按照图 19 进行钢结构抗火设计，需解决的关键理论问题为：

①火灾升温特性；

②火灾下钢构件的升温；

③高温钢材性能；

④高温钢构件极限承载力；

⑤火灾下结构所需承受的荷载；

下面介绍我们对以上问题的研究成果[5～98]。

7.6.1　火灾升温特性

讨论建筑室内火灾升温特性应区分小室火灾与大空间火灾。

7.6.1.1　小室火灾升温

小室指一般建筑办公室、住宅房间大小，火灾发生后，会经历初期增长阶段、全盛阶段和衰退阶段（见图 20）。在初期增长阶段和全盛阶段间，会经历一个短暂的轰燃过程。

所谓“轰燃”，是指混合可燃物中火焰波的快速传播或在有限空间的气相着火现象（见图 21）。一旦发生轰燃，室内所有可燃物表面都开始剧烈燃烧。

火灾发生轰燃的条件与着火房间地面上的热辐射强度有关，如火灾发生后达不到轰燃条件，将产生不了轰燃而熄灭（见图 20），这种火灾的危害一般不大。能产生危害的火灾是会发生轰燃的火灾，故一般仅关心轰燃后的小室火灾升温特性。

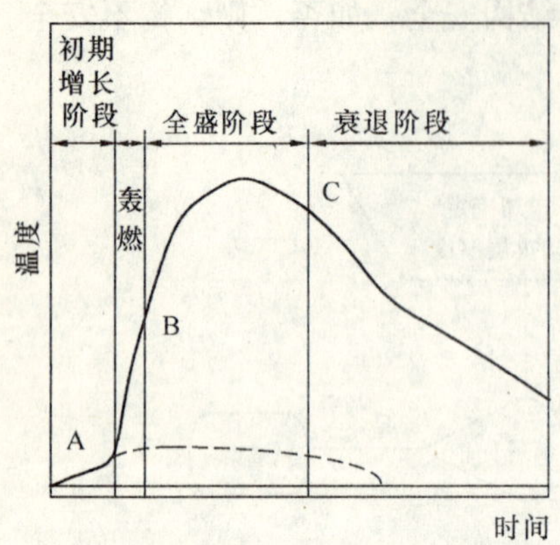

图 20　小室火灾温度的发展过程

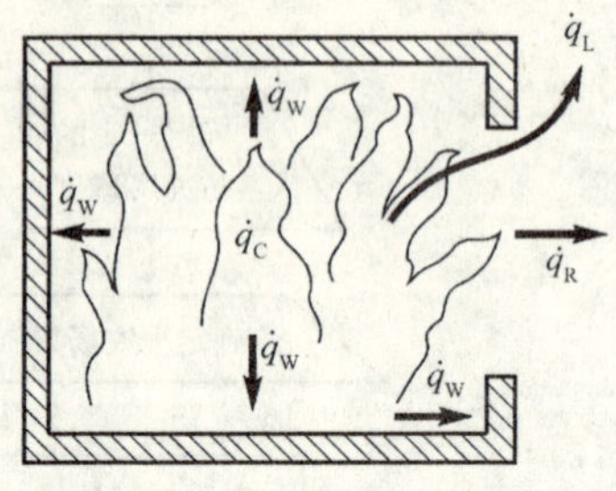

图 21　小室火灾的轰燃

轰燃后的小室火灾升温主要与下列几个因素有关：

①火灾荷载密度；

②房间通风条件；

③房间墙体的导热性。

火灾荷载密度定义为：

$$q=\frac{1}{A_{fl}}\Sigma M_{\rm V}H_{\rm V} \tag{2a}$$

或

$$q_{\rm t}=\frac{1}{A_{\rm t}}\Sigma M_{\rm V}H_{\rm V} \tag{2b}$$

式中，q(或 $q_{\rm t}$) 是火灾荷载密度（MJ/m^2）；$M_{\rm V}$ 是室内可燃材料 V 的总质量（kg）；$H_{\rm V}$ 是室内可燃材料 V 的热值（MJ/kg）；A_{fl}是室内地面面积（m^2）；$A_{\rm t}$ 是房间所有壁面、地面及顶面面积之和（m^2）。

可见，火灾荷载密度实际上是一个单位面积上的可燃物释放热量值的概念。

房间的通风条件可采用通风因子（也称开口因子）定义，即

$$F_{\rm V}=\frac{A_{\rm w}\sqrt{h_{\rm w}}}{A_{\rm t}} \tag{3}$$

式中，$F_{\rm V}$ 是通风因子（m$^{1/2}$）；$A_{\rm w}$ 是房间所有开口（门、窗等）的总面积（m^2）；$A_{\rm w}=\Sigma A_i$；A_i 是房间各开口的面积（m^2）；$h_{\rm w}$ 是房间所有开口平均高度（m），$h_{\rm w}=\Sigma(A_ih_i)/A_{\rm w}$；$h_i$ 是房间各开口的高度（m）；$A_{\rm t}$ 是房间所有壁面、地面及顶面面积之和（m^2）。

一般情况下，火灾荷载密度越大，火灾升温的最高温度越高，高温持续时间越长，房间墙体的导热性越好，火灾升温越慢，达到的最高温度越低。

房间升温条件对火灾升温的影响要复杂些。当房间开口面积很小时，火灾时进入房间的空气受到限制，可燃物由于空气量不足而燃烧不充分，此时，随着房间开口面积增大，进入空气量增多，会使可燃物燃烧率增加，因而火灾升温速度会增大，达到的最高温度会提高。这种受房间开口空气流量影响的火灾称为通风控制型火灾。

当房间开口面积进一步增大到空气供给足以满足可燃物燃烧所需时，则空气量的继续增加不会提高燃烧速率，反而会带走高温烟气的热量，降低火灾升温速度和达到的最高温度（见图 22）。此时，可燃物的燃烧速率将由可燃物的性质和可燃物的分布决定，将这种

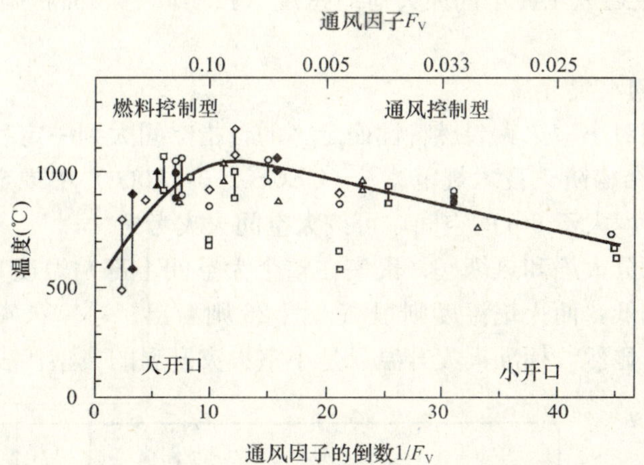

图 22　通风因子与小室火灾最高温度的关系

特性的火灾称为燃料控制性火灾。

研究表明，小室火灾的以上两种类型，可采用通风因子判别：

当 $F_V \geqslant 0.07$ 时，为燃料控制型火灾；

当 $F_V < 0.07$ 时，为通风控制型火灾。

图 23 和图 24 分别为恒定火灾荷载密度（按房间总表面积确定）不同通风条件的火灾

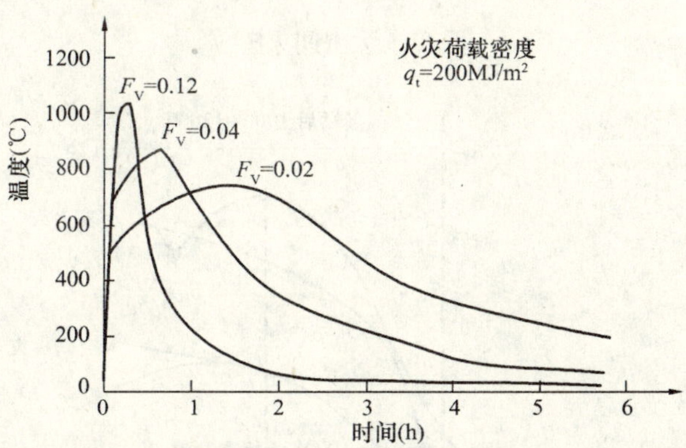

图 23　恒定火灾荷载密度不同通风条件的火灾升温-时间曲线

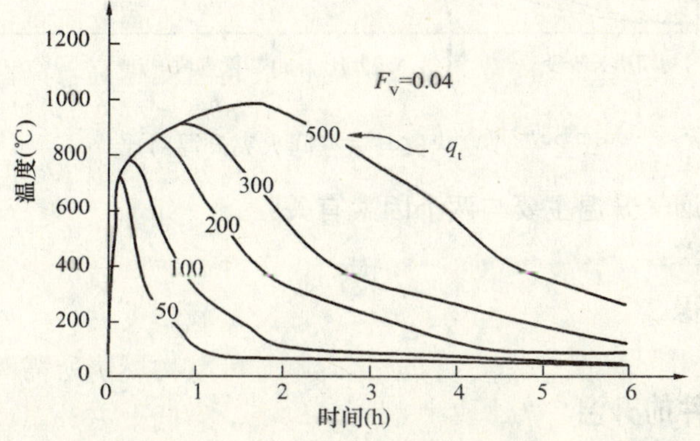

图 24　恒定通风条件不同火灾荷载密度的火灾升温-时间曲线

升温-时间曲线和恒定通风条件不同火灾荷载密度（按房间总表面积确定）的火灾升温-时间曲线。

7.6.1.2 大空间火灾升温

大空间火灾是相对小室火灾轰燃特性而定义的，指空间大到一定程度而不可能产生轰燃的火灾。《建筑钢结构防火技术规范》[99]（CECS 200：2006）中规定，单体建筑面积不小于 500m^{2}且净空高度大于 6m 的空间，可按大空间火灾考虑。

大空间火灾应区分火焰和热烟气（聚集在整个大空间上部）温度（见图 25)，大空间火灾热烟气温度会较低，而火焰温度则很高。图 26 则给出了小室火灾与大空间火灾（热烟气）升温的对比，显然，标准火灾升温只是小室火灾升温的典型代表。

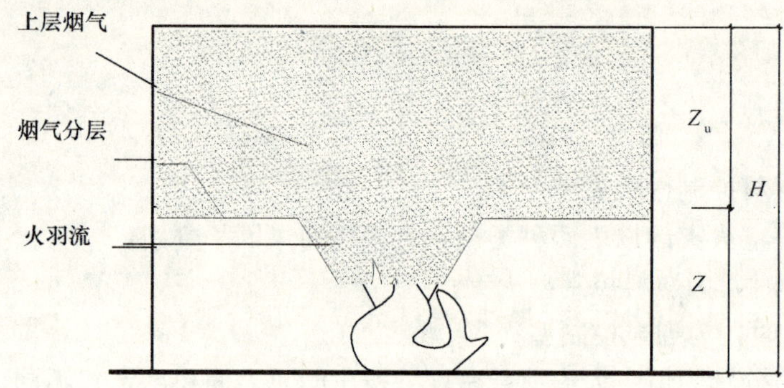

图 25 大空间火灾

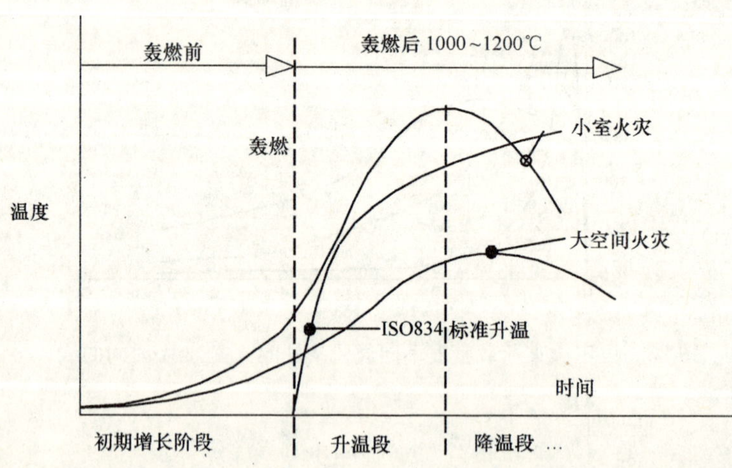

图 26 小室火灾与大空间火灾升温对比

大空间火灾的热烟气升温主要与两个因素有关：

(1) 火灾热释放率；

(2) 大空间的体积。

7.6.2 火灾下钢构件的升温

火灾下可燃物由燃烧所释放的热量，可通过热烟气以辐射、对流、传导三种方式传递

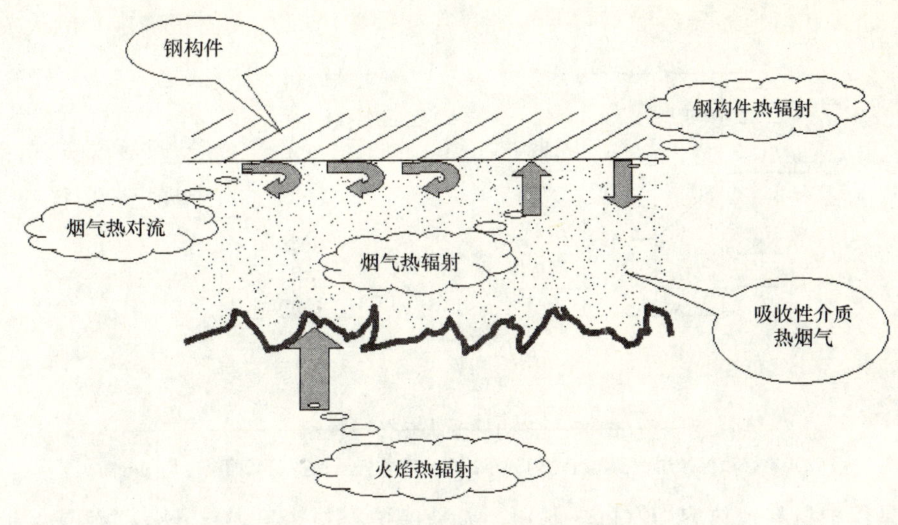

图 27 钢构件受火传热模型

给钢构件（见图 27），钢构件则以升温的方式吸收这部分热量。

由于钢材热传导性好，可认为火灾中任一钢构件的内部温度分布是均匀的，则可采用如下一维传热模型计算钢构件的受火升温：

$$B \cdot (T_g - T_s) = \rho_s \cdot c_s \cdot \frac{dT_s}{dt} \tag{4}$$

式中：B 是构件单位长度综合传热系数[W/(m^2 · ℃)]；T_s 是钢构件温度(℃)；T_g 是热烟气的温度(℃)；c_s 是钢的比热容[J/(kg · ℃)]；ρ_s 是钢的密度(kg/m^3)；t 是时间(s)。

当构件无防火保护时，火灾热烟气主要以热辐射和热对流的方式传热，此时

$$B = (\alpha_r + \alpha_c)\frac{F}{V} \tag{5}$$

式中：V 是单位长度构件的体积(m^3/m)；F 是单位长度构件的受火表面积(m^2/m)；α_c 是对流传热系数，对于纤维类燃烧火灾，可取 α_c＝25W/(m^2 · ℃)；α_r 是以辐射方式由空气向构件表面传热的传热系数[W/(m^2 · ℃)]。

$$\alpha_r = \frac{2.041}{T_g - T_s} \cdot \left[\left(\frac{T_g + 273}{2}\right)^4 - \left(\frac{T_s + 273}{2}\right)^4\right] \tag{6}$$

当构件有防火保护时，火灾热烟气主要通过防火保护层以传导的方式传热。如果防火保护层本身吸收的热量不大，对火灾下钢构件升温的影响不大，则这样的保护层成为轻质保护层（图 28*a*），否则为非轻质保护层（图 28*b*）。有防火保护层时，钢构件综合传热系数可按下列公式计算：

$$B = \frac{\lambda_i}{d_i} \cdot \frac{F_i}{V} \qquad \text{(轻质保护层)} \tag{7}$$

$$B = \frac{1}{1 + \frac{c_i \rho_i d_i F_i}{2 c_s \rho_s V}} \cdot \frac{\lambda_i}{d_i} \cdot \frac{F_i}{V} \qquad \text{(非轻质保护层)} \tag{8}$$

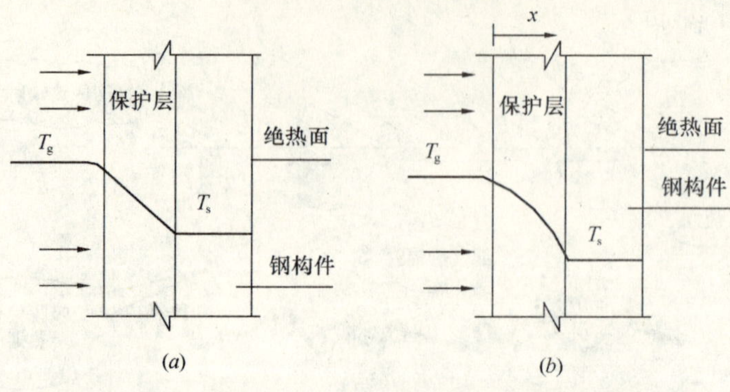

图 28 钢构件受火热传导模型

(a) 有轻质保护层截面温度均匀分布；(b) 有非轻质保护层截面温度均匀分布

式中：c_i 是保护材料比热容[J/(kg·K)]；ρ_i 是保护材料密度(kg/m³)；d_i 是保护材料厚度(m)；λ_i 是保护材料 500℃时的导热系数或等效导热系数[W/(m·K)]；F_i是构件单位长度防火保护材料内表面积(m²/m)。

对于任意火灾，若热烟气温度时程 $T_g(t)$是确定的，则采用差分方法求解方程式(4)，可解得构件受火的升温时程 $T_s(t)$。

如构件受特定的标准火灾，则构件升温可很简便地按下式计算：

$$T_s(t) = (\sqrt{0.044 + 5.0 \times 10^{-5} B} - 0.2)t + 20 \tag{9}$$

式中：t 是受火时间 (s)。

对于大空间火灾中无防火保护的钢构件，由于火灾热烟气温度与火焰温度不同，则对于距火焰较近的钢构件，除同样需考虑热烟气通过辐射与对流传递热量给钢构件外，还应考虑火焰通过热辐射传递热量给钢构件 (图 27)。

大空间火灾中，考虑火焰辐射对无防火保护钢构件升温的影响（与不考虑火焰辐射相比）与下列因素有关：

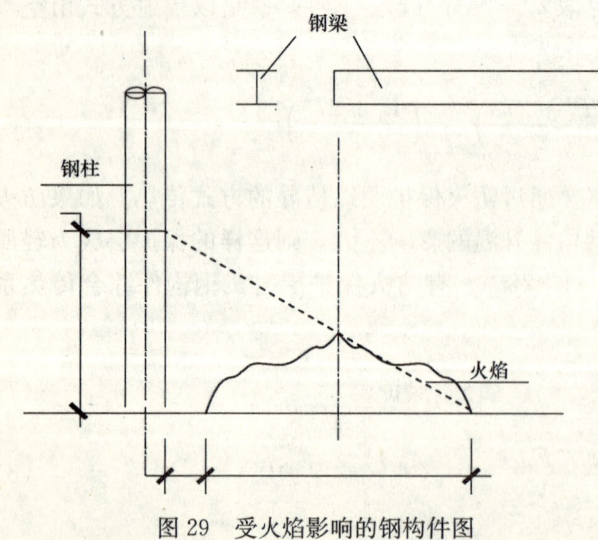

图 29 受火焰影响的钢构件图

(1) 构件距火焰的距离，距离越近，影响越大；

(2) 构件表面与火焰中心的连线与构件表面的夹角（见图 29)，此夹角为 90°时，影响最大，夹角越小，影响越小；

(3) 受火焰辐射的表面积与构件表面积的比值，此比值越大，影响越大；

(4) 火焰温度与火灾热烟气温度的差值，此差值越大，影响越大。在小室火灾中，热烟气温度就是火焰温度，故没有额外火焰辐射问题。

大空间火灾中，对于有防火保护的钢构件，无需考虑火焰辐射对钢构件升温的影响。

应该指出，在大空间火灾中，如果钢构件被火焰直接接触，则钢构件应按受火焰辐射计算构件温度。

7.6.3 高温钢材性能

高温下钢材的强度，弹性模量等受力性能将发生变化，而这些性能对钢构件抗火能力极为重要，是钢结构进行抗火计算与设计的基础。

高温钢材力学性能参数可通过试验确定。试验方法主要有两种，即恒温加载试验（又称稳态试验，Isothermal or Steady-state Test）和恒载升温试验（又称瞬态试验，Anisothermal or Transient-state Test）。钢材高温材性试验大多采用恒温加载试验，图 30 是一组低碳钢各温度下的应力-应变曲线试验结果。

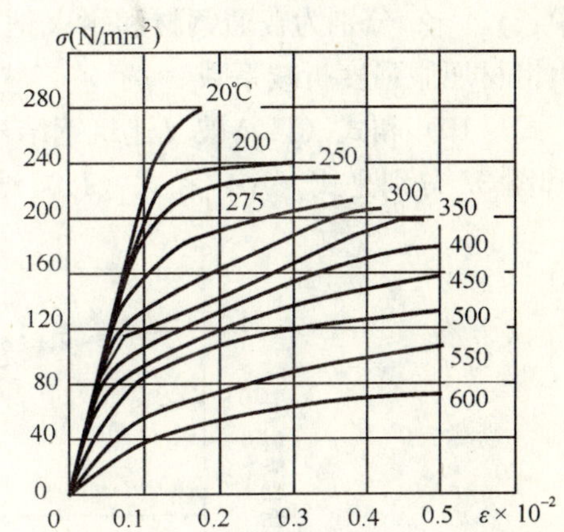

图 30　低碳钢各温度下的应力-应变曲线

为获得各种钢材的高温性能，同济大学进行了大量试验，包括：

(1) 19 个 Q345 钢（钢结构常用钢）试件试验；

(2) 29 个 SM41 钢（轻型建筑钢结构常用钢）试件试验；

(3) 33 个 20MnTiB 钢（高强度螺栓常用钢）试件试验；

(4) 47 组 Q345 钢板间的摩擦系数（高强度螺栓连接常用）试件试验；

(5) 94 个耐火钢（新型结构钢）试件试验；

(6) 27 个高强钢绞线（索结构用钢）试件试验。

依据上述试验的结果，获得了普通钢材和耐火钢材弹性模量和屈服强度随温度变化的折减系数

$$\chi_T = \begin{cases} \dfrac{7T_s - 4780}{6T_s - 4760} & 20℃ \leqslant T_s < 600℃ \\[2ex] \dfrac{1000 - T_s}{6T_s - 2800} & 600℃ \leqslant T_s \leqslant 1000℃ \end{cases} \tag{10a}$$

$$\chi_T = \begin{cases} 1 - \dfrac{T_s - 20}{2520} & 20℃ \leqslant T_s < 650℃ \\[2ex] 0.75 - \dfrac{7(T_s - 650)}{2500} & 650℃ \leqslant T_s < 900℃ \\[2ex] 0.5 - 0.0005T_s & 900℃ \leqslant T_s \leqslant 1000℃ \end{cases} \tag{10b}$$

$$\eta_{\mathrm{T}}=\begin{cases}1.0 & 20℃\leqslant T_{\mathrm{s}}\leqslant 300℃\\ 1.24\times10^{-8}T_{\mathrm{s}}^{3}-2.096\times10^{-5}T_{\mathrm{s}} & \\ \quad +9.228\times10^{-3}T_{\mathrm{s}}-0.2168 & 300℃<T_{\mathrm{s}}<800℃\\ 0.5-T_{\mathrm{s}}/2000 & 800℃\leqslant T_{\mathrm{s}}\leqslant 1000℃\end{cases}\tag{11a}$$

$$\eta'_{\mathrm{T}}=\begin{cases}\dfrac{6(T_{\mathrm{s}}-768)}{5(T_{\mathrm{s}}-918)} & 20℃\leqslant T_{\mathrm{s}}<700℃\\ \dfrac{1000-T_{\mathrm{s}}}{8(T_{\mathrm{s}}-600)} & 700℃\leqslant T_{\mathrm{s}}\leqslant 1000℃\end{cases}\tag{11b}$$

式中：χ_{T}，χ'_{T}分别为普通钢材和耐火钢材弹性模量折减系数；η_{T}，η'_{T}分别为普通钢材和耐火钢材屈服强度折减系数。

式（10）和式（11）被《建筑钢结构防火技术规范》[99]（CECS 200：2006）采用。图31和图32分别是式（10）和式（11）结果与国外规范或试验结果的比较。

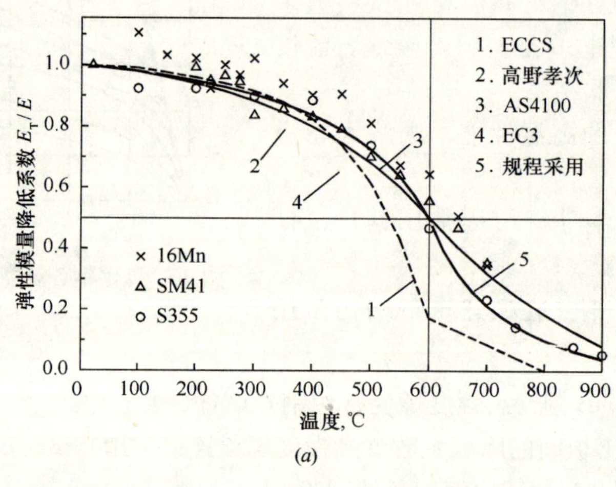

(*a*)

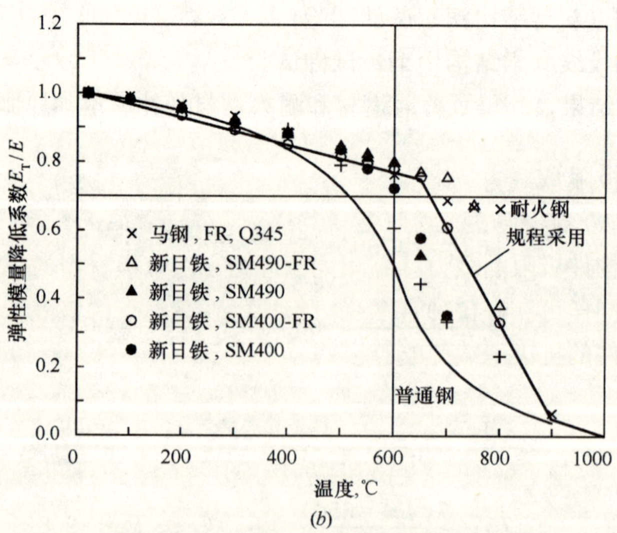

(*b*)

图31 钢材高温弹性模量

（*a*）普通钢；（*b*）耐火钢

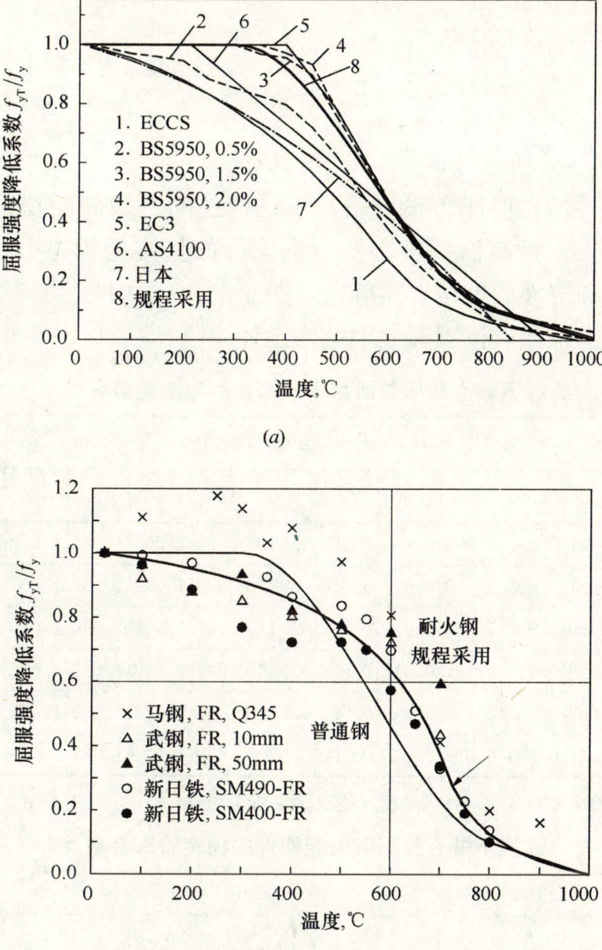

图 32　钢材高温屈服强度

(*a*) 普通钢；(*b*) 耐火钢

7.6.4　高温钢构件极限承载力

7.6.4.1　基本受力钢构件

采用与常温下钢构件极限承载力计算相同原理与方法，但材料的强度和弹性模量取值考虑温度的影响，则同样可获得各种基本受力钢构件在高温下的极限承载力，进而获得如下各种基本受力钢构件的抗火承载力验算公式：

1. 轴心受拉构件

$$\frac{N}{A_n} \leqslant \eta_T \gamma_R f \tag{12}$$

式中：N 是火灾下构件的轴向拉力或轴向压力设计值；A_n 是构件的净截面面积；η_T 是高温下钢材的强度折减系数；γ_R 是钢材的抗力系数，近似取 $\gamma_R = 1.1$；f 是常温下钢材的强度设计值。

2. 轴心受压构件

$$\frac{N}{\varphi_{\mathrm{T}}A}\leqslant\eta_{\mathrm{T}}\gamma_{\mathrm{R}}f \tag{13a}$$

$$\varphi_{\mathrm{T}}=\alpha_{\mathrm{c}}\varphi \tag{13b}$$

式中：N 是火灾时构件的轴向压力设计值；A 是构件的毛截面面积；φ_{T} 是高温下轴心受压钢构件的稳定系数；α_{c} 是高温下轴心受压钢构件的稳定验算参数，对于普通结构钢构件，根据构件长细比和构件温度按表 4 确定，对于耐火钢构件，按表 5 确定；φ 是常温下轴心受压钢构件的稳定系数，按现行《钢结构设计规范》[100]（GB 50017）确定。

高温下轴心受压普通结构钢构件的稳定验算参数 α_{c} **表 4**

$\lambda\sqrt{\frac{f_y}{235}}$	温　度（℃）															
	⩽50	100	150	200	250	300	350	400	450	500	550	600	650	700	750	800
⩽10	0.999	0.998	0.997	0.995	0.993	0.990	0.989	0.991	0.996	1.001	1.002	1.002	0.996	0.995	1.000	1.000
50	0.998	0.995	0.991	0.986	0.980	0.973	0.970	0.977	0.990	1.002	1.007	1.007	0.989	0.986	1.001	1.000
100	0.996	0.988	0.979	0.968	0.955	0.939	0.933	0.947	0.977	1.013	1.046	1.050	0.976	0.969	1.005	1.000
150	0.994	0.983	0.970	0.955	0.937	0.915	0.906	0.926	0.967	1.019	1.063	1.069	0.965	0.955	1.008	1.000
200	0.994	0.982	0.968	0.952	0.933	0.910	0.902	0.922	0.965	1.023	1.075	1.082	0.963	0.952	1.009	1.000
⩽250	0.994	0.981	0.968	0.951	0.932	0.909	0.900	0.920	0.965	1.024	1.081	1.088	0.962	0.952	1.009	1.000

注：温度在 50℃及以下时 α_{c} 取 1.0，其他温度 α_{c} 按线性插值确定。

高温下轴心受压耐火钢构件的稳定验算参数 α_{c} **表 5**

$\lambda\sqrt{\frac{f_y}{235}}$	温　度（℃）															
	⩽50	100	150	200	250	300	350	400	450	500	550	600	650	700	750	800
⩽10	1.000	0.999	0.998	0.998	0.998	0.998	0.998	1.000	1.000	1.001	1.002	1.004	1.006	1.008	1.011	1.012
50	0.999	0.997	0.995	0.994	0.994	0.994	0.996	0.999	1.001	1.004	1.008	1.014	1.023	1.030	1.044	1.050
100	0.997	0.993	0.989	0.987	0.986	0.987	0.990	0.998	1.008	1.023	1.054	1.105	1.188	1.245	1.345	1.378
150	0.996	0.989	0.984	0.980	0.979	0.980	0.986	0.997	1.012	1.035	1.073	1.136	1.250	1.350	1.589	1.722
200	0.996	0.989	0.983	0.979	0.978	0.979	0.985	0.996	1.014	1.041	1.087	1.164	1.309	1.444	1.793	1.970
⩽250	0.996	0.988	0.983	0.979	0.977	0.979	0.985	0.996	1.015	1.045	1.094	1.179	1.341	1.497	1.921	2.149

3. 压弯构件

绕强轴 x 轴弯曲：

$$\frac{N}{\varphi_{\mathrm{xT}}A}+\frac{\beta_{\mathrm{mx}}M_{\mathrm{x}}}{\gamma_{\mathrm{x}}W_{\mathrm{x}}(1-0.8N/N'_{\mathrm{ExT}})}+\eta\frac{\beta_{\mathrm{ty}}M_{\mathrm{y}}}{\varphi'_{\mathrm{byT}}W_{\mathrm{y}}}\leqslant\eta_{\mathrm{T}}\gamma_{\mathrm{R}}f \tag{14a}$$

$$N'_{\mathrm{ExT}}=\pi^{2}E_{\mathrm{T}}A/(1.1\lambda_{\mathrm{x}}^{2}) \tag{14b}$$

绕弱轴 y 轴弯曲：

$$\frac{N}{\varphi_{\mathrm{yT}}A}+\eta\frac{\beta_{\mathrm{tx}}M_{\mathrm{x}}}{\varphi'_{\mathrm{bxT}}W_{\mathrm{x}}}+\frac{\beta_{\mathrm{my}}M_{\mathrm{y}}}{\gamma_{\mathrm{y}}W_{\mathrm{y}}(1-0.8N/N'_{\mathrm{EyT}})}\leqslant\eta_{\mathrm{T}}\gamma_{\mathrm{R}}f \tag{15a}$$

$$N'_{\mathrm{EyT}} = \pi^2 E_{\mathrm{T}} A / (1.1\lambda_{\mathrm{y}}^2) \tag{15b}$$

式中：N 是火灾时构件的轴向压力设计值；M_x，M_y 分别为火灾时所计算构件段范围内对强轴（x）和弱轴（y）的最大弯矩设计值；A 是构件的毛截面面积；W_x，W_y 分别为对强轴和弱轴的毛截面模量；N'_{ExT}，N'_{EyT} 分别为高温下绕强轴弯曲和绕弱轴弯曲的参数；λ_x，λ_y 分别为对强轴和弱轴的长细比；φ_{xT}，φ_{yT} 是高温下轴心受压钢构件的稳定系数，分别对应于强轴失稳和弱轴失稳，按式（16）计算；

$$\varphi_{\mathrm{T}} = \alpha_{\mathrm{c}} \varphi \tag{16}$$

φ'_{bxT}，φ'_{byT} 是高温下均匀弯曲受弯钢构件的稳定系数，分别对应于强轴失稳和弱轴失稳，按式（17）计算；

$$\varphi'_{\mathrm{bT}} = \begin{cases} \alpha_{\mathrm{b}}\varphi_{\mathrm{b}} & \alpha_{\mathrm{b}}\varphi_{\mathrm{b}} \leqslant 0.6 \\ 1.07 - \dfrac{0.282}{\alpha_{\mathrm{b}}\varphi_{\mathrm{b}}} \leqslant 1.0 & \alpha_{\mathrm{b}}\varphi_{\mathrm{b}} > 0.6 \end{cases} \tag{17}$$

γ_x，γ_y 分别为绕强轴弯曲和绕弱轴弯曲的截面塑性发展系数；对于工字型截面，$\gamma_x=1.05$，$\gamma_y=1.2$；对于箱形截面，$\gamma_x=\gamma_y=1.05$；对于圆钢管截面，$\gamma_x=\gamma_y=1.15$；η 是截面影响系数，对于闭口截面，$\eta=0.7$，对于其它截面，$\eta=1.0$；β_{mx}，β_{my} 是弯矩作用平面内的等效弯矩系数，按现行《钢结构设计规范》[100]（GB 50017）确定；β_{tx}，β_{ty} 是弯矩作用平面外的等效弯矩系数，按现行《钢结构设计规范》[100]（GB 50017）确定。

7.6.4.2　约束钢梁的悬链线效应

在英国 Cardington 进行的一个 8 层足尺钢框架结构的抗火试验中发现，在结构中受到周围构件约束的钢梁的抗火性能可比一个独立钢梁的抗火性能要好的多，受约束的框架梁在火灾中的挠度可非常大，但梁不坍塌。见图 33，其原因是，随着梁的挠度的加大，梁中将产生拉力，梁的承载由原来的单纯的受弯机制，转变为受弯和受拉的联合机制。梁中的拉力将也提供抵抗梁横向荷载的承载力，这种效应称为梁的悬链线效应，见图 34。可见悬链线效应将对梁的抗火承载力有利。

图 33　Cardington 试验中钢框架梁的变形

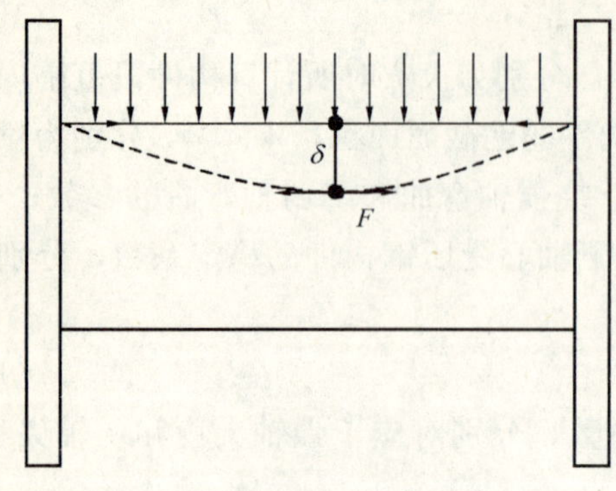

图 34 梁的悬链线效应

图 35 是一在端部受有轴向与转动约束钢梁的分析模型，火灾下约束钢梁轴力与挠度的变化如图 36 所示。可见，在受火初始阶段，由于热膨胀，梁内将产生轴压力。但随着温度的继续增加，梁挠度的增大，梁的悬链线效应将在梁内产生轴拉力，而抵消部分轴压力。当梁挠度达到一定值，悬链线效应产生的轴拉力与热膨胀效应产生的轴压力相等时，梁内轴压力降至零。当梁挠度继续增加时，梁内拉力将超过压力使梁受拉，最终，梁在拉力和弯矩作用下超过截面承载力而破坏。

显然，对于约束钢梁，以轴压力和轴压力平衡（即轴力为零）时的状态进行抗火设计，是偏于安全的。

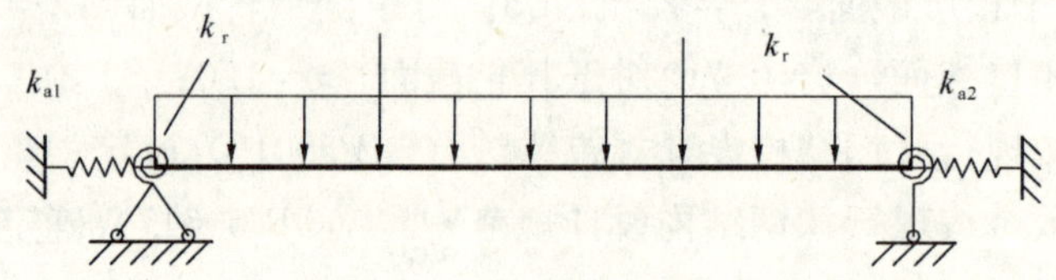

图 35 约束钢梁模型

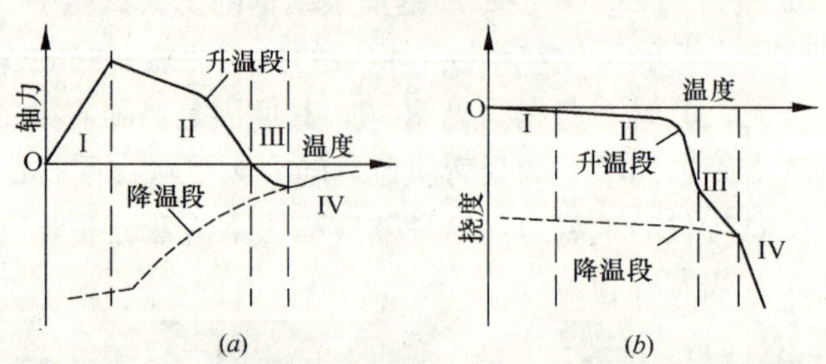

图 36 约束钢梁受火反应

（a）轴力变化；（b）挠度变化

7.6.4.3 楼板的薄膜效应

在 2001 年台北东方科技园区高层钢结构建筑火灾中，楼板产生了很大的挠曲线变形（见图 9），然而楼板并未坍塌，其原因是楼板的薄膜效应。薄膜效应是指薄板在大挠度下产生的一种张拉力，可以承受较大的板上横向荷载。

在钢结构中的楼板，在板面位置一般会配温度钢筋，见图 37，以防在常温下由于温度变化而引起的热胀缩，使板面开裂。这种温度钢筋会做成钢筋网的形式，可利用这种温度钢筋网承受楼板在大挠度下由薄膜效应产生的张拉力。图 38 显示了楼板随着挠度的增加，其承受荷载机制由弯曲形式逐渐转变为薄膜张力

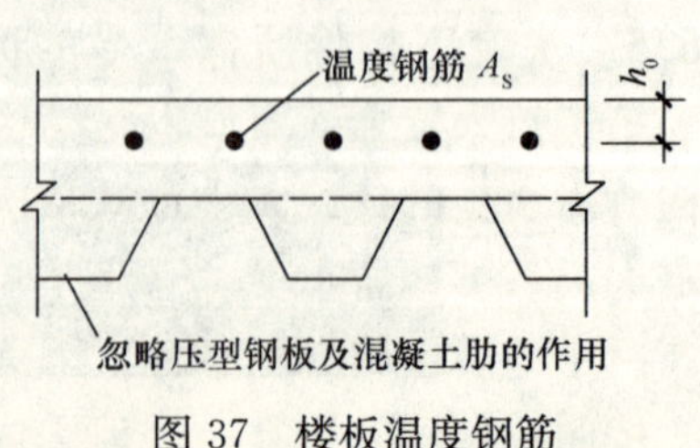

图 37 楼板温度钢筋

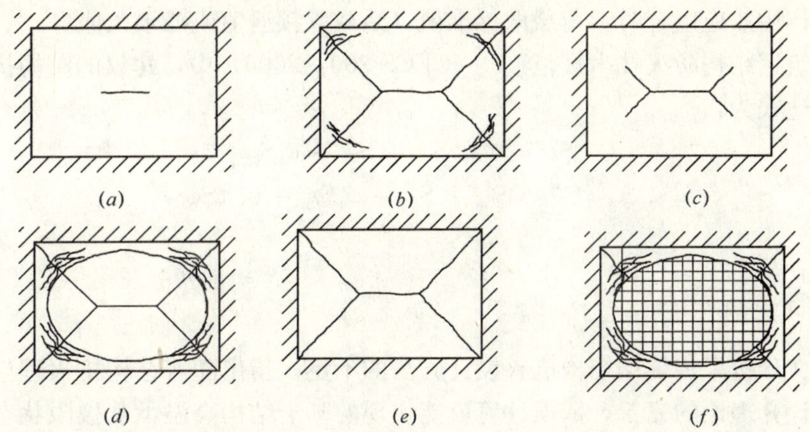

图 38　楼板内薄膜效应的形成过程

(a) 开始屈服；(b) 形成破坏机构；(c) 薄膜效应充分发展；
(d) 屈服线进一步发展；(e) 薄膜效应的产生；(f) 薄膜效应的极限状态

形式的过程，最终在楼板四周会形成一个混凝土压力环，承受钢筋网内的拉力。

利用楼板薄膜效应，我们有可能只对主梁进行防火保护，见图 39，只要所保护的钢梁所支撑的楼板在火灾下通过利用薄膜效应不坍塌，则板内的一些次梁可不进行防火保护。这样钢结构的防火成本可减少。

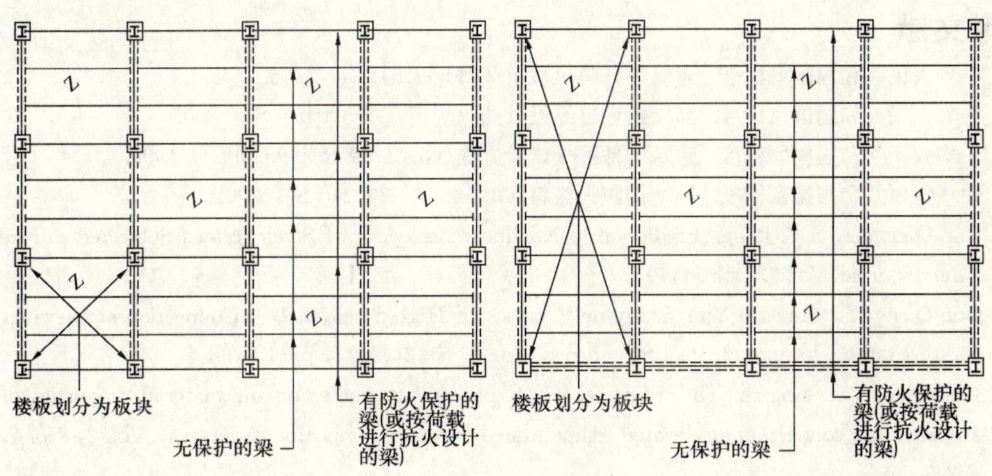

图 39　支撑楼板的钢梁防火保护方式

7.6.5　火灾下构件所需承受的荷载

结构抗火安全，实质上是结构在火灾条件下足够承受火灾发生时的荷载，因此火灾下结构所需承受的荷载大小，对结构抗火设计影响很大。由于火灾属于小概率偶然事件，在确定火灾下结构所承受的荷载时，应取火灾发生时荷载、楼面或屋面活载最可能出现的值。由于地震也属偶然事件，因此，火灾条件下可不考虑地震作用。

按照我国国家标准《工程结构可靠度设计统一标准》[101]（GB 50153）的规定，确定偶然设计状况作用效应组合时，荷载取标准值，活载取频遇值或准永久值。

在《建筑钢结构防火技术规范》[99]（CECS 200：2006）中，建议的结构抗火验算作用效应组合表达式为：

$$S_m = \gamma_{0T}(S_{Gk} + S_{Tk} + \psi_f S_{Qk}) \tag{18a}$$

$$S_m = \gamma_{0T}(S_{Gk} + S_{Tk} + \psi_q S_{Qk} + 0.4S_{Wk}) \tag{18b}$$

7.7 总结

（1）本文分析了传统钢结构抗火设计方法的不足，指出随着结构抗火理论研究的深入和结构抗火计算理论的完善，依据计算就能实现的基于结构高温承载极限状态的现代钢结构抗火设计方法必将得到推广应用。

（2）尽管国内外钢结构抗火研究取得了较大进展，但仍有一些涉及钢结构抗火设计的问题没有很好的解决，这些问题主要有：

①建筑火灾实际升温的精确且实用模型；

②结构中约束构件温度效应与约束效应的综合影响；

③火灾下结构整体非线性反应分析与倒塌判别准则；

④结构抗火可靠分析与抗火可靠度设计。

参考文献

[1] 中华人民共和国国家标准. 建筑设计防火规范(GB 50016)[S]，2006.

[2] 中华人民共和国国家标准. 高层民用建筑设计防火规范(GB 50045)[S]，2005.

[3] 中华人民共和国国家标准. 建筑构件耐火试验方法(GB/T 9978—1999)[S]，1999.

[4] 中华人民共和国国家标准. 钢结构防火涂料(GB 14907—2002) [S]，2002.

[5] Guo-Qiang Li，S. C. Jiang. Prediction to Nonlinear Behaviour of Steel Frames Subjected to Fire，Fire Safety Journal，Vol. 32，No. 4，1999，pp347-368

[6] Guo-Qiang Li，Ying-zhi Yin. Study on Behavior of High-Strength Bolt Connection at Elevated Temperatures，Int. Journal of Steel Structures，Vol 1，No 2，2001，pp113-122

[7] Guo-Qiang Li，Ying-zhi Yin，Ming-fei Li. Experimental studies on the material properties of high-strength bolt connection at elevated temperatures，Steel & composite structures，Vol. 2，No. 4，2002，pp247-258

[8] Guo-Qiang Li，S. C. Jiang，Ying-zhi Yin，Kai Chen，Ming-fei Li. Experimental Structural on the Properties of Constructional Steel at Elevated Temperatures，Journal of Structural Engineering，Vol 129，No. 12，2003，pp1717-1721

[9] Jun Ding，Guo-Qiang Li，Y. Sakumoto. Parametric studies on fire resistance of fire-resistant steel members，Journal of Constructional steel research，(60)2004，PP1007-1027

[10] Guo-Qiang Li and Jun Ding，Y. Sakumoto：A Practical approach for fire safety design of fire-resistant steel members，Steel and Composite Structures，Vol. 5. No. 1(2005)，PP71-86

[11] G. Q. Li，S. X. Guo and S. C. Jiang. On the Chinese Code on Fire Safety Design of Steel Building Structures，Steel and Composite Structures，Vol. 5. No. 5(Oct. 2005)，pp395-405

[12] Guo-Qiang Li, Pei-jun Wang. Spline Finite Element Methods for Analysis of Axially Restrained Steel Beams Subjected to Elevated Temperatures in Fire, Advances in Structural Engineering, Vol. 10 No. 2(April 2007), pp111-120

[13] Guo-Qiang Li, Shi-xiong Guo, Hao-sheng Zhou. Modeling of membrane action in floor slabs subjected to fire, Engineering Structure, Vol. 29 No. 6(June 2007), pp880-887

[14] Wei-Yong Wang, Guo-Qiang Li, Yu-Li Dong. Experimental study and spring-component modeling of extended end-plate joints in fire, Journal of Constructional Steel Research, 63(2007), pp1127-1137

[15] Guo-Qiang Li, Pei-jun Wang, S. C. Jiang. Non-linear finite element analysis of axially restrained steel beams at elevated temperatures in a fire, Journal of Constructional Steel Research, Vol. 63 No. 9(Sep. 2007), pp1175-1183

[16] 李国强，沈祖炎．钢框架结构非线性实用分析方法，结构工程师，1990，4．pp14-18

[17] 李国强，沈祖炎．建筑火灾损失的预测模型，四川建筑科学研究，1992，1.

[18] 李国强，沈祖炎．高温下轴心受压钢构件的极限承载力，建筑结构，1993，9．pp23-25，p36

[19] 李国强．火灾下钢框架结构的极限状态分析，土木工程学报，1994，1．pp49-56

[20] 李国强．钢梁抗火计算和设计的实用方法，工业建筑，1994，7．pp43-46

[21] 李国强，沈祖炎．钢柱抗火计算和设计的实用方法，工业建筑，1995，2．pp31-37，p61

[22] 蒋首超，李国强．高温下结构钢的材料特性，钢结构，1996，2．PP49-57

[23] 蒋首超，李国强．火灾下钢结构构件的升温，钢结构，1996，2．PP19-24

[24] 李国强，蒋首超．非均布高温钢结构梁单元切线刚度方程，同济大学学报，1999，1．PP1-5

[25] 李国强，殷颖智，蒋首超．火灾下组合楼板的温度场分析，工业建筑，1999，12．PP47-49

[26] 陈凯，李国强．火灾下变截面门式刚架结构整体抗火性能分析，结构工程师，1999，增刊 PP155-161

[27] 李国强，贺军利，蒋首超．约束钢梁的抗火试验与验算，土木工程学报，2000，4．PP23-26

[28] 李国强，贺军利，蒋首超．钢柱的抗火试验与验算，建筑结构，2000，9．PP12-15

[29] 李国强．钢结构抗火设计的发展，钢结构，2000，3．PP47-49

[30] 蒋首超，李国强．局部火灾下钢框架温度内力的实用计算方法，工业建筑，2000，9．PP56-61

[31] 李国强，陈凯，蒋首超，殷颖智．高温下 Q345 钢的材料性能试验研究，建筑结构，2001，1. pp53-55

[32] 李国强，贺军利，韩林海．钢管混凝土轴压构件抗火承载力的计算，建筑结构，2001，1. pp60-62

[33] 李国强，张晓进，蒋首超，殷颖智．高温下 SM41 钢的材料性能试验研究，工业建筑，2001，6. PP57-59

[34] 李国强，陈凯．门式钢刚架结构实用抗火设计方法，建筑结构，2001，6．PP14-18

[35] 李国强，李明菲，殷颖智，蒋首超．高温下高强度螺栓 20MnTiB 钢的材料性能试验研究，土木工程学报，2001，5．PP100-104

[36] 丁军，李国强，蒋首超．火灾下钢结构构件的温度分析，钢结构，2002，2．pp53-56

[37] 李国强．现代钢结构抗火设计方法，消防科学与技术，2002，1．pp8-11

[38] 蒋首超，李国强，李明菲．高温下压型钢板-混凝土粘结强度的试验，同济大学学报，2003，3. pp273-276

[39] 丁军，李国强．火灾下钢结构楼板的薄膜作用，建筑钢结构进展，2003，2．pp17-23

[40] 李国强，殷颖智．钢结构高强度螺栓连接抗火性能的有限元分析，土木工程学报，2003，6. PP18-25

[41] 李国强，丁军. 耐火钢梁的抗火性能参数分析与抗火设计，钢结构，2003，5. pp52-55

[42] 李国强，李兆治. 钢结构性能化抗火设计的初步设想，消防科学与技术，2004，1. pp46-48

[43] 蒋首超，李国强，楼国彪，孙元杰. 钢-混凝土组合楼盖抗火性能的数值分析方法，建筑结构学报，2004，3. PP38-44

[44] 蒋首超，李国强，周宏宇，王琦. 钢-混凝土组合楼盖抗火性能的试验研究，建筑结构学报，2004，3. PP45-50

[45] 丁军，李国强. 耐火钢柱的抗火性能参数分析与抗火临界温度估计，建筑钢结构进展，2004，3. PP19-22

[46] 李明菲，李国强，王银志. 墙板对火灾下钢柱截面温度分步的影响，建筑钢结构进展，2004，3. PP45-50

[47] 蒋首超，李国强，叶绮玲. 火灾下钢框架结构非线性反应分析与试验研究，工业建筑，2004，8. PP66-69

[48] 李国强，杜咏. 大空间建筑顶部火灾空气升温的参数分析，消防科学与技术，2005，1. pp19-21

[49] 李国强，李明菲，王银志. 考虑墙板约束影响的轴心受压钢柱抗火设计方法，建筑钢结构进展，2005，1. pp41-46

[50] 李国强，郭士雄，蒋首超，杜咏. 上海南站钢屋盖结构性能化抗火安全评估，建筑钢结构进展，2005，2. pp31-36

[51] 李国强，王银志，李明菲. 高温下墙板约束对轴心受压钢柱承载力的影响，钢结构，2005，3. pp85-88

[52] 王银志，李国强. 火灾中考虑整体性的钢梁破坏形态研究，结构工程师，2005，3. pp25-29

[53] 黄珏倩，李国强，杜咏. 对运用双区域模型计算大空间建筑火灾温度方法的修正，消防科学与技术，2005，3. pp279-282

[54] 李国强，杜咏. 实用大空间建筑火灾空气升温经验公式，消防科学与技术，2005，3. pp283-287

[55] 李国强，丁军，张韧. 耐火钢梁抗火计算与设计的方法，工业建筑，2005，7. pp80-82

[56] 丁军，李国强. 耐火钢构件抗火性能非线性有限元分析，建筑结构，2005，8. pp61-63

[57] 李国强，黄珏倩. 双区域模拟大空间火灾烟气下降和升温规律的分析，消防科学与技术，2005，5. pp527-531

[58] 周焕廷，李国强. 双曲索网结构抗火极限承载力影响因素分析，钢结构，2005，5. PP86-89

[59] 王银志，李国强. 火灾中约束钢梁破坏过程分析，钢结构，2005，5. PP79-81

[60] 王佳，李国强. 防火涂料局部破损对钢构件在火灾下的温度分布影响，结构工程师，2005，5. PP30-35

[61] 郭士雄，李国强. 火灾下约束钢梁的受力性能及抗火设计方法，建筑结构，2005，12. PP59-61

[62] 蒋东红，李国强，王世伟，张彬. 钢骨混凝土轴压柱抗火极限承载力计算，钢结构，2005，6. PP87-91

[63] 周宏宇，李国强，王银志. 简支组合梁抗火性能参数研究，钢结构，2005，6. PP92-96

[64] 李国强，吴波，韩林海. 结构抗火研究进展与趋势，建筑钢结构进展，2006，1. PP1-13

[65] 李国强，杜咏，王银志，蒋首超. 杭州国际会议中心钢结构抗火性能评估与设计，建筑钢结构进展，2006，1. PP14-22

[66] 李国强，蒋首超，陆立新，杜咏，郭士雄. 某综合楼钢结构抗火安全设计与评估(I)，防灾减灾工程学报，2006，1. PP13-20

[67] 陆立新，李国强，蒋首超，沈俊昶，吴保桥，吴结才. 马钢耐火钢梁的抗火试验与设计验算，建筑结构，2006，4. PP59-60 转 76

[68]　李国强，郭士雄，蒋首超．某综合楼钢结构抗火安全设计与评估(II)，防灾减灾工程学报，2006，2. PP123-128

[69]　楼国彪，李国强，雷青．钢结构高强度螺栓端板连接研究现状(I)，建筑钢结构进展，2006，2. PP8-21

[70]　蒋东红，李国强，张彬．钢骨混凝土偏压抗火极限承载力分析研究，建筑钢结构进展，2006，2. PP55-62

[71]　楼国彪，李国强，雷青．钢结构高强度螺栓端板连接研究现状(II)，建筑钢结构进展，2006，3. PP16-23

[72]　杜咏，李国强．基于场模型的大空间建筑火灾钢构件升温的简化计算方法，消防科学与技术，第 25 卷．3. PP299-303

[73]　李国强，陆立新，蒋首超，沈俊昶，吴保桥，吴结才．马钢耐火钢柱的抗火试验与设计验算，建筑钢结构进展，2006，4. PP12-16

[74]　蒋首超，李国强，周昊圣，吕毅．钢-混凝土组合楼板实用抗火设计方法，建筑结构，2006，8. PP87-89 转 86

[75]　周宏宇，李国强，王银志．影响组合梁抗火性能的两个因素分析，建筑钢结构进展，2006，5. PP40-45

[76]　李国强 郭士雄．受火约束钢梁在升温段和降温段行为的理论分析(Ⅰ)，防灾减灾工程学报，2006，3. PP241-250

[77]　李国强，郭士雄．约束钢梁高温下大变形状态分析，同济大学学报(自然科学版)，2006，7. PP853-857

[78]　蒋首超，陆立新，李国强，沈俊昶，吴保桥，吴结才．马钢耐火钢高温下材料性能试验研究，土木工程学报，2006，8. PP72-75

[79]　沈俊昶，杨才福，马鸣图，蒋首超，陆立新，李国强．耐火钢综合性能及构件抗火试验分析，钢结构，2006，4. PP87-91

[80]　王卫永，李国强，于克强．外伸端板节点在火灾下的转角-温度关系，钢结构，2006，3. PP92-94

[81]　李国强，王银志，王孔藩．考虑结构整体的组合梁极限抗火性能分析，力学季刊，2006，4. PP726-731

[82]　靳飞，李国强．全盛期室内火灾参数化模型的参数随机性，建筑科学与工程学报，2006，4. PP44-48

[83]　郭士雄，李国强．受火约束钢梁在升温段和降温段行为的理论分析(Ⅱ)，防灾减灾工程学报，2006，4. PP359-368

[84]　王卫永，董毓利，李国强．外伸端板节点火灾行为的试验研究和理论分析，哈尔滨工业大学学报，2006，12. PP2125-2128

[85]　李国强，王培军．轴向约束梁柱在火灾引起温度沿截面分布不均匀下的大变形分析，力学季刊，2007，2. PP246-255

[86]　黄珏倩，李国强．门式钢刚架结构不需防火保护层的条件，消防科学与技术，2007，4. pp371-374

[87]　刘玉姝，李国强，蒋首超，楼国彪．北京奥林匹克公园会议中心钢管摇摆柱性能化抗火分析与设计，土木工程学报，2007，9. pp1-7

[88]　李国强，周昊圣，郭士雄．火灾下钢结构建筑楼板的薄膜效应机理及理论模型，建筑结构学报，2007，5. pp40-47

[89]　李国强，郭士雄，周昊圣．火灾下钢结构建筑楼板的薄膜效应模型验证及实用方法，建筑结构学

报，2007，5. pp48-53

[90] 李国强，周宏宇. 钢-混凝土组合梁抗火性能试验研究，土木工程学报，2007，10. pp19-26

[91] 黄珏倩，李国强，包盼其，刘克. 门式钢刚架实用抗火临界温度计算方法，建筑钢结构进展，2007，6. pp42-48

[92] 李国强，王银志，郝坤超. 约束组合梁抗火性能试验研究，自然灾害学报，2007，6. pp93-98

[93] 杜咏，李国强，黄珏倩. 大空间建筑火灾中烟气温度计算模型的比较，自然灾害学报，2007，6. pp99-103

[94] 李国强，周焕廷. 火灾(高温)下索网结构计算的连续化方法，力学季刊，2007，4. pp638-646

[95] 李国强，周焕廷. 索网结构抗火实用设计方法，空间结构，2007，4. pp43-50

[96] 王培军，李国强. 弹性轴向约束平面钢梁火灾下非线性分析的弧线坐标法，工程力学，2008，1. pp137-144

[97] 李国强，韩林海，楼国彪，蒋首超. 钢结构及钢混凝土组合结构抗火设计. 北京：中国建筑工业出版社，2006

[98] 李国强，蒋首超，林桂祥. 钢结构抗火计算与设计. 北京：中国建筑工业出版社，1999

[99] 中国工程建设标准化协会标准. 建筑钢结构防火技术规范(CECS200：2006)[S]，2006

[100] 中华人民共和国国家标准. 钢结构设计规范(GB 50017)[S]，2003

[101] 中华人民共和国国家标准. 工程结构可靠度设计统一标准(GB 50153)[S]，1992

第 8 章　Chapter 8

充分利用结构健康监测体系:香港的经验

Making Good Use of Structural Health Monitoring Systems: Hong Kong's Experience

Y. L. Xu(徐幼麟)

Department of Civil and Structural Engineering, The Hong Kong Polytechnic University, China

E-mail: ceylxu@polyu. edu. hk

Abstract: Many innovative tall buildings, long span bridges and large space structures emerge in recent years but expose to harsh conditions such as strong winds and severe earthquakes. The installation of long-term comprehensive structural health monitoring systems (SHMS) to important structures becomes a trend to monitor their loading conditions, to assess their performance, to detect their damage, and to guild their maintenance with the utmost goals of ensuring the structures function properly during their long service life and preventing them from catastrophic failure under extreme events. Nevertheless, it is still not clear how to make good use of SHMS toward these goals although SHMS have been installed in some large structures. This paper takes the Tsing Ma Bridge in Hong Kong as an example to manifest how the SHMS installed in the bridge has been utilized since 1997. The SHMS installed in the Tsing Ma Bridge is briefly introduced first. How to use the SHMS for investigating highway loading, railway loading, wind characteristics, and temperature effects is then presented. Identification of time varying natural frequencies and modal damping ratio of the bridge under strong winds in terms of the SHMS is also demonstrated. Toward the performance assessment and damage detection of the bridge, SHMS-based computer simulation and damage assessment are targeted and some typical examples are given. The establishment of SHMS-based bridge rating system for bridge maintenance is commented as an on-going research work. The difficulties encountered and future work needed in this field are finally highlighted for both academic research and practical application.

Keywords: Structural health monitoring system; long span bridges; highway loading; railway loading; wind; temperature; computer simulation; performance assessment; bridge rating system; future work

8.1 Introduction

The increasing globalization of world economy significantly impacts the engineering profession and intensifies the competition of construction industries worldwide. Such competition also brings many innovative large civil structures made of new materials and implemented by new technologies. For instance, in addition to a number of newly-built long span cable-supported bridges, Hong Kong and Mainland China are constructing the world longest cable stayed bridge with a main span over 1000 m and Italy is planning to build the world longest suspension bridge with a main span over 3000m. These super long span bridges surrounded by harsher environment than ever before engender many challenges to professionals on how to ensure these structures function properly during their long service period and how to prevent them from sudden failure or fatal disaster during strong winds, severe earthquakes, terrorist attacks, and other abnormal events.

Recently-developed structural health monitoring technology provides a better solution for the problems concerned. Structural health monitoring technology is based on a comprehensive sensory system and a sophisticated data processing system implemented with advanced information technology and supported by cultivated computer algorithms. The main objectives of the structural health monitoring are to monitor the loading conditions of a structure, to assess its performance under various service loads, to verify or update the rules used in its design stage, to detect its damage or deterioration, and to guide its inspection and maintenance. In Hong Kong and Mainland China, structural health monitoring systems (SHMS) have been installed in more than 40 long span bridges and a few tall buildings and large space structures. The early-installed SHMS are simple with a few sensors, but the newly-installed SHMS are comprehensive with a variety of sensors located at different parts of the structure. Nevertheless, many key issues remain unsolved in pursuing the utmost goals. It is also not clear how to make good use of data recorded by SHMS toward the utmost goals.

The Tsing Ma Bridge in Hong Kong is the longest suspension bridge in the world carrying both highway and railway. The Tsing Ma Bridge is also located in one of the most active typhoon prone regions in the world. The Hong Kong Highways Department installed a comprehensive Wind And Structural Health Monitoring System (WASHMS) and a Global Positioning System-On-Structure Instrumentation System (GPS-OSIS) in the Tsing Ma Bridge in 1997 and 2000, respectively (Wong et al. 2001*a*, 2001*b*). This paper takes the Tsing Ma Bridge as an example to manifest how the SHMS (WASHMS + GPS-OSIS) installed in the bridge has been used since 1997. The SHMS installed in the Tsing Ma Bridge is briefly introduced first. How to use the SHMS for investigating highway loading, railway loading, strong wind characteristics, and temperature effects is then presented. Identification of time varying natural frequencies and modal damping ratio of the bridge under strong winds in terms of the SHMS is also demonstrated. Toward the performance assessment and damage detection of the bridge, SHMS-based computer simulation and damage assessment are targeted and some typical examples are given. The establishment of SHMS-based bridge rating system for bridge maintenance is commented as an on-going research work. The difficulties encountered and future work needed in this field are finally highlighted for both academic research and practical application.

8.2 Shms in tsing ma bridge

8.2.1 Tsing Ma Bridge

The Tsing Ma Bridge has an overall length of 2,132 m and a main span of 1,377 m between the Tsing Yi tower in the east and the Ma Wan tower in the west (see Fig. 1). The height of the two reinforced concrete towers is 203 mPD. The two main cables of 1.1 m diameter and 36 m apart in the north and south are accommodated by the four saddles located at the top of the tower legs. The bridge deck is a hybrid steel structure consisting of Vierendeel cross frames supported on two longitudinal trusses acting compositely with stiffened steel plates. The bridge deck is suspended by suspenders in the main span to allow a sufficiently large navigation channel and in the Ma Wan side span to minimize the number of substructures found in the sea. Due to the highway layout requirement, on the Tsing Yi side the deck is supported by three piers rather than suspenders. The bridge deck carries a dual three-lane highway on the upper level of the deck and two railway tracks and two carriageways on the lower level within the bridge deck. The further information on the bridge can be found in Xu et al. (1997a), Xu et al. (1997b), and Ko et al. (1998).

8.2.2 WASHMS in Tsing Ma Bridge

The Wind And Structural Health Monitoring System (WASHMS) in the Tsing Ma Bridge is composed of five sub-systems, namely, sensory system, data acquisition system, data processing and analysis system, computer for system operation and control, and fibre optic cabling network system (Wong et al, 2001*a*). The sensory system consists of about 300 sensors and associated interfacing units installed at different locations of the bridge (see Fig. 1). They include anemometers, temperature sensors, accelerometers, strain gauges, level sensing stations, displacement transducers, weigh-in-motion sensors, signal amplifiers, and interfacing equipment. The data acquisition system refers to the computer controlled data acquisition outstation units with appropriate data acquisition interfaces and the software for parameter configuration. The data acquisition outstation units are installed on/inside the bridge and their major functions are to collect and digitize signals received from the sensory system and to deliver them to the data processing and analysis system through the fiber optical cabling network system installed along the bridge alignment. The data processing and analysis system is located at the Tsing Yi administrative building and it is a

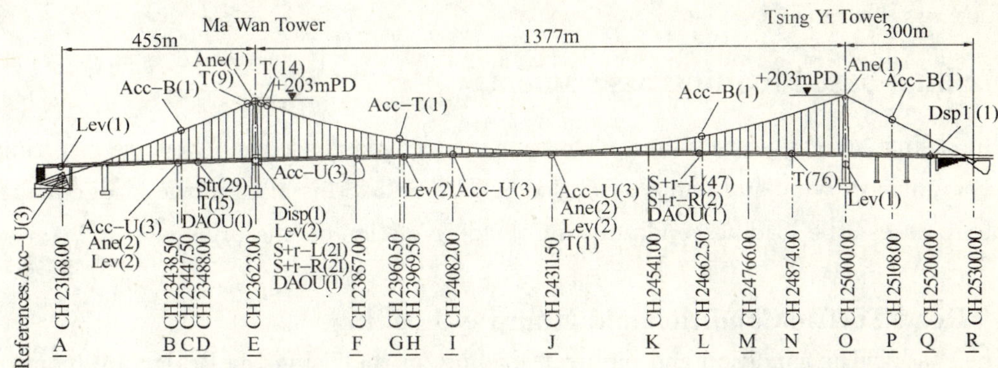

Fig. 1　Tsing Ma Bridge and layout of sensory system

workstation for overall data collection, transmission, storage, control, and post-processing. The computer for system operation and control is also a workstation located at the Tsing Yi administrative building and is equipped with appropriate software for graphical inputs and outputs of structural modeling and analysis works.

8.2.3 GPS-OSIS in Tsing Ma Bridge

The WASHMS in the Tsing Ma Bridge, however, has some weak points in bridge displacement response monitoring. The level sensing stations provide the real time monitoring of displacements at typical stiffening deck sections but in vertical direction only. To improve the efficiency and accuracy of the WASHMS in displacement monitoring, the Hong Kong Highways Department further installed a Global Positioning System-On-Structure Instrumentation System (GPS-OSIS) in December 2000 with on-going updating until September 2002 to monitor the absolute displacements of the cables, the stiffening deck and the bridge towers (Wong et al, 2001*b*).

The GPS-OSIS installed on the Tsing Ma Bridge consists of five sub-systems, namely, GPS sensory system, local data acquisition system, global data acquisition system, GPS computer system, and optical fiber network system. The GPS sensory system is installed on locations where the maximum bridge displacements are expected. There are a total of 14 fixed measuring points for the bridge located at the edges of deck sections at mid-span, the main cables at mid-span, and the tower-tops. Two additional GPS sensory systems installed adjacent to the Tsing Yi administrative building are used as the base reference stations. The GPS-OSIS of the bridge adopts a highly effective and stable optical fiber network system for data transmission between the local data acquisition system and the global data acquisition system. The GPS computer system consists of two workstations. One workstation is used for processing, archiving, storage and graphical treatment of data and for displaying the real-time motions of the bridge. It is also responsible for the operation and control of the whole GPS-OSIS. The other workstation is used for post-processing of the GPS data and to work with the measured data from the WASHMS.

8.3 Shms-based loading assessment

The Tsing Ma Bridge is subjected to four major types of loads. They are referring to highway, railway, wind and temperature loads. The SHMS installed in the bridge makes it possible to assess the loading conditions and loading effects on the bridge.

8.3.1 Road Traffic Condition and Highway Loading

The road traffic condition and highway loading on the Tsing Ma Bridge are monitored by dynamic weigh-in-motion (WIM) stations for two carriageways (the Airport bound way and the Kowloon bound way) at the approach to the Lantau Toll Plaza near the Lantau administration building as shown in Fig. 2. Each WIM station is composed of two bending path pads and two magnetic loop detectors. The two bending path pads are placed on the left and right sides respectively of each station for detecting vehicle weight. The two magnetic loop detectors are installed in the front and rear lines respectively of each station at a definite distance for measuring axle numbers, axle spacing, vehicle speed and others.

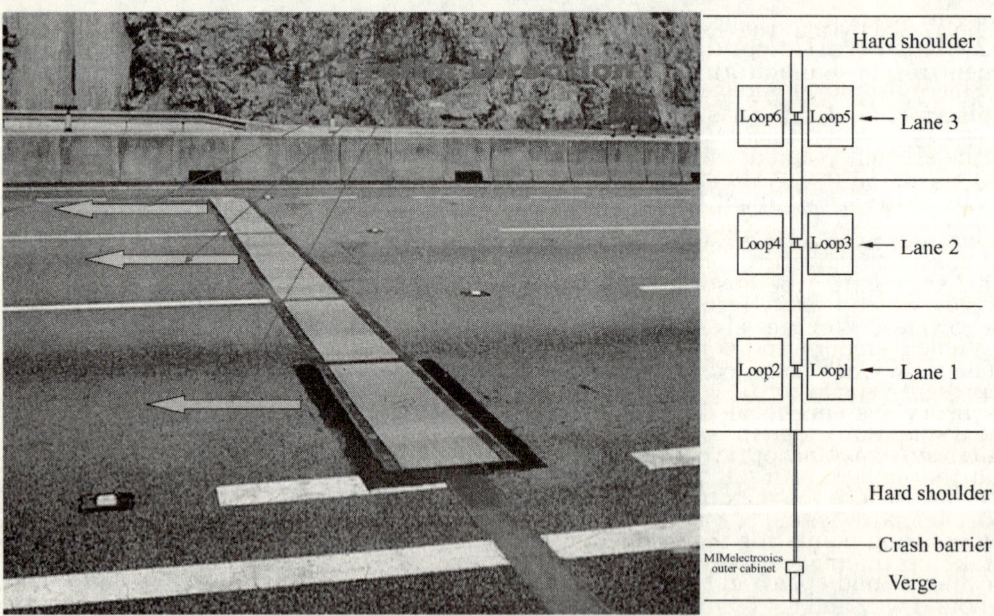

Fig. 2 Dynamic weigh-in-motion stations at Lantau Toll Plaza

The M-class vehicle classification system is used to classify types of vehicles running on the bridge according to their main features such as the number of axles, axle distance and gross weight. According to the M-class vehicle classification system (see Table 1), the vehicles running on the bridge can be classified into 8 categories based on the information from the WIM stations.

M-class vehicle classification system　　Table 1

Class (1-8)	Vehicle category	Short string	No. of axle	Magnetic vehicle length	Axle distance (1. to 2.)	Gross weight
1	Motor cycles	MC	2	<2 m plus magnetic contour analysis		<0.6T
2	Cars, Vans or Taxi	C/V	2, 3, 4	<6.3m		< 3T
3	Public service vehicles	PSV		Magnetic contour analysis	< 5.5m	< 4T
4	Light goods vehicles	LGV	2, 3, 4	≥6.3m <10m	<3.5m	<5.5T
5	Medium goods vehicles	MGV	2,3,4,5,6,7,8		≥3.5m<7.8m	<24T
6	Rigid heavy goods vehicles	RHGV	2,3,4,5,6,7,8		≥3.5m<7.8m	<38T
7	Articulated heavy goods vehicles	AHGV	2,3,4,5,6,7,8		≥3.5m<7.8m	<44T
8	Buses and Coaches	BUS	2,3,4	Magnetic contour analysis	≥7.8m	<24T

8.3.2 Road traffic condition

The WIM stations have recorded the traffic information since August 1998. Owing to the need to conduct major updates for the entire system, both the Airport and Kowloon bound stations were in suspension from July 2004 to March 2005. Furthermore, the traffic information detailed to each lane was not available until April 2005. Therefore, to have a detail analysis of traffic condition for each lane, the WIM data for the whole year 2006 are analyzed, which best reflects the current traffic condition and provides traffic lane information. Vehicle amount and composition of different vehicle types are two important indexes to represent vehicle traffic condition. The traffic volume and traffic composition in each lane of the Tsing Ma Bridge in 2006 are shown in Fig. 3 for 8-class vehicle classification. It is observed that in 2006, a total of 8.5 million vehicles run through the Tsing Ma Bridge for the Airport bound way and 8.8 million vehicles for the Kowloon bound way. The car, vans and taxi take up the biggest percentage among all types of vehicles: about 69.8% for the Airport bound way and 69.0% for the Kowloon bound way. The percentage of heavy goods vehicles, including rigid heavy goods vehicles and articulated heavy goods vehicles, is about 5% to 6% of the total vehicles. More vehicles run on the middle lane than other lanes for both the Airport and Kowloon bound ways, which account for 49.1% and 55.5% respectively of the total vehicles. Most of heavy goods vehicles use the slow lane for both the Airport and Kowloon bound ways, accounting for 95.1% and 89.1% respectively of the total heavy vehicles. Most of medium goods vehicles, including medium goods vehicles and buses, use the slow lane for both the Airport and Kowloon bound ways, accounting for

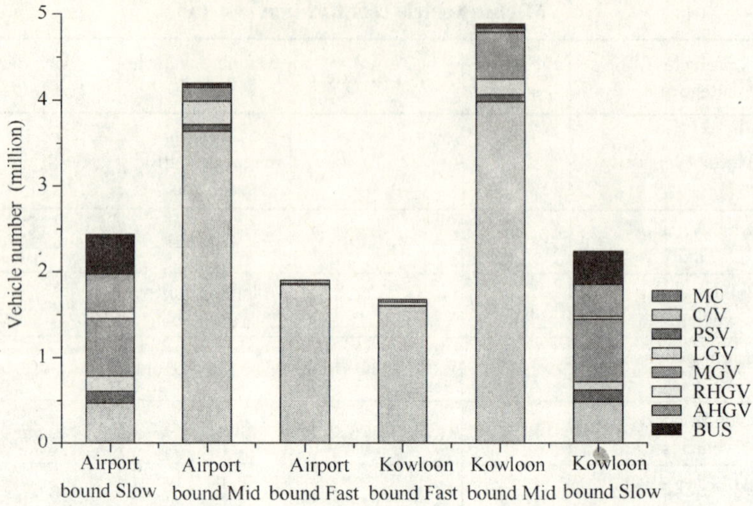

Fig. 3 Vehicle count by 8-class vehicle classification

85.7% and 64.4% respectively of the total medium vehicles.

8.3.3 Highway loading

Axle load is an important parameter for bridge pavement and bridge structure. The percentage of axle load in different loading ranges with 1 ton interval is determined according to the aforementioned vehicle classification system. Based on the statistical WIM data of year 2006, the distributions of axle load are displayed in terms of vehicle category in Fig. 4 and bridge lane in Fig. 5. It is observed that the axle number decreases with increasing axle load: only 12.5% of the total vehicle axles have an axle load more than 5 tonnes. Most of vehicle axles have an axle load less than 1 tonne, which account for 54.1% and 48.6% of the total axles respectively for the Airport and Kowloon bound ways. The second most axle number corresponds to the axle load between 1 to 2 tons. More axles of vehicles run on the

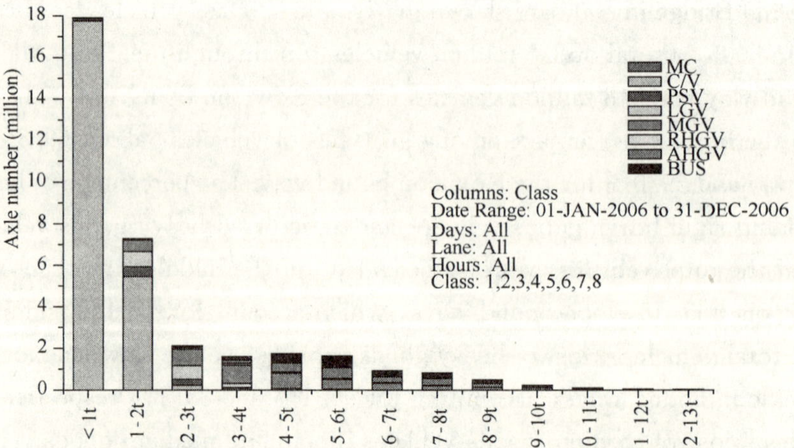

Fig. 4 Axle loading distribution against axle number and vehicle category

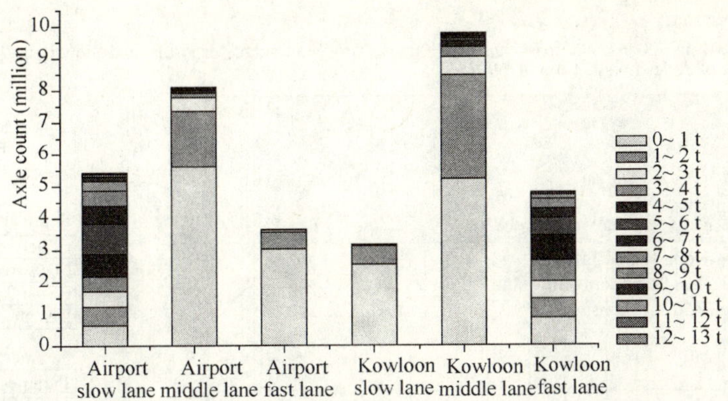

Fig. 5　Axle loading distribution against axle number and bridge lane

middle lane than the slow lane and fast lane, which account for about 47. 2% and 55. 1% of the total axles respectively for the Airport and Kowloon bound ways. Most of the axles with axle load more than 5 tonnes run on the slow lane, accounting for about 95. 5% for the Airport bound way and 84. 2% for the Kowloon bound way.

Gross vehicle weight (GVW) is another important parameter for bridge pavement and bridge structure. The percentage of vehicles in different GVW ranges with 4 tone interval is determined based on the WIM data of year 2006. Because of space limitation, the results are not presented here. For traffic loading determination, the derived numbers of axles and vehicles falling into their corresponding individual load ranges shall be presented in ascending series in a probabilistic way to form an axle load spectrum and/or a GVW spectrum. These spectra can then be used as a basis for deriving the equivalent number of passage of standard fatigue vehicle for estimating fatigue life of the bridge. A typical example of the axle load spectrum from the recorded WIM data in 2006 for the slow lane of the Airport bound way is presented in Fig. 6 together with the two reference load spectra: the standard vehicle axle load histogram specified in BS5400 Part 10 and the design vehicle axle load histogram.

In deriving the axle load spectrum plotted in Fig. 6, an effective axle load is defined based on axle loads in all the axle load ranges and their corresponding proportions. Some typical points in the axle load spectrum are also specified in Fig. 6. For instance, according to the Hong Kong road traffic regulations and considering the overloaded axle, the upper limit of the allowable maximum axle load is set as 12. 5 tons. The percentage of overloaded axles based on the measurement data in 2006 is therefore given in Fig. 6. In addition, some important exceeding possibilities are also provided in Fig. 6 for the sake of clarification. It is observed that the axle load spectra from the measured WIM data and based on BS5400 both are on the safer side compared with the design spectrum. The axle load spectrum from the measured WIM data is quite close to that based on BS5400, but the axle load spectrum from the measured WIM data is said to be safer than the reference axle load spectrum based on BS5400 because the former effective axle load is smaller than the latter one. Further-

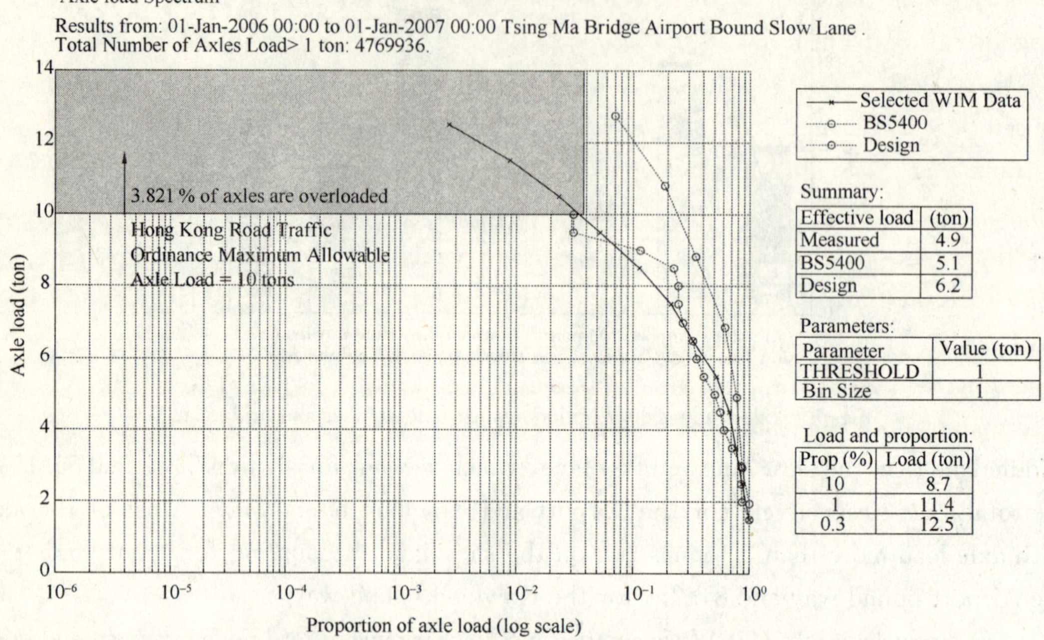

Fig. 6 Axle load spectrum for airport bound slow lane

more, there is a significant proportion of overloaded axles (axle loads more than 10 tons) on the slow lane with the percentage of 3.8% of all the vehicle axles concerned.

8.3.4 Train Traffic Condition and Railway Loading

The railway connecting the Lantau Island including the Hong Kong International Airport to the existing commercial centres of Hong Kong Island and Kowloon via the Tsing Ma Bridge is managed by the Mass Transportation Railway (MTR) Corporation Hong Kong Limited. The railway operation began in June 1998. Railway tracks supported on the bottom chords of the cross frames of the bridge deck are constituted of track plates, rail waybeams and tee diaphragms. The rail waybeams are the main longitudinal members which exist in pair in each railway track. The rail waybeams are stiffened by the diaphragms which are the T-shape members interconnecting the two parallel rail waybeams in orthogonal direction. On top of the stiffened rail waybeams, 20 mm thick steel track plates are laid over them. To monitor train traffic flow and identify bogie load distribution, a set of strain gauges were installed on the inner waybeam of each pair of waybeams under the two rail tracks at chainage 24662.5. Through a proper calibration, the signals from the strain gauges can be converted to the bogie load data, by which the requested information on train traffic flow and bogie load distribution can be obtained. This special measurement system was put into operation in 2000.

8.3.5 Train traffic condition

Traffic volume and composition of trains are two important indexes of train traffic condition. Based on the bogie load records, the annual number of trains passing through the bridge is calculated in terms of the number of bogies and shown in Fig. 7 for 6 years from 2000 to 2005. It is observed that there is a significant increase in annual train count in 2003 and afterwards the annual train count becomes stable around 150,000. Before 2003, 14-bogie trains (7-car trains) are dominant with about 96% of all the trains passing through the bridge annually. Since 2003, the number of 16-bogie trains (8-car trains) becomes more than that of 14-bogie trains. After the opening of the Hong Kong Disneyland in September 2005, 16-bogie trains become dominant. In November 2005, the percentage of the 16-bogie trains is already 90%. November and December of 2005 can be considered as the months representing the current train traffic condition. The monthly train count is about 12000 and more than 90% of the trains are 16-bogie trains.

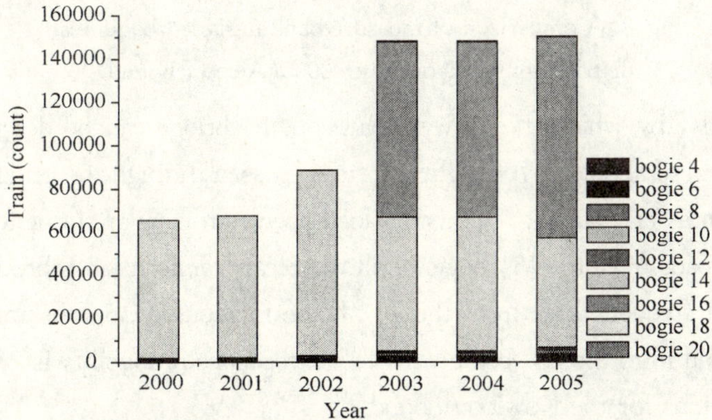

Fig. 7　Annual train count (2000～2005)

8.3.6 Railway loading

The bogie load data recorded in November 2005 are chosen as a database to investigate bogie load distribution. Because about 90% of the trains, that is, a total of 10705 trains, passing through the Tsing Ma Bridge in November 2005 are 16-bogie trains, the investigation of bogie load distribution focuses on 16-bogie trains. The bogie load distribution in a 16-bogie train is actually the distribution of 16 bogie loads along the train. For each 16-bogie train passing through the bridge in either the Airport bound or the Kowloon bound, one bogie load distribution can be obtained. The bogie load distributions with the maximum bogie load and the minimum bogie load in November 2005 are shown in Fig. 8 for the Airport bound. The mean of the bogie load distributions is also plotted together with the design bogie load distribution provided by the MTR Corporation Hong Kong Limited. It can be seen that that the pattern of the measured bogie load distributions follows the design pattern. The bogie load distribution with the maximum bogie load is below the design bogie load distribution for both the Airport and Kowloon bounds.

Bogie load spectrum is defined as either the probability density function or the probability dis-

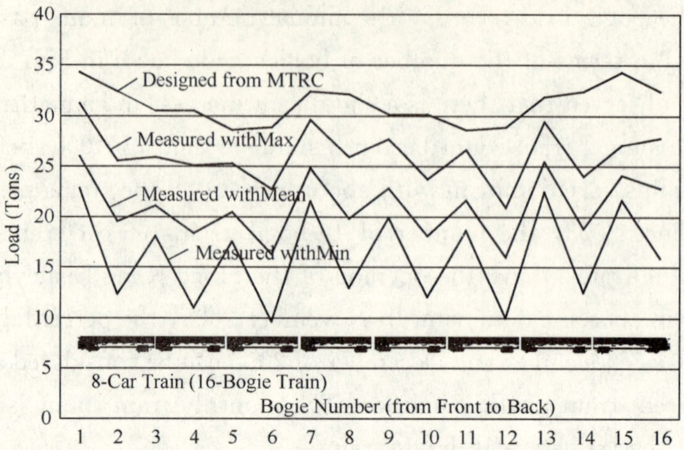

Fig. 8 Comparison of measured and designed bogie load distributions in November 2005(Airport bound)

tribution of bogie loads, by which the railway loads on the bridge can be described statistically. In November 2005, a total of about 10,705 eight-car trains passed through the bridge. By taking the 13th bogie in an 8-car train as an example, the bogie load spectrum is figured out to see the probability distribution of bogie loads from the 13th bogie of all the 8-car trains passing through the bridge. It can be seen that the most frequent load from the 13th bogie is about 25.5 tons for the Airport bound (Fig. 9). The bogie load from the 13th bogie with 1% exceedance probability is 28.5 tons for the Airport bound and 26.4 tons for the Kowloon bound.

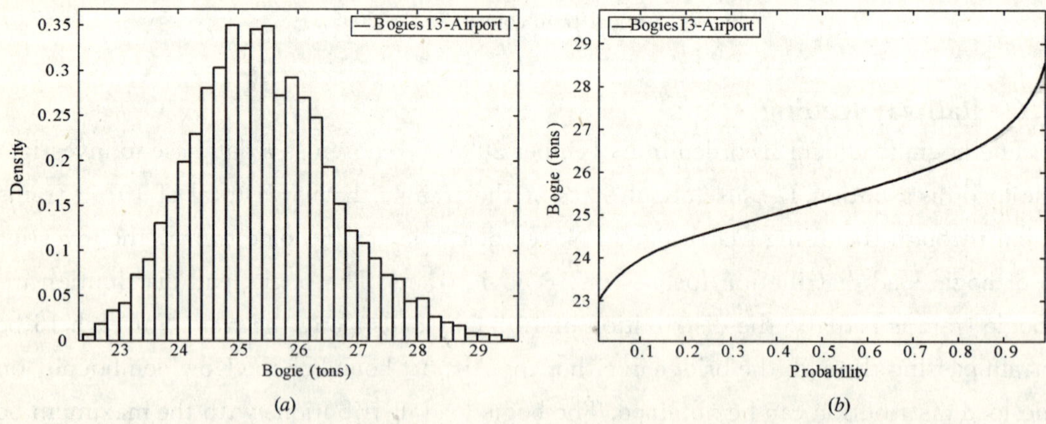

Fig. 9 Bogie load spectrum of the 13th Bogie (Airport bound)

(*a*) Probability density distribution; (*b*) Probability distribution

Gross train weight (GTW) spectrum is defined as either the probability density function or the probability distribution of gross train weights. In November 2005, a total of about 10,705 eight-car trains passed through the Tsing Ma Bridge. Fig. 10 shows the gross train weight spectrum for the Airport bound from all the 8-car trains passing through the Tsing Ma Bridge in November 2005. It can be seen that the most frequent GTW among all

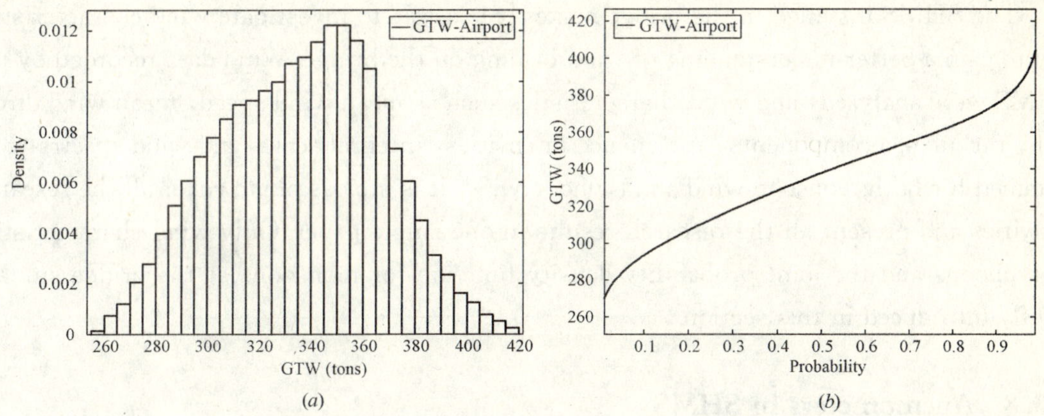

Fig. 10　GTW spectrum (Airport bound)

(*a*) Probability density distribution; (*b*) Probability distribution

the 8-car trains in November 2005 is about 350 tons for the Airport bound. The GTW with the exceedance probability of 1% is 405 tons for the Airport bound and 412 tons for the Kowloon bound, which is below the design maximum load of 498.4 tons.

8.3.7 Wind Characteristics

Hong Kong is situated at latitude N22.2° and longitude E114.1° and it is just on the south eastern coast of China facing the South China Sea. The weather system of Hong Kong is influenced by the land mass to its north as well as by the ocean to its south and east. Two types of wind conditions dominate Hong Kong: monsoon wind prevailing in the months from November to April and typhoon wind predominating in the summer. The local topography surrounding the bridge is quite unique and complex, which includes sea, islands, and mountains of 69 to 500m high (see Fig. 11). The alignment of bridge deck deviates for 17° in counterclockwise from the east-west axis (see Fig. 12). The complex topography makes wind characteristics at the bridge site very complicated.

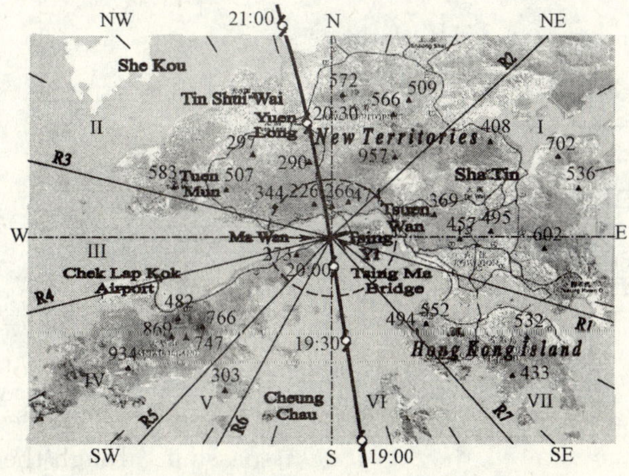

Fig. 11　Local topography of Hong Kong

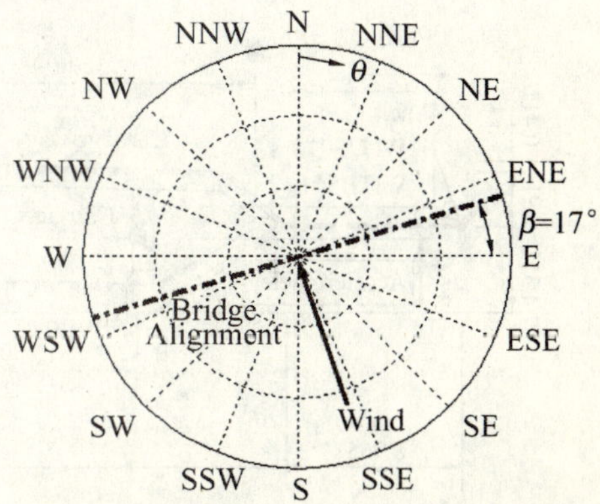

Fig. 12　Alignment of Tsing Ma Bridge

The SHMS installed in the bridge makes it possible to investigate wind characteristics and to gain a better understanding of wind loading on the bridge. Wind data recorded by the SHMS were analyzed, and wind characteristics such as mean wind speed, mean wind direction, turbulence components, turbulence intensities, integral scales and wind spectra were obtained for both monsoon wind and typhoon wind. It is impossible to cover all the research activities and present all the research results in one single paper. Only wind characteristics of typhoons and the joint probability density function for monsoons at the bridge site are briefly introduced in this section.

8.3.8 Anemometers in SHMS

The SHMS of the bridge includes a total of six anemometers with two at the middle of the main span; two at the middle of the Ma Wan side span; and one of each on the Tsing Yi Tower and Ma Wan Tower (see Fig. 13). To prevent disturbance from the bridge deck, the anemometers at the deck level were respectively installed on the north side and south side of the bridge deck via a boom of 8.965 m long from the leading edge of the deck (see Fig. 14).

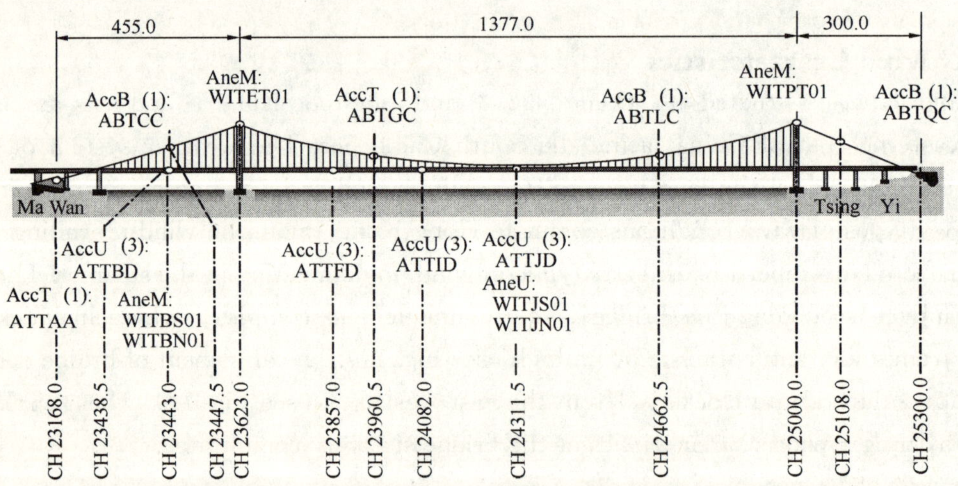

Fig. 13 Locations of anemometers and accelerometers

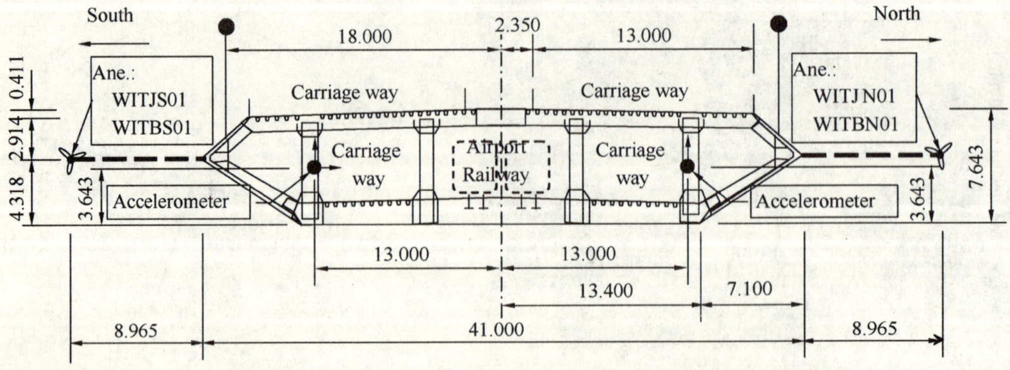

Fig. 14 Deck cross-section and sensor positions

The anemometers installed on the north side and south side of the bridge deck at the middle of main span, respectively specified as WI-TJN-01 and WI-TJS-01, are the digital type Gill Wind Master ultrasonic anemometers. The anemometers located at the two sides of the bridge deck near the middle of the Ma Wan approach span, specified as WI-TBN-01 on the north side and WI-TBS-01 on the south side, are the analogue mechanical anemometers. Each analogue anemometer consists of a horizontal component (RM Young 05106) with two channels, giving the horizontal resultant wind speed and its azimuth, and a vertical component (RM Young 27106) with one channel, providing the vertical wind speed. Other two analogue mechanical anemometers of a horizontal component only are arranged at the level of 11 m above the top of each bridge tower. They are specified as WI-TPT-01 for the Tsing Yi tower and WI-TET-01 for the Ma Wan tower. The sampling frequency of measurement of wind speeds was set as 2. 56 Hz.

8. 3. 9 Typhoon Victor

On 31 July 1997, less than three months after the opening of the Tsing Ma Bridge, the tropical depression Victor formed in the middle of the South China Sea. The tropical depression Victor became a real typhoon when it entered the region of 250 km south of Hong Kong at 8:00 on 2 August 1997 HKT (Hong Kong Time). The SHMS installed on the bridge timely recorded wind speed and bridge response time histories. The measured wind data of seven-hour duration from 17:00 to 24:00 HKT were used in the analysis to investigate typhoon wind characteristics. The 10-minute averaged mean wind speed and mean wind direction at the anemometers WITJN01 and WITJS01 were computed using the traditional approach and plotted against time in Figures 15a and 15b, respectively. It can be seen that there is a sudden change of wind direction from north-east to south-west and the mean wind speed becomes very small around 20:00 HKT. This indicates that the eye of Typhoon Victor just crossed over the bridge at that time. The measurement results also showed that turbulence intensities of winds during Typhoon Victor were higher than those due to monsoon winds and that some of wind samples were not stationary as assumed traditionally. The further information can be found in the literature (Xu et al, 2000*a*).

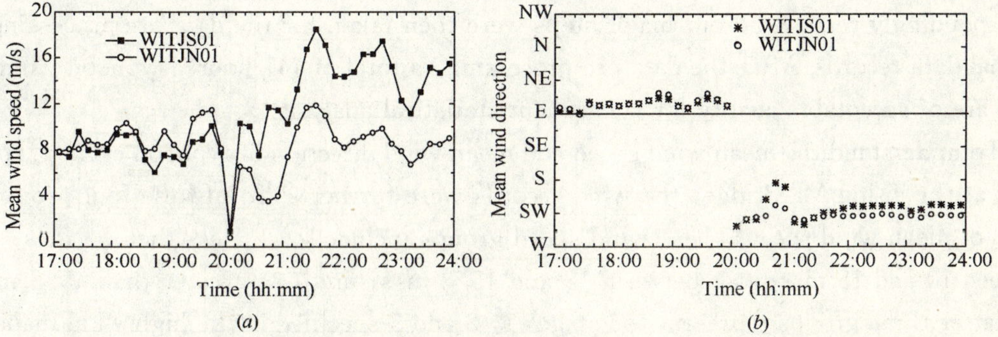

Fig. 15 Variations of 10-min mean wind speed and direction

(*a*)Mean wind speed; (*b*)Mean wind direction

8.3.10 Non-stationary wind model

The above findings motivated the author and his colleagues to propose a new approach for characterizing non-stationary wind speed (Xu and Chen, 2004). In the new approach, a measured wind sample of one-hour duration is decomposed into the sum of a deterministic time-varying mean wind speed plus a stationary fluctuating wind component by using the empirical mode decomposition (EMD) (Huang et al, 1998). The time-varying mean wind speed identified by the EMD at a designated intermittency frequency level is more natural than the traditional time-averaged mean wind speed over the certain time interval. The field measurement wind data recorded from the anemometers installed in the bridge during Typhoon Victor were used to test the new approach. It was found that the new approach could be used to check whether a wind sample is stationary or not. Most of the non-stationary wind samples recorded during Typhoon Victor could be decomposed into a time-varying mean wind speed plus a well-behaved fluctuating wind speed admitted as a stationary random process. The fluctuating wind components obtained by the EMD from the non-stationary wind samples complied with the standard Gaussian distribution very well while those gained from the traditional approach departed significantly from the Gaussian distribution. The spectral density functions of fluctuating wind components obtained by the EMD method had slightly lower amplitude in the low frequency range than those gained by the traditional approach. The turbulence intensities calculated by the EMD were more appropriate than the traditional approach. The gust factors obtained by the EMD were similar to those obtained by the traditional approach. It can be concluded that the proposed approach is more appropriate than the traditional approach for characterizing wind speed. The further information can be found in the literature (Chen et al, 2004).

8.3.11 Typhoon wind characteristics

To understand typhoon wind characteristics at the bridge site, typhoons with signal No. 3 and above hoisted by Hong Kong Observatory (HKO) during the period from July 1997 to September 2005 were targeted. A total of 247 hourly typhoon data records were correspondingly retrieved. Four major steps were then taken for the data pre-processing of original data records. After the data pre-processing, a total of 147 hourly typhoon wind records are of acceptable quality for subsequent statistical analysis.

To understand the mean wind speed and mean wind direction of typhoon events experienced at the Tsing Ma Bridge, the wind records were further split into the four groups in terms of mean wind speed. These wind speed groups included: (1) less than 10 m/s; (2) between 10 and 18 m/s; (3) between 18 and 45.8 m/s; and (4) greater than 45.8 m/s. The latter three groups represent the stages 1, 2 and 3 specified in the high wind management system for the bridge. Fig. 16 shows the polar plot of the 10-minute mean wind direction for typhoon events. It can be seen that almost 23% records are taken from the N direction. This observation is consistent with that made by the HKO.

Fig. 17 displays the polar plot of the 10-minute mean wind speed for typhoon events. It can be observed that wind speeds within the range from 10 to 18 m/s are dominant for typhoon events. This indicates that the stage 1 of the high wind management system was hoisted during most of typhoons for the bridge during the period concerned. Amongst the 16 direction sectors, the SW direction shows the maximum 10-minute mean wind speed of 22.67m/s.

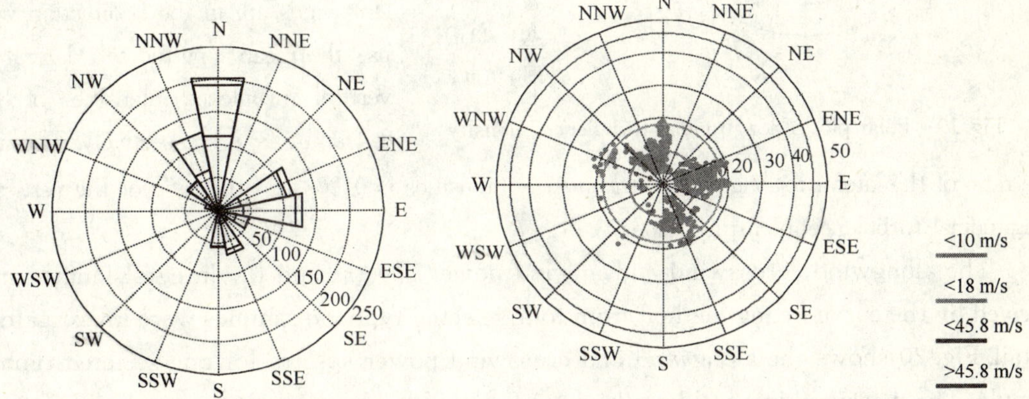

Fig. 16　Polar plot of mean wind direction　　Fig. 17　Polar plot of mean wind speed

The mean wind incidence is defined as the angle between the mean wind velocity and the horizontal plane. The positive mean wind incidence means the wind blowing upward. Fig. 18 displays the polar plot of the 10-minute mean wind incidence recorded at the deck level of the bridge for typhoon events. All the wind incidences are within ±10° at the 95% upper and lower limits of wind incidences under a mean wind speed of 20 m/s. However, the wind incidences measured in the easterly directions are much more scattered than those in other directions. This may be because easterly wind directions are almost parallel to the longitudinal direction of the bridge deck. There is a possibility that the flow of air could be disrupted by the bridge deck. It can also be seen that wind incidences tend to be approximately zero for the open-sea area. The mean values of 10-minute mean wind incidences were recorded as 0.33°, 0.09° and 0.52° in the SE, SSE and S directions, respectively.

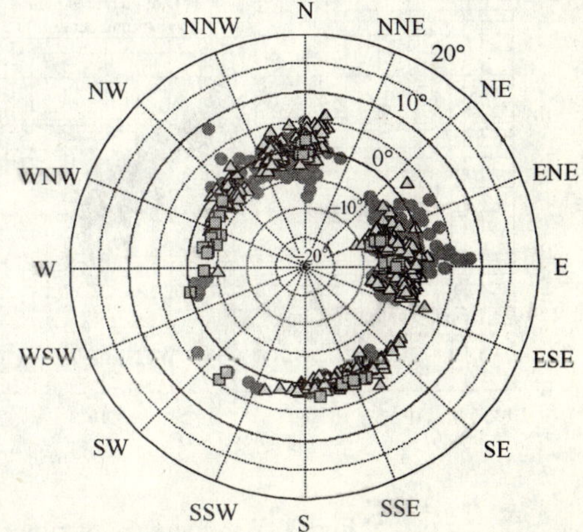

Fig. 18　Polar plot of mean wind incidence

Fig. 19 shows the polar plot of the 10-minute turbulence intensity in the longitudinal direction at the deck level of the bridge for typhoon events. It can be seen that the turbulence intensities measured in the north-easterly and easterly directions vary within a larger range than those measured in other di-

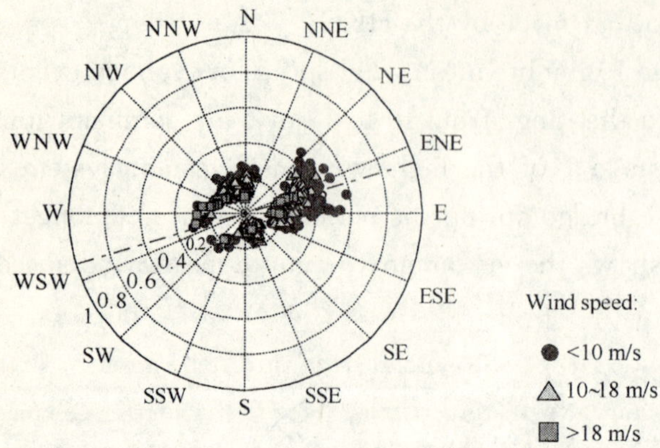

Fig. 19 Polar plot of longitudinal turbulence intensity

rections. The NE direction shows the most turbulent winds for typhoon events. The mean values of the longitudinal, lateral and vertical turbulence intensities in this direction are 38. 6, 36. 2 and 26. 1%, respectively. Contrarily, the least turbulent winds are in the S direction which has the mean longitudinal, lateral and vertical turbulence intensities of 8. 6, 8. 4 and 5. 3%, respectively. The average ratio of the lateral to longitudinal turbulence intensities is 0. 903 and the ratio of the vertical to longitudinal turbulence intensities is 0. 703.

The alongwind, crosswind and upwind power spectra and the integral length scales derived by the curve fitting method from four selected typhoon samples were investigated in detail. Fig. 20 shows the alongwind and crosswind power spectra for one selected typhoon sample. The measured integral length scales for the over-land exposure vary between 94. 81 m and 188. 52m (the average value being 136. 75 m) forL_u^x, between 31. 34m and 54. 55m (the average value being 44. 96m) forL_v^x, and between 33. 18m and 43. 04 m (the average value being 37. 93 m) forL_w^x. Contrarily, for the open-sea fetch the measured integral length scales range between 110. 66m and 539. 66m (the average value being 280. 76 m) forL_u^x, between 43. 74m and 149. 18m (the average value being 83. 85m) forL_v^x, and between 29. 15m and 60. 04m (the average value being 40. 91m) forL_w^x. It seems that the integral length scales of L_u^x, L_v^x and L_w^x from the open-sea fetch are larger than those from the over-land fetch. This observation appears to be consistent with the comment in the literature that the length scales is a decreasing function of terrain

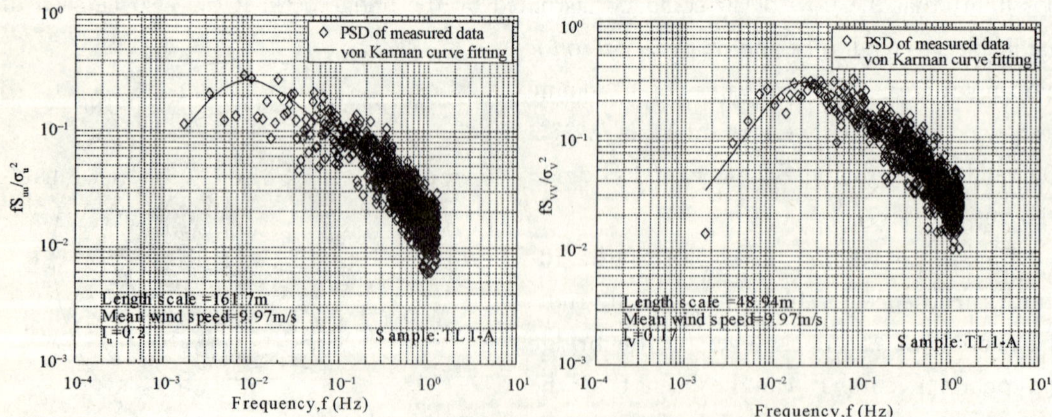

Fig. 20 Wind spectra of typhoon sample 1 from over-land fetch

(*a*) Alongwind spectrum; (*b*) Crosswind spectrum

roughness (Simiu and Scanlan, 1996).

8.3.12 Monsoon wind and joint probability density function

The joint probability density function of wind speed and wind direction is essential when assessing wind-induced fatigue damage to the bridge. A practical joint probability distribution function was adopted for a complete population of wind speed and wind direction based on two assumptions: (1) the distribution of the component of wind speed for any given wind direction follows the Weibull distribution; and (2) the interdependence of wind distribution in different wind directions can be reflected by the relative frequency of occurrence of wind.

$$P_{u,\theta}(U,\theta) = P_\theta(\theta)\left(1-\exp\left[-\left(\frac{U}{c(\theta)}\right)^{k(\theta)}\right]\right)=\iint f_\theta(\theta) f_{u,\theta}(U,k(\theta),c(\theta))\,\mathrm{d}u\mathrm{d}\theta \quad (1)$$

$$f_{u,\theta}(U,k(\theta),c(\theta)) = \frac{k(\theta)}{c(\theta)}\left(\frac{U}{c(\theta)}\right)^{k(\theta)-1}\exp\left[-\left(\frac{U}{c(\theta)}\right)^{k(\theta)}\right] \quad (2)$$

$$P_\theta(\theta) = \int_0^\theta f_\theta(\theta)\mathrm{d}\theta \quad (3)$$

in which $0\leqslant\theta<2\pi$; and $P_\theta(\theta)$ is the relative frequency of occurrence of wind in wind direction θ. The occurrence frequency $P_\theta(\theta)$ as well as the distribution parameters $k(\theta)$ and $c(\theta)$ can be estimated using wind data recorded at the bridge site. Wind records of hourly mean wind speed and direction within the period between 1 January 2000 and 31 December 2005 from the anemometer installed on the top of the Ma Wan tower were used to ascertain the joint probability density function of hourly mean wind speed and direction. The height of the anemometer is 214 m above the sea level. Wind records having an hourly mean wind speed lower than 1 m/sec were removed in order to avoid any adverse effect on the statistics. As a result, 19775 hourly monsoon records were available for calculation of the joint probability density function of wind speed and direction. The number of hourly typhoon wind records during the concerned period was so small that the corresponding joint probability density function could not be obtained at present.

All the monsoon records were classified into 16 sectors of the compass with an interval of $\Delta\theta=22.5°$ according to the hourly mean wind direction (see Fig. 12). In each sector, mean wind speed was further divided into 16 ranges from zero to 32 m/sec with an interval of $\Delta U=2$m/sec. This leads to a total of 256 cells, and the relative frequency of hourly mean wind speed and wind direction in each cell was calculated. Based on the relative frequencies of wind speed and wind direction calculated, the theoretical expression of joint probability density function was deduced based on Equation (1). The Weibull function was used to fit the histogram of hourly mean wind speed for each wind direction, and the typical results in the east, south and west directions are depicted in Figures 21a to 21c, respectively. The Weibull function was also applied to the complete wind records without considering wind direction, as shown in Fig. 21d. The results show that the Weibull function also fits the

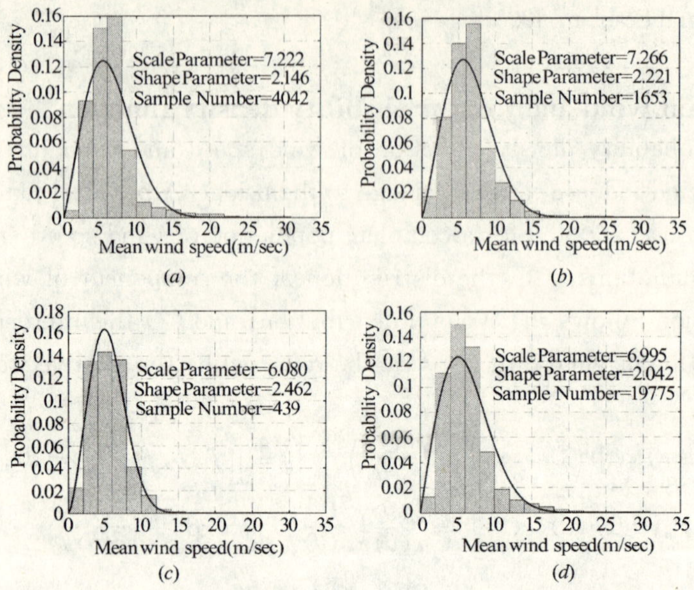

Fig. 21 Weibull distribution of hourly mean wind speed

(*a*) East direction; (*b*) South direction; (*c*) West direction; (*d*) All directions

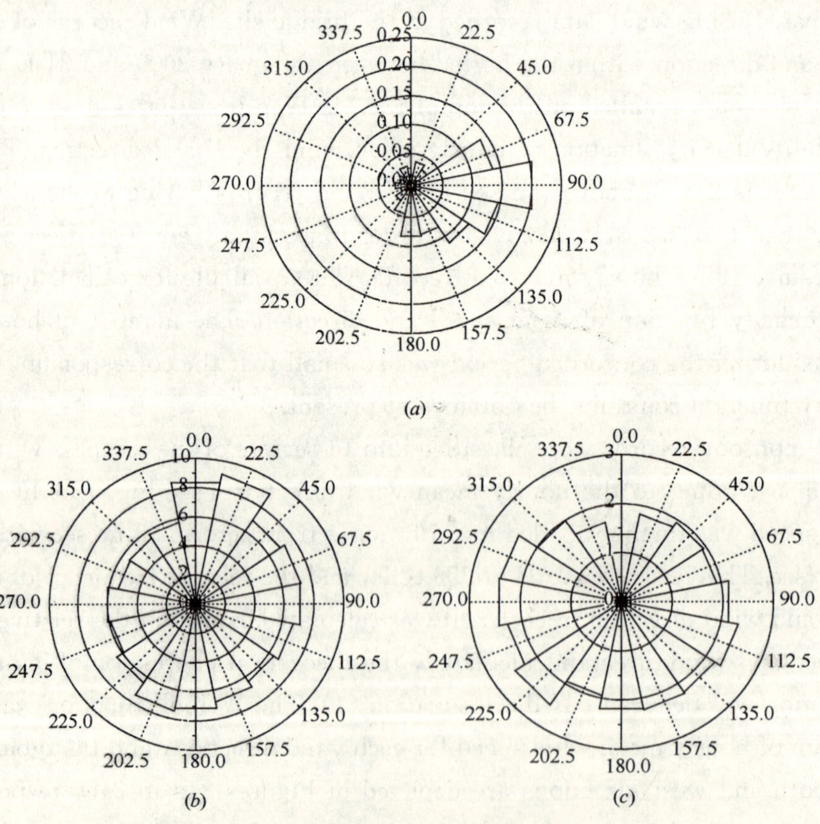

Fig. 22 Relative frequency of wind direction and Weibull scale and shape parameters

(*a*) Relative frequency of wind direction; (*b*) Weibull scale parameter; (*c*) Weibull shape parameter

complete wind data adequately. The relative frequency of wind direction and the scale and shape parameters of the Weibull function obtained are given in polar plot in Figures 22a to 22c, respectively. It can be seen that the dominant monsoon direction is the east, and the scale and shape parameters do not vary significantly with wind direction. The further details can be found in the literature (Xu et al. 2008*a*).

8.3.13 Temperature Effects

Long suspension bridges are subjected to continuous temperature variations primarily due to solar radiation and ambient air temperature. The temperature variations will then cause the bridge to expand and contract in the longitudinal direction and to bend in the vertical plane. Since the movements of the bridge are often accommodated by the bearings and expansion joints, large forces may develop in the bridge structure if any component of the movements is restrained. These forces sometimes may cause damage to the bridge. This section introduces the temperature effects on the displacement responses of the Tsing Ma Bridge using the long-term measurement data from the SHMS and provides a basis for the real-time monitoring of temperature effect. To this end, the statistical details of ambient air temperature, bridge effective temperature, and bridge displacement responses are analysed. The statistical relationships between the bridge effective temperature and the displacement responses of the bridge are established.

8.3.14 Temperature sensors and global positioning systems

The ambient temperature and bridge member temperature are obtained from temperature sensors installed at six different locations along the bridge longitudinal axis (see Fig. 23). The ambient temperature, which is denoted as T1 in Fig. 23 with numbers of sensors marked in parenthesis, is measured at two locations of the bridge. One of the locations installed with one temperature sensor is approximately at the middle section of the main span. There are 5 more sensors measuring the ambient air temperature inside the bridge deck section of the main span near the Tsing Yi tower. The total numbers of sensors implemented for the temperature measurement of the bridge members, named T2 in Fig. 23, are 109. 23 out of 109 sensors are thermocouple embedded inside the main cables. The remaining 86 sensors are distributed at two different locations. A larger amount of sensors, to-

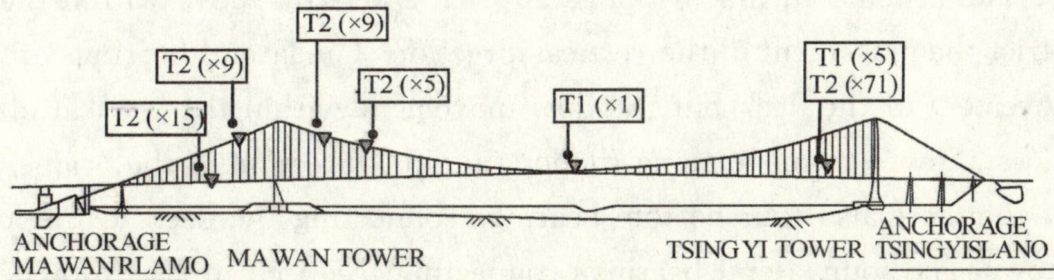

Fig. 23　Distribution of temperature sensors in Tsing Ma Bridge

tally 71, are installed to members of the main span deck section nearby the Tsing Yi tower for the determination of bridge effective temperature. The sampling frequency of all the temperature sensors is 0.07 Hz.

After September 2002, the displacements of the bridge in longitudinal, lateral and vertical directions are mainly monitored by GPS. There are totally 14 GPS receivers installed in the mentioned three major components and their distribution is shown in Fig. 24. The Ma Wan tower and the Tsing Yi tower are respectively installed with a pair of GPS receivers and they are mounted at the top of saddles on each of the tower leg. The displacement of the main cables is only monitored in the middle of the main span and correspondingly a pair of GPS receivers is located at the mid-span of the cables. The bridge deck sections in the main span and the Ma Wan side span are the other two regions mounted with four pairs of GPS receivers. The mid-span of the Ma Wan side span is noted with one pair of receivers. The other three pairs of GPS receivers are located at one quarter, one half and three quarter of the main span of the bridge deck with reference from the Ma Wan tower. The sampling frequency of the GPS receivers is 10 Hz.

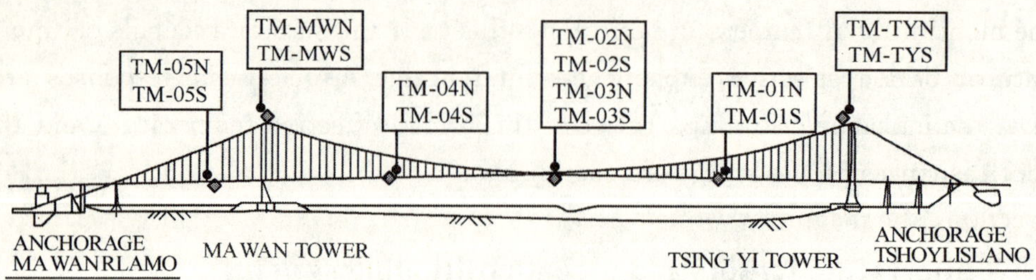

Fig. 24 Distribution of GPS receivers in Tsing Ma Bridge

8.3.15 Temperature effects

The translational movements of the bridge deck at the Ma Wan abutment are restricted in three directions, but the deck is free for rotation about the lateral axis. At the Tsing Yi abutment, only the vertical and lateral movements of the bridge deck are restrained; the longitudinal movement and the rotations about the lateral axis are allowable. At the Ma Wan tower, the bridge deck is connected to the bottom cross beam of the tower through four articulated link bearings (or rockers) and to the tower legs through four lateral bearings (rollers). The articulated link bearings allow the deck to move within the horizontal plane but restrict the movement in the vertical direction. The lateral bearings are to restrain the lateral movement of the deck but to allow movement within the vertical plane. Therefore, the deck is allowed to move along the longitudinal direction of the bridge. At the Tsing Yi tower, there are also four bottom bearings connecting the deck to the lowest cross beam of the tower and four lateral bearings connecting the deck to the tower legs. All the piers allow the bridge deck to move in the longitudinal direction.

To closely understand the effects of temperature on the displacement responses of the Tsing Ma Bridge, the relationship between bridge temperature and displacement is examined using the long-term measurement data after having studied bridge member temperature and bridge displacement. The temperature data adopted in the analysis is the effective deck temperature. The daily bridge mean displacement is utilized to explore the temperature-displacement relationship. All the temperature and displacement data used in the analysis are recorded during 2003 to 2005. After each set of the temperature and displacement data is obtained, the following linear analytical function is adopted to fit the temperature-displacement relationship using the least-squares method to determine the two factors in the following equation.

$$D = a \cdot T + b \tag{4}$$

where D is the displacement at a certain location and in a certain direction of the bridge; T denotes the effective deck temperature (degree Celsius); a is the slope factor which actually depicts the variation rate of displacement to temperature; b is a constant factor. The factor "a" actually reflects the variation rate of displacement response to temperature change: if the temperature increase is 1 °C, the change in displacement is "a" in mm. Fig. 25 displays the distribution of variation rates of the longitudinal displacement in the bridge based on daily mean data. It is observed that the magnitude of displacement variation rate of deck sections at the main span is much larger than that of other parts of the bridge. As far as the bridge deck is concerned, owing to the accumulation of bridge displacement in the longitudinal direction, the magnitude of displacement variation rate gradually increases in the longitudinal direction from the Ma Wan side span to the Tsing Yi abutment.

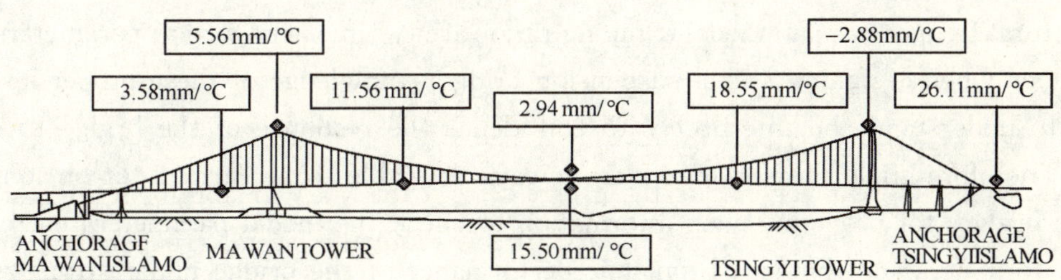

Fig. 25　Distribution of variation rates of longitudinal displacement

Fig. 26 displays the distribution of variation rate of vertical displacement of the bridge. It was found that only the vertical displacements of deck sections and cable section

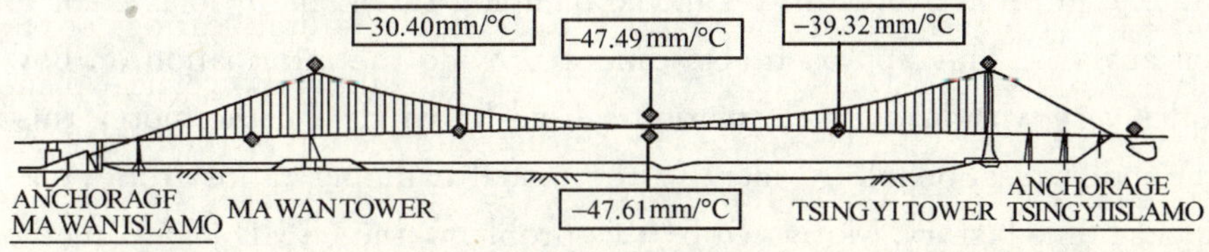

Fig. 26　Distribution of variation rates of vertical displacement

at the main span are well correlated with the temperature. Therefore, the variation rates of vertical displacement of three deck sections and one cable section are analyzed and displayed in Fig. 26. Observed in Fig. 26, the displacement variation rates of deck sections in the middle of the main span is almost the same as that of the cable section. The temperature-induced vertical displacement of the deck section in the middle of the main span is thus partially controlled by the deformation of the main cables. It can also be seen that the variation rates of vertical displacement of deck sections at one quarter span and three quarter span of the main span from the Ma Wan tower are smaller to some extent in comparison with that of the deck section in the middle of the main span.

Based on the linear temperature-displacement relationships determined through curve fitting, the measured maximum variation ranges of displacement are compared with those estimated using the linear function expressed by Equation (4). The corresponding results demonstrate that the estimating errors of longitudinal displacement are on average about 5% and the maximum absolute error is no more than 13%. Regarding the vertical displacement, the maximum absolute error is no more than 11% during 2003 to 2005. Therefore, it is concluded that the statistical relationship between effective deck temperature and displacement can effectively estimate the bridge movements subjected to the variation of temperature, which can be used for real-time monitoring of temperature effects. The further information can be found in the literature (Xu et al. 2007*a*).

8.4 Shms-based system identification

Natural frequencies and modal damping ratios are two most important parameters to be determined when designing a long suspension bridge. Knowledge of these properties is essential to understand and interpret with confidence the response of the bridge to strong winds. Therefore, field measurements are always desirable to be carried out on long suspension bridges for providing such information to check the modal parameters used in the design, to understand the actual dynamic performance of the bridge under strong winds, and to develop better design theories for future bridges. For a long suspension bridge, the most popular approach for identifying natural frequency and modal damping ratio is to carry out ambient vibration measurements and then to apply the FFT-based method to the measured bridge response time histories. Though this approach can identify natural frequencies quite accurately, it often overestimates modal damping ratios due to bias error involved in the spectral analysis. This approach also cannot provide the information on how dynamic characteristics vary with bridge deformation caused by winds. Furthermore, this approach may become more questionable to identify the modal damping ratios from non-stationary bridge response time history. Motivated by this problem, the (EMD+HT) method, which consists of mainly the empirical mode decomposition (EMD) and the Hilbert transform

(HT), is used and applied to the field measurement results recorded by the SHMS of the bridge during Typhoon Victor.

8.4.1 EMD+HT Method

The EMD+HT method is a two-step data analyzing method. The first step is the EMD by which a measured structural response time history can be decomposed into a series of intrinsic mode functions (IMF) that admit well-behaved Hilbert transforms (Huang et al., 1998). This decomposition is based on the direct extraction of the energy associated with various intrinsic time scales of the time history itself. Furthermore, to avoid the mode mixing during the sifting process, the intermittency check is used to separate the waves of different periods into different modes based on the period length to obtain modal response time histories. The random decrement technique (RDT) is then applied to each of the modal response time histories to obtain the free modal response time histories. The second step of the EMD+HT method is implemented by performing the Hilbert transform to each of the free modal response time histories to find the corresponding natural frequency and modal damping ratio.

8.4.2 Natural Frequencies and Modal Damping Ratios

The EMD+HT method and the FFT method are applied to the measured acceleration response time histories of one hour duration of the bridge during Typhoon Victor for identifying the natural frequencies and modal damping ratios of Lat1, Lat2, Ver1, Ver2, and Tor1, which stand for the first and second lateral vibration modes, the first and second vertical vibration modes, and the first torsional vibration mode, respectively (Chen et al., 2004). The resulting natural frequencies from six hour-long records are listed in Table 2 and the resulting modal damping ratios are shown in Table 3. It is noted that the damping identified here is the total damping that consists of the net structural damping and aerodynamic damping.

It is encouraging to see from Table 2 that for a given mode of vibration, the natural frequencies identified by the EMD+HT method from the six records are very close to each other. It is also interesting to see that the natural frequencies identified by the EMD+HT method are very close to those obtained by the FFT-based method. The modal damping ratios identified by the EMD+HT method are smaller than those obtained from the FFT-based method (see Table 3). It was well known that the bias error in the FFT-based method always leads to an overestimation of damping because of the limitation on the frequency resolution, and hundreds of stationary samples should be used in the damping ratio estimation to reduce the random error. Obviously, it is almost impossible in reality to obtain a stationary record during typhoon long enough for damping estimation. Considering all these factors and the adaptive nature of the EMD+HT method performed in the frequency-time domain, one may say that the EMD+HT method is superior to the FFT-based method for modal damping estimation.

Comparison of natural frequencies **Table 2**

Lat1		Lat2		Ver1		Ver2		Tor1	
EMD	FFT	EMD	FFT	EMD	FFT	EMD	FFT	EMD	FFT
0.0686	0.0688	0.1602	0.1625	0.1137	0.1141	0.1365	0.1359	0.2652	0.2656
0.0677	0.0672	0.1608	0.1609	0.1138	0.1149	0.1373	0.1359	0.2653	0.2656
0.0683	0.0688	0.1610	0.1609	0.1139	0.1141	0.1387	0.1375	0.2655	0.2641
0.0677	0.0672	0.1598	0.1594	0.1133	0.1141	0.1362	0.1359	0.2641	0.2641
0.0681	0.0688	0.1587	0.1594	0.1141	0.1141	0.1376	0.1375	0.2645	0.2641
0.0677	0.0672	0.1601	0.1594	0.1145	0.1141	0.1373	0.1359	0.2645	0.2641

Comparison of modal damping ratios **Table 3**

Lat1		Lat2		Ver1		Ver2		Tor1	
EMD	FFT	EMD	FFT	EMD	FFT	EMD	FFT	EMD	FFT
0.80%	2.54%	1.27%	1.39%	0.35%	1.24%	1.01%	1.68%	0.29%	0.45%
1.52%	2.09%	0.91%	1.02%	1.27%	1.61%	0.66%	0.98%	0.75%	0.81%
0.82%	1.97%	0.81%	1.27%	1.03%	0.97%	0.78%	0.93%	0.56%	0.54%
0.55%	1.67%	0.58%	1.09%	0.68%	1.09%	0.49%	0.92%	0.35%	0.48%
1.31%	2.22%	1.07%	1.61%	0.73%	1.14%	0.94%	1.62%	0.47%	0.62%
1.58%	2.20%	1.33%	1.24%	0.89%	1.55%	0.62%	0.89%	0.47%	0.54%

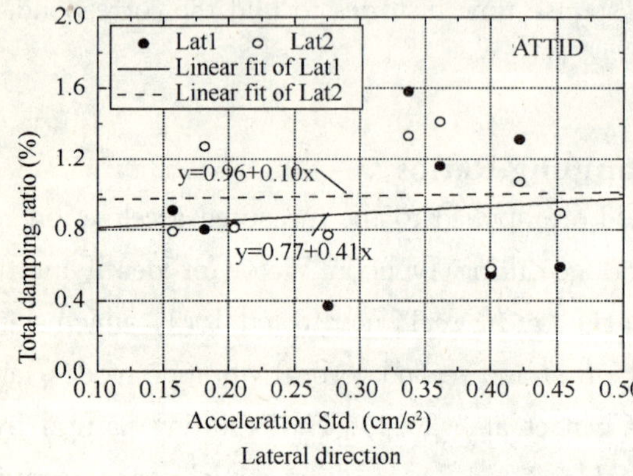

Fig. 27 Variations of total damping ratio

Furthermore, the variations of natural frequency, total modal damping ratio, and structural modal damping ratio with vibration amplitude and mean wind speed are examined. The variation of structural modal damping ratio with modal frequency is also investigated. The results demonstrated that the natural frequencies of the bridge decreased very slightly with the increase in either mean wind speed or vibration amplitude (see Fig. 27). The total modal damping ratios and the structural damping ratios both exhibited an increasing trend with increasing vibration amplitude or increasing mean wind speed (see Fig. 28). In consideration that most of the bridge response time his-

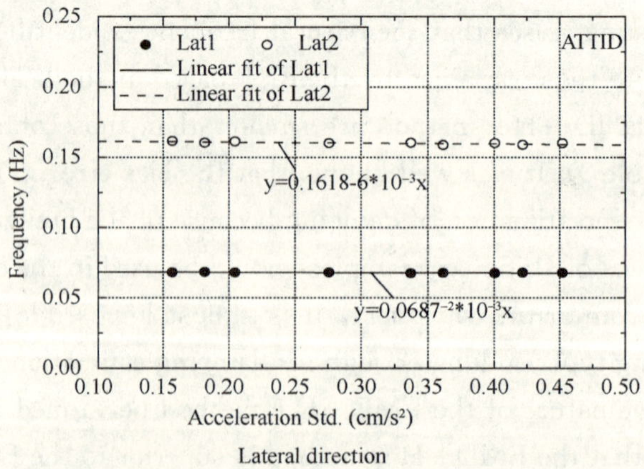

Fig. 28 Variations of natural frequency

tories recorded during Typhoon Victor are non-stationary, it may be concluded that the EMD+HT method is superior to the FFT-based method for damping identification of large civil structures under strong typhoon.

8.5 Shms-based computer simulation

Real-time monitoring of bridge behavior by SHMS is the best way to examine the currently-used design rules and to develop and verify new analytical methods if necessary. Only two research projects conducted by the author and his colleagues are presented in this paper: wind effects on train-bridge systems and buffeting response of long span bridge to skew winds.

8.5.1 Wind Effects on Train-Bridge Systems

Heavy trains running on a long suspension bridge may significantly change the dynamic characteristics and affect the serviceability of the bridge. Long suspension bridges are often very slender and low damped. If a long suspension railway bridge is built in wind prone area, the bridge will experience considerable vibrations due to aerodynamic effects. The vibration of the bridge due to both strong winds and running trains may in turn deform the railway track laid on the bridge deck and affect the running safety of trains and the comfort of passengers. Therefore, the understanding of dynamic behavior and the prediction of dynamic response of long suspension bridges under both high winds and running trains becomes an important task. Xu et al. (2003) presented a framework for predicting the dynamic response of a long suspension bridge to high winds and running trains. Xu et al. (2004) then extended their work to investigate fully dynamic interaction of a long span cable-stayed bridge with running trains subjected to crosswinds using the most up-to-date information in the areas of wind-bridge interaction, bridge-train interaction, and wind-train interaction. However, the aforementioned two studies are numerical studies only. The rationality and feasibility of the proposed framework and the accuracy of dynamic responses of the system predicted from the framework are needed to be verified before it can be used for bridge health assessment.

On 16 September 1999, Typhoon York, which is the strongest typhoon since 1983 and the typhoon of longest duration on record, passed by Hong Kong. All vehicles except trains were prohibited from running on the Tsing Ma Bridge for two and a half hours. This event provided a distinctive opportunity to examine the proposed framework. The field measurement data recorded by the SHMS during Typhoon York were analyzed and the four particular cases were identified (Xu et al 2007*b*; Guo et al, 2007). The four particular cases identified included the bridge without any vehicles (Case 1), the bridge with one train (Case 2), the bridge with two trains running in opposite direction (Case 3), and the bridge with three running trains (Case 4). For each case, wind characteristics, bridge acceleration responses, and bridge displacement responses were analyzed using the measurement data from anemometers, accelerometers, and level sensing systems, respectively. The number, speed and location of trains running on the bridge were also identified using the measurement data from strain gauges. The field measurement results were finally used to verify the proposed framework; wind characteristics and train information identified from the measurement data were used as inputs; and the acceleration and displacement responses of the bridge under cross winds and running trains were computed and compared with the responses measured from the field for each case. The dynamic responses of trains running on the bridge were also computed to evaluate their safety and serviceability. Because of the limitation of space, only Case 2 is presented in this paper.

Wind data of two and a half hours duration recorded during Typhoon York were analyzed. The wind data were evenly divided into segments of 3-minute long each with 1. 5-minute overlap between neighboring segments. The duration of 3 minutes was chosen based on the fact that it took about 1. 5 to 2. 0 minutes for a train to completely pass through the whole bridge and the fact that some anemometers at the deck level were out of order during Typhoon York. A complete set of field measurement data of 3 minutes duration were extracted from original data base and subsequently analyzed. The strain data recorded by the strain gauges (SSTLN04 and SSTLS04) stuck on the main span cross-frame at Section L (CH 24662. 5 in Fig. 29) were analyzed to identify the number of trains running on the bridge, train speed, and train location relative to the bridge.

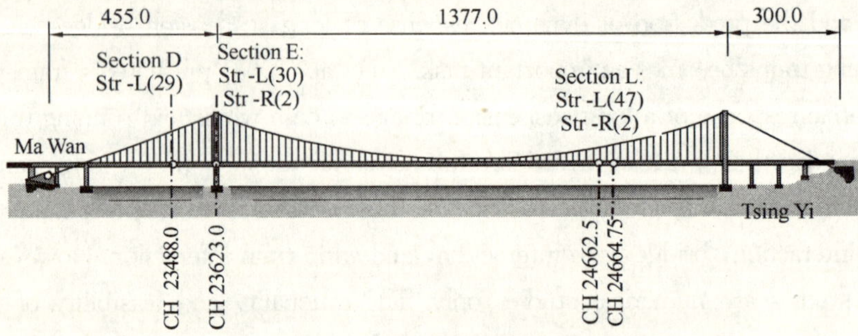

Fig. 29 Distribution of strain gauges in Tsing Ma Bridge

Figures 30a and 30b illustrate the strain time histories of 3 minutes duration from the strain gauges on the north side (SSTLN04) and the south side (SSTLS04), respectively, in Case 2 during Typhoon York. It can be seen that there are a cluster of strain peaks around 90 second in either time history, but the peak strain values in the time history of the north strain gauge are much larger than those in the time history of the south strain gauge. The 12 seconds portion of the time history of the north strain gauge, containing a cluster of strain peaks, is expanded in Fig. 30c to permit a close look at signal features due to running trains. Eight strain peaks, corresponding to 7 coaches, clearly appear in the strain time history. Based on these results, one may conclude that in Case 2, there was only one train running on the north track heading the Tsing Yi Island. The total time needed for the train

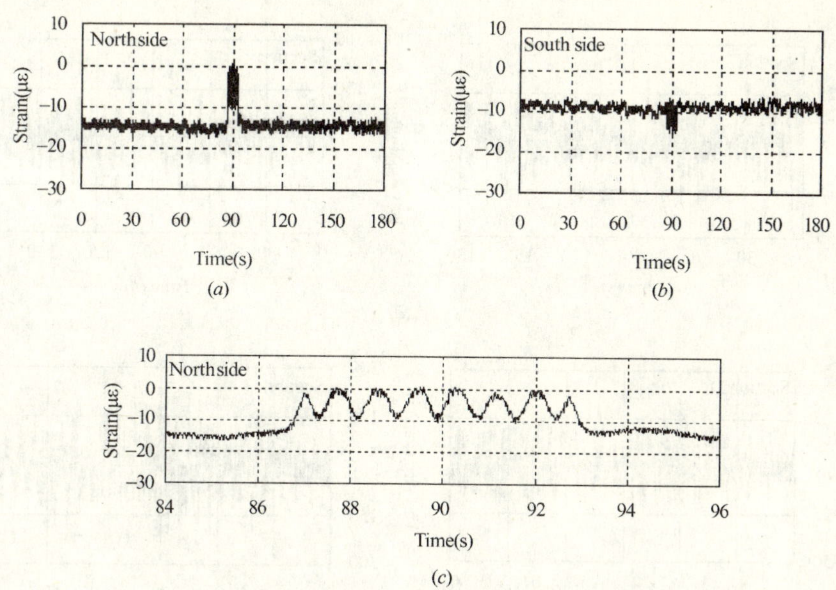

Fig. 30　Strain time histories in Case 2

(*a*) North side; (*b*) South side; (*c*) Zoom of strain on north side

to completely pass through Section L is 6.6 seconds. The moment of the train just passing Section L is at 86.5 second. By assuming that the train ran on the bridge at a constant speed, the train speed can be deduced as 99km/h (27.5 m/s). The location of the train at zero time instant is deduced as 2380 m away from Section L in the west direction. The measured mean wind speed in this case is 18.9 m/s and the measured turbulent intensity is 14.9% in the horizontal direction and 7.9% in the vertical direction. The measured lateral, vertical and torsional acceleration responses of the bridge deck were obtained directly from the accelerometers of the SHMS.

The fluctuating wind components and modal buffeting forces were simulated based on the measured results. The responses of the bridge to both train and cross winds were calculated in the time domain. Fig. 31 illustrates the time histories of both the measured and computed deck acceleration responses of 3 minutes duration in the main span. Displayed in Fig. 31a are the time histories of the measured lateral acceleration response at Section F (around 1/6 of the main span) and the measured vertical acceleration response at Section J (around 1/2 of the main span). Fig. 31b displays the computed lateral and vertical acceleration responses at the same sections. It can be seen that the computed acceleration time history in the lateral direction is compatible with the measured one but the computed acceleration time history in the vertical direction is higher than the measured one in terms of response amplitude. Displayed in Fig. 32 are the computed and measured maximum acceleration responses along the bridge deck in the lateral and vertical direction. It can be seen that the computed maximum acceleration responses of the bridge deck are all close to the measured results. The absolute relative discrepancies are less than 8% for lateral response and

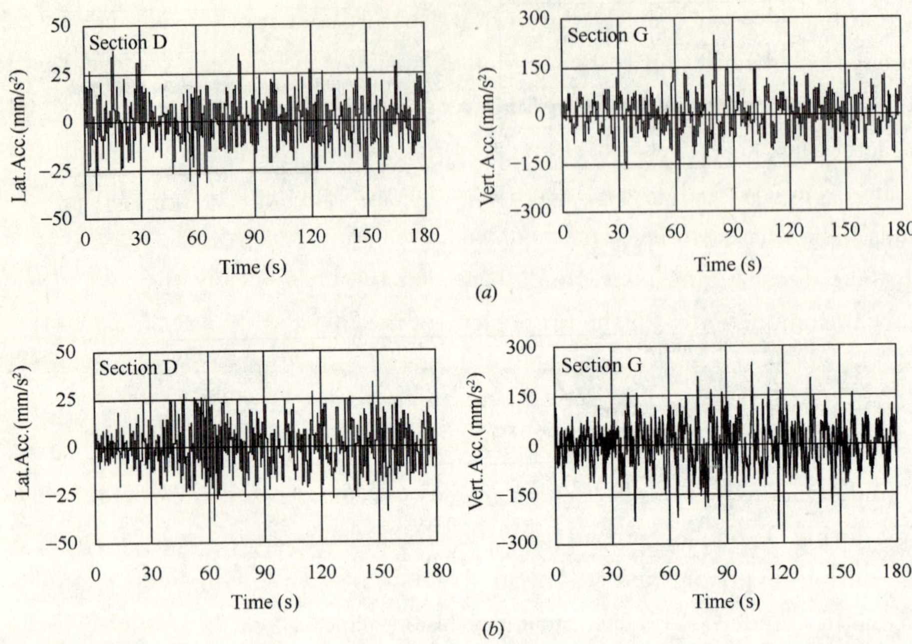

Fig. 31 Comparison of measured and computed deck accelerations in main span (Case 2)

(*a*) Filtered measured deck acceleration responses; (*b*) Computed deck acceleration responses

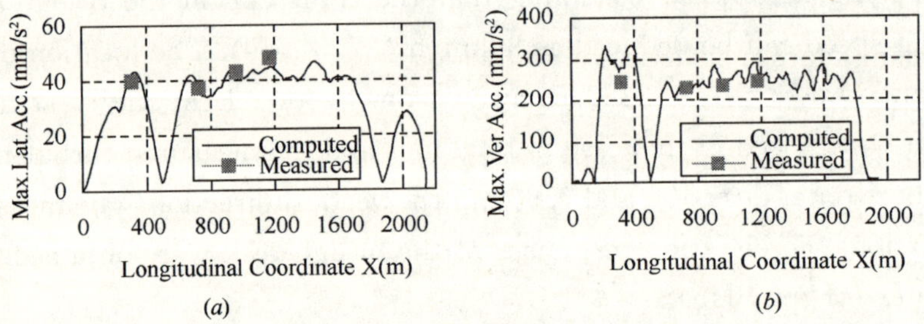

Fig. 32 Comparison of measured and computed maximum deck accelerations (Case 2)

(*a*) Lateral acceleration response; (*b*) Vertical acceleration response

less than 18% for vertical response. Further comparison showed that not only the computed acceleration responses but also displacement responses of the bridge deck agreed well with the measured ones. The comparison was thus found to be satisfactory and the proposed framework for predicting dynamic response of coupled train-bridge systems in cross winds is feasible. The further information on this topic can be found in the literature (Xu et al. 2007*b*; Guo et al. 2007).

8.5.2 Buffeting Response of Long Span Bridge to Skew Winds

The field measurement data recorded by the SHMS of the Tsing Ma Bridge during typhoons manifested that mean wind directions often deviated from the normal of the bridge longitudinal axis (Xu et al. 2000*a*). Wind characteristics, including mean wind speed and turbulence intensity, varied along the bridge deck. The currently-used analytical methods for the prediction of buffeting response of long suspension bridges, however, assume that mean wind speed is coming at a right angle to the longitudinal axis of the bridge deck and wind characteristics keep constant along the bridge deck. This leads to some difficulties to examine the accuracy of the currently-used analytical methods for the bridge subject to skew winds. The currently-used method was thus improved by the author and his colleagues to take into account skew winds based on the quasi-steady theory and the oblique strip theory in conjunction with the finite element method and the pseudo excitation method (Zhu and Xu, 2005; Xu and Zhu 2005). A series of wind tunnel tests were performed to measure the aerodynamic coefficients and flutter derivatives of the bridge deck under skew winds using innovative test rigs developed for this particular study (Zhu et al. 2002*a*; 2002*b*). Wind structures and buffeting responses measured by the WASHMS of the Tsing Ma Bridge during Typhoon Sam were analyzed. The comparison of buffeting response of the bridge during Typhoon Sam was carried out between the computed results using the improved method and those measured results. The comparison was found satisfactory in general. The detail information on this topic could be found in the literature (Liu et al. 2004; Zhu and Xu, 2005; Xu and Zhu, 2005).

It should be pointed out that in this study, some key information regarding the modeling of aerodynamic forces such as admittance functions was not quantified. The field measurement data regarding wind characteristics were not comprehensive, such as the lack of spatial correlation of turbulent winds. Thus, more filed measurements with improved SHMS and more comparisons having measured aerodynamic admittance functions should be carried out in the future.

8.6 Shms-based damage assessment

As discussed above, the recent trend is to install a comprehensive SHMS in a long suspension bridge to monitor its performance and its safety through the analysis and synthesis of real-time measurement data. However, many key issues remain unsolved as how to take full advantage of the real-time data for effective and reliable health assessment of the bridge. Moreover, the number of sensors is always limited for a long suspension bridge and the locations of structural defects or degradation may not be at the same positions as the sensors. Possibility exists that the worst structural condition may not be directly monitored by sensors. Therefore, for the accomplishment of the structural health assessment of the bridge, SHMS-based computer simulation and damage assessment are necessary and imperative. This section introduces one example in the structural health monitoring based damage assessment: wind-induced fatigue damage assessment of the bridge.

8.6.1 Background

When a long suspension bridge is built in a wind-prone region, the bridge will suffer considerable buffeting-induced vibration which appears within a wide range of wind speeds and lasts for almost the whole design life of the bridge. As a result, the frequent occurrence of buffeting response of relatively large amplitude may cause fatigue damage to steel structural members of a long suspension bridge. Although many works have been conducted on traffic-induced fatigue damage of steel bridges, there has been very limited research on buffeting-induced fatigue damage of long suspension bridges. A systematic framework for assessing long-term buffeting-induced fatigue damage to a long suspension bridge is proposed in this paper using the structural health monitoring system and by integrating a few important wind/structural components with the continuum damage mechanics (CDM)-based fatigue damage assessment method.

By taking the Tsing Ma Bridge as an example, a joint probability density function of wind speed and direction is first established based on wind data recorded by the WASHMS as described before. The numerical procedure for buffeting-induced stress analysis of the bridge based on its structural health monitoring-orientated finite element model is then established and used to identify stress characteristics at hot spots of critical steel members under different wind speeds and directions. The accumulative fatigue damage to the critical members at their hot-spots during the bridge design life is finally evaluated using a CMD-based fatigue damage model taking into consideration of long-term effects of buffeting forces.

8.6.2 SHMS-Oriented Finite Element Model

The sensors for strain measurement in the WASHMS for the bridge are always limited: not all the stress responses of all the local components can be directly monitored. To facilitate an effective assessment of stress-related bridge safety, a structural health monitoring orientated finite element model (FEM) is needed for a long suspension bridge so that stresses/strains in all of the important bridge components can be directly computed and some of them can be compared with the measured ones for verification. However, the currently-conducted buffeting analyses of long span bridges are often based on a simplified spine beam FEM of equivalent sectional properties (Xu et al. 2000*b*). Such simplified model is effective to capture the dynamic characteristics and global structural behaviour of the bridge under strong winds without heavy computational effort. However, local structural behaviour linked to stress and strain, which is prone to cause local damage, could not be estimated directly. On the other hand, with the rapid development of information technology, the improvement of speed and memory capacity of personnel computers (PC) has made it possible to establish a structural health monitoring orientated finite element model for a long suspension bridge. In this regard, a complex structural health monitoring (SHM) oriented finite element model (FEM) has recently been established by the author and his colleagues for the Tsing Ma Bridge with significant modelling features of the bridge deck included for the good replication of geometric details of the as-built complicated deck. The proposed SHMS oriented FE model has also been updated using the measured natural frequencies and mode shapes of the bridge with the updated parameters being material properties only because the geometric features and supports of bridge deck have been modelled in a great detail in the proposed SHM orientated FE model. It turns out that the updated complex FE model could provide comparable and credible structural dynamic modal characteristics. The detail information on this topic can be found in the literature (Fei, et al. 2007; Zhang et al. 2007).

8.6.3 Wind-Induced Stress Analysis

Based on the established structural health monitoring orientated FEM, a numerical procedure for wind induced stress analysis of long suspension bridges has been proposed by the author and his colleagues. Significant improvements of the proposed procedure are that the effects of the spatial distribution of both buffeting forces and self-excited forces on a bridge deck structure are taken into account, as opposed to lumping all buffeting forces and self-excited forces at the centre of elasticity as in the case of an equivalent beam finite element model. Local strains and stresses in structural members of the bridge deck, which are prone to cause local damage, are predicted directly using the mode superposition technique in the time domain. The field measurement data including wind, acceleration and stress recorded by the anemometers, accelerometers, and strain gauges in the WASHMS installed in the Tsing Ma Bridge during Typhoon York have been analyzed. The buffeting-induced acceleration responses at the locations of 12 accelerometers installed in the bridge has been computed using the mode superposition method and compared with the measured results. The wind-induced stress responses at the locations of 9 strain gauges installed in the bridge have been computed through the modal stress analysis and compared with the measured ones. The comparative results show that the computed stress time histories are similar in both pattern and magnitude with the measured ones. The detail information on this topic can be found in the literature (Liu et al. 2007).

8.6.4 Wind-Induced Fatigue Damage Assessment

In the health monitoring orientated finite element model of the bridge, there are a total of 15,904 beam elements used to model the bridge deck. To find the most critical beam elements and the corresponding most critical stresses in the bridge deck, a buffeting-induced stress analysis is carried out by considering a 15 m/s mean wind perpendicular to the bridge axis from the south for one hour. By comparing the maximum values and the standard deviations of all stress time histories, the cross section of the bridge deck at the Ma Wan tower is identified as the most critical section (CH23623 in Figure1), in which the six elements of no. 34111 and 38111, 40881 and 48611, 58111 and 59111 are identified as the most critical elements. The elements 34111 and 38111 are the bottom chords of the outer north and south longitudinal trusses, respectively, on the main span side (see Fig. 33). The elements 40881 and 48611 are the bottom chords of the inner north and south longitudinal trusses, respectively, on the main span side. The elements 58111 and 59111 are the bottom chords in the middle of the cross frame close to the north and south inner longitudinal trusses respectively. The hot spot stress is then determined by multiplying the nominal stress by the corresponding stress concentration factors.

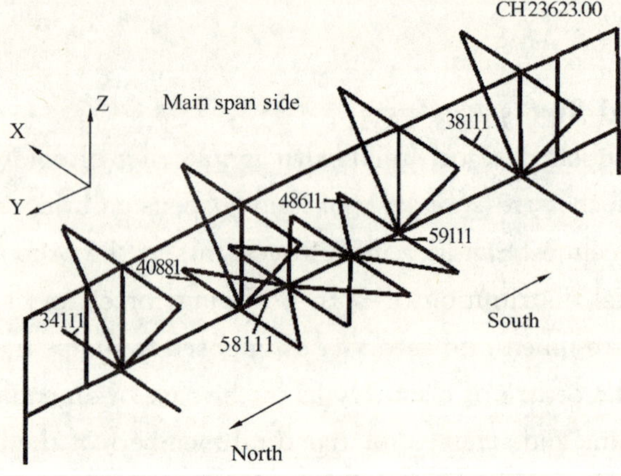

Fig. 33 Critical deck section and critical elements identified

For the subsequent buffeting-induced fatigue damage assessment of the bridge deck at the six hot spot stress locations for a wind return period of 120 years, the preceding exercise has to be repeated for the mean wind speeds from 5m/s to 30m/s with an interval of 5m/s for winds over the over-land fetch and from 5m/s to 20m/s at an interval of 5m/s for winds over the open-sea fetch, respectively. This yields a total of 60 one-hour time histories of the hot spot stresses for the bridge deck. For each of 60 one-hour time histories, the Rainflow counting method is applied to obtain the hop spot stress characteristics within one hour for different wind speeds and directions. The hop spot stress characteristics include the total number of stress cycles (N_r), the stress range (σ_r), the mean value (σ_m) of each stress cycle and the maximum stress range ($\sigma_{r,max}$). The mean value of each stress cycle includes the mean stress caused by the mean wind speed. As a result, a total of 60 data sets of hop spot stress characteristics are produced.

Based on the continuum damage mechanics (CDM) and some mathematical manipulation, the damage evolution model for fatigue damage assessment of a long suspension bridge can be given as

$$D_k = 1 - \left\{ (1 - D_{k-1})^{\alpha_{e,k}+1} - \frac{(\alpha_{e,k}+1)}{B(\beta+3)} \left(\sum_{j=1}^{m_{b,k}} \left[(\sigma_{r,jk} + 2\sigma_{m,jk}) \sigma_{r,jk} \right]^{\frac{\beta+3}{2}} (1 - D_{k-1})^{\alpha_{e,k} - \alpha_{j,k}} \right) \right\}^{\frac{1}{1+\alpha_{e,k}}} \quad (5)$$

where α_e is determined by the maximum stress range $\sigma_{r,max}$ through the function $\alpha_j = f(\sigma_{r,j})$; and k denotes the kth block. To assess the buffeting induced fatigue damage to the bridge deck at the identified six hot spot stress locations for a wind return period of 120 years, the occurrence sequences of 1,051,200 (=24×365×120) blocks of one hour duration should be determined in consideration of different wind speeds and directions. Because only monsoon wind effect on the bridge is considered in this study and the monsoon wind in Hong Kong normally blows from south in summer and from north in winter, it is assumed that monsoon wind blows over the open-sea fetch in summer and over the over-land fetch in winter. Two random permutation sequences of uniform distribution are then generated according to the total number of wind records in summer and in winter, respectively, in a particular year. The first random permutation sequence actually brings out the occurrence sequence of wind records over the open-sea fetch while the second random permutation sequence leads to the occurrence sequence of wind records over the over-land fetch for that particular year. Each wind record is then converted to each wind block according to it wind direction and wind speed.

The hot spot stress characteristics corresponding to each wind block can be best found from one of the 60 data sets of the hot spot stress characteristics obtained in advance. The fatigue damage accumulation of the bridge deck at each hot-spot stress location can finally be processed using the damage evolution model (Equation 5) one year after another until 120 years by assuming the zero damage at the beginning of fatigue damage accumulation. Fig. 34 shows the damage evolution of the bridge deck at the six hot spot stress locations during 120 years period. It can be seen that damage index increases with time. Slight nonlinear relationship between the damage index and time can be observed for the hot spot stress of the element 48611, which indicates the nonlinear nature of fatigue initiation and growth

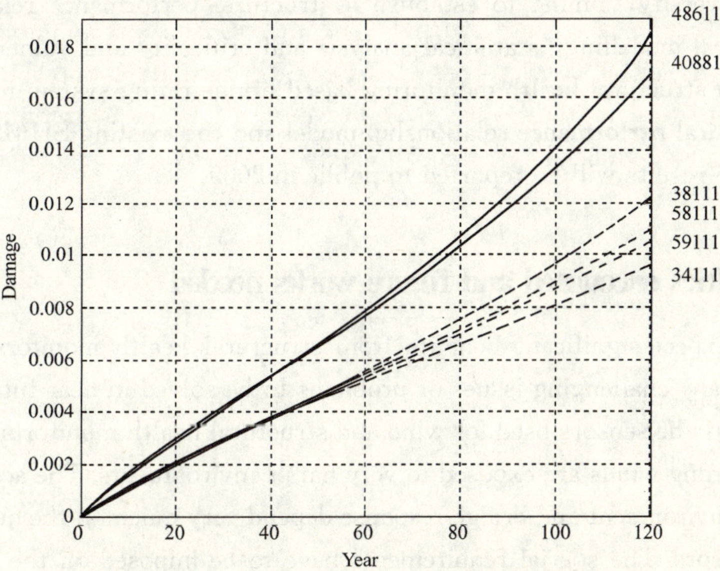

Fig. 34　Damage Evolution during 120-year Return Period

and the capability of the damage evolution model used in this study. It can be seen that monsoon wind induced fatigue damage to the bridge deck is not significant. The further information on this topic can be found in the literature (Xu et al. 2008*b*). It should be noticed that this study does not take typhoon effects and traffic effects into account. It will be done when long-term field measurement data on typhoons and traffic are available.

8.7 Shms-based bridge rating system

Although the SHMS installed in the Tsing Ma Bridge has demonstrated various degrees of successes in providing useful field measurement data for identification of various loads on the bridge, verification of the rules used in the design, and assessment of the bridge performance, the complete structural health monitoring functions have not been reached. Since the number of sensors is always limited for such a large structure and the locations of structural defects or degradation may not be at the same positions as the sensors, possibility exists in the condition that the worst structural condition may not be directly monitored. Therefore, the development of a structural performance relationship model for relating the structural performance conditions of the Tsing Ma Bridge to the measurement results at limited locations from the current SHMS becomes an imperative task. Furthermore, there is no an effective and scientific bridge rating system to provide a rational basis for rating risk of major bridge structural components utilizing the existing SHMS and for selecting types and frequencies of inspection and maintenance. The development of an effective bridge rating system based on the SHMS is another imperative task. A collaborative research project was initiated by the Hong Kong Highways Department and The Hong Kong Polytechnic University, aiming to establish a structural performance relationship model through numerical modelling, statistical analysis and criticality and vulnerability analyses and to develop a structural health monitoring-based bridge rating system for the bridge by using the structural performance relationship model and the existing SHMS. It is expected that some of the results will be reported to public in 2009.

8.8 Difficulties encounted and future works needed

While we expect significant benefits from structural health monitoring technology, there are still many challenging issues or problems to be solved in near future.

(a) Sensors: the sensors used for wind and structural health monitoring of a long span bridge during strong winds are exposed to very harsh environments. The accurate measurements of wind environment and bridge response depend very much on the quality and sensitivity of the sensors. The special requirements have to be imposed on the sensors used in WASHMS, and the advanced and reliable sensors have to be developed.

(b) Data transmission: the current SHMS often consists of sensors directly connected to data acquisition units through hard-wire or optical fiber connections. When monitoring a long span bridge with a large amount of sensors, long lines may attenuate data signals, particularly high frequency signals, due to the resistance and capacitance of cables. Signals may also become noisy and degrade due to coupled noise sources near cable path. Wireless data acquisition systems may be a solution for overcoming the problem in data transmission.

(c) Data management: the long-term wind and structural health monitoring of a long span bridge will produce a huge amount of data. Data storage, archive, maintenance, process, and presentation pose a challenging problem to professionals. The post-processing, related to improper use of complex software and selection of the algorithm, may introduce errors to the measurement results. It will be better if we have intelligent systems that can automatically switch the monitoring system from a slow-speed monitoring to a high-speed monitoring or vice versa.

(d) Complete wind environment monitoring: although the SHMS installed in the Tsing Ma Bridge is quite advanced, there exist some problems to be solved. For instance, more anemometers are required in order to catch complete pictures of both mean wind speed profile and turbulence intensity profile. More anemometers are also needed in order to explore the spatial correlation of fluctuating winds.

(e) Wind loading monitoring: there are no sensors in the SHMS installed in the Tsing Ma Bridge to monitor wind pressures and wind loading. How to monitor wind pressure/loading and how to identify aerodynamic admittance and flutter derivatives by using the SHMS is a very exciting and challenging task.

(f) Damage detection: another challenging issue is how to detect structural damage using the measured data from the SHMS. Long span bridges using innovative materials and structural systems under extreme wind loading often exhibit nonlinearities. Strong typhoons attacking the bridge are often non stationary. These factors make system identification and damage detection more difficult.

(g) Inspection and maintenance works: there is no an effective and scientific bridge rating system at present to provide a rational basis for rating risk of major bridge structural components and for selecting types and frequencies of inspection and maintenance by utilizing the SHMS. The development of an effective bridge rating system based on the SHMS is another imperative task. A collaborative research project was initiated by the Hong Kong Highways Department and The Hong Kong Polytechnic University, aiming to establish a structural health monitoring-based bridge rating system for the Tsing Ma Bridge. It is expected that some of the results will be reported to public in 2009.

(h) Computer simulation platform for extreme events: one of the tasks of SHMS is to prevent a bridge from catastrophic collapse. An advanced computer simulation platform is thus required to develop extreme loadings models by sufficiently utilizing the measurement data from the SHMS, to model the bridge including geometric and material nonlinearities, to perform load path and redundancy analysis, to carry out progressive collapse analysis of the bridge under the extreme wind loading, and finally to evolve an appropriate early warning system.

The last but not least is the economic and organizational challenge. Long-term wind and structural health monitoring is costly and time consuming. It is often difficult to sustain due to the lack of continuous financial supports. New concepts, ideas and technologies for a successful wind and structural health monitoring also require mutual respects and effectively coordinated collaboration between professional engineers, academic researchers, government agencies, contractors, managers, and owners.

8.9 Conclusions

This paper has demonstrated how to make good used of structural health monitoring systems for important large structures by taking the Tsing Ma Bridge in Hong Kong as an example. The structural health monitoring system installed in the Tsing Ma Bridge has been used for the identification of highway loading, railway loading, wind loading, and temperature loading in terms of either the existing methods or the methods developed from this study. The structural health monitoring system has also been used to identify time-varying modal properties of the bridge during strong wind and to verify the newly-developed computer simulations of the bridge under complicated loading conditions toward a better understanding of bridge behavior. The structural health monitoring system has also utilized for a better assessment of bridge performance. The idea of the establishment of a structural health monitoring system-based bridge rating system has also been commented as an on-going research work. Although a significant progress has been made in this new frontier, there are many challenging issues to be solved for a complete wind environment monitoring, wind loading identification, damage detection, inspection and maintenance works, and decision making after extreme events. Continuing efforts should also be paid to the quality of sensors, data transmission, data storage, data management, and data analysis to enhance the efficiency of structural health monitoring systems.

Opportunities created by modern technologies and new engineering methods are abundant, and their frontiers are without limits. It is important for researchers, governmental officials, practitioners, and other professionals to stay close to the new technology developments, to explore their potential of applications, and to integrate them fully into the decision making and strategy implementation for multi-hazards mitigation.

Acknowledgements

The work described in this paper was financially supported by the Research Grants

Council of Hong Kong through several CERG grants, The Hong Kong Polytechnic University through its niche area project on performance-based structural health monitoring of large civil engineering structures, the Hong Kong Highways Department through a contract research on bridge health and engineering. The support from the Lantau Fixed Crossing Project Management Office, the Hong Kong Highways Department, to allow the author to access the measured data for academic purpose only is particularly appreciated. Sincere thanks should also go to many co-workers and students of the author. Without their helps, it is impossible to come up with this paper. Any opinions and conclusions presented in this paper are entirely those of the author.

References

[1] Chen, J., Xu, Y. L. and Zhang, R. C. (2004). "Modal parameter identification of Tsing Ma suspension bridge under Typhoon Victor: EMD-HT method," *Journal of Wind Engineering and Industrial Aerodynamics*, Vol. 92, pp. 805-827.

[2] Chen, J. and Xu, Y. L. (2004). "On modelling of typhoon-induced non-stationary wind speed for tall buildings," *The Structural Design of Tall and Special Buildings*, Vol. 13. No. 2, pp. 145-163.

[3] Fei, Q. G., Xu, Y. L., Ng, C. L., Wong, K. Y., Chan, W. Y. and Man, K. L (2007). "Structural health monitoring oriented finite element model of Tsing Ma bridge tower," *International Journal of Structural Stability and Dynamics*, Vol. 7, No. 4, 647-668.

[4] Guo W. W., Xu Y. L., Xia H, Zhang W. S. and Shum K. M. (2007). "Dynamic response of suspension bridge to typhoon and trains. II: numerical results", *Journal of Structural Engineering*, Vol. 133(1), 12-21.

[5] Huang N. E., Shen Z., Long S. R., Wu M. C., Shi H. H., Zheng Q., Yen N. C., Tung C. C. and Liu H. H. (1998). "The Empirical mode decomposition and the Hilbert spectrum for nonlinear and non-stationary time series analysis," *Proceedings of Royal Society London A*, Vol. 454, 903-995.

[6] Ko, J. M., Xue, S. D. and Xu, Y. L. (1998). " Modal analysis of suspension bridge deck at erection stage", *Engineering Structures*, Vol. 20, No. 12, Dec., pp. 1102-1112.

[7] Liu G., Xu, Y. L. and Zhu, L. D. (2004) "Time domain buffeting analysis of long suspension bridges under skew winds" *Wind & Structures-An International Journal*, Vol. 7, No. 6, 421-447.

[8] Liu, T. T., Xu, Y. L., and Zhang, W. S. (2007). "Buffeting-induced stresses in a long suspension bridge II: Structural health monitoring orientated stress analysis", Submitted to *Wind and Structures*

[9] Simiu E., and Scanlan, R. H. (1996), "Wind effects on structures", New York: John Wiley & Sons.

[10] Wong, K. Y., Man K. L, and Chan W. Y. K. (2001*a*). "Monitoring of wind load and response for cable-supported bridges in Hong Kong", *Proceedings of SPIE 6th International Symposium on NDE for Health Monitoring and Diagnostics*, *Health Monitoring and Management of Civil Infrastructure Systems* (*Chase and Aktan eds.*), *Newport Beach*, *California*, Vol. 4337, 292-303.

[11] Wong, K. Y., Man K. L, and Chan W. Y. K. (2001*b*). "Application of global positioning system to structural health monitoring of cable-supported bridges", *Proceedings of SPIE 6th International Symposium on NDE for Health Monitoring and Diagnostics*, *Health Monitoring and Management of Civil Infrastructure Systems* (*Chase and Aktan eds.*), *Newport Beach*, *California*, Vol. 4337,

390-401.

[12] Xu, Y. L., Ko, J. M., and Yu, Z. (1997a). 'Modal analysis of tower-cable system of Tsing Ma long suspension bridge', *Engineering Structures*, Vol. 19, No. 10, pp. 857-867.

[13] Xu, Y. L., Ko, J. M and Zhang, W. S. (1997b). "Vibration studies of Tsing Ma long suspension bridge", *Journal of Bridge Engineering*, Vol. 2 (4), 149-156.

[14] Xu Y. L., Zhu L. D., Wang K. Y., Chan K. W. Y. (2000*a*). Field measurement results of Tsing Ma suspension Bridge during typhoon Victor. *Journal of Structural Engineering and Mechanics*, 2000, Vol. 10, No. 6, 545-559.

[15] Xu Y. L., Sun D. K., Ko J. M. and Lin J. H. (2000*b*). "Fully coupled buffeting analysis of Tsing Ma suspension bridge", *Journal of Wind Engineering and Industrial Aerodynamics*, Vol. 85, 97-117.

[16] Xu Y. L., Xia H. and Yan Q. S. (2003). "Dynamic response of suspension bridge to high wind and running train", *Journal of Bridge Engineering*, Vol. 8(1), 46-55.

[17] Xu Y. L., Zhang N, and Xia H. (2004). "Vibration of coupled train and cable-stayed bridge systems in cross winds", *Engineering Structures*, Vol. 26(10), 1389-1406.

[18] Xu Y. L. and Chen J. (2004). "Characterizing non-stationary wind speed using empirical mode decomposition", *Journal of Structural Engineering*, ASCE, 2004, Vol. 130(6), 912-920.

[19] Xu Y. L. and Zhu L. D. (2005). "Buffeting response of long span cable-supported bridges under skew winds-Part II: case study". *Journal of Sound and Vibration*, 281(3-5), 675-697.

[20] Xu, Y. L., Chen, B., Ng, C. L., Wong, K. Y. and Chan, W. Y. (2007*a*). "Monitoring temperature effect on a long suspension bridge in Hong Kong," *The* 5th *Workshop on Nondestructive Evaluation of Civil Infrastructural System*, Taipei, Taiwan, pp. 171-194.

[21] Xu Y. L., Guo W. W., Chen J., Shum K. M. and Xia H. (2007*b*), "Dynamic response of suspension bridge to typhoon and trains. I: field measurement results", *Journal of Structural Engineering*, Vol. 133(1), 3-11.

[22] Xu, Y. L., Chen, J., Ng, C. L. and Zhou, H. J. (2008*a*). "Occurrence probability of wind-rain-induced stay cable vibration", Advances in Structural Engineering-An International Journal, Vol. 11, No. 1, 53-69.

[23] Xu, Y. L. Liu, T. T. and Zhang, W. S. (2008*b*), "Buffeting-induced fatigue damage assessment of a long suspension bridge," *International Journal of Fatigue* (in press).

[24] Zhang, W. S., Wong, K. Y., Xu, Y. L. and Liu T. T. (2007). "Buffeting-induced stresses in a long suspension bridge I: structural health monitoring oriented finite element model", Submitted *to Wind and Structures*.

[25] Zhu, L. D., Xu, Y. L., Zhang, F. and Xiang, H. F. (2002*a*). "Tsing Ma bridge deck under skew winds. I: aerodynamic coefficients". *Journal of Wind Engineering and Industrial Aerodynamics*, Vol. 90, No. 7, 781-805.

[26] Zhu, L. D., Xu, Y. L., and Xiang, H. F. (2002*b*). "Tsing Ma bridge deck under skew winds. II: flutter derivatives". *Journal of Wind Engineering and Industrial Aerodynamics*, Vol. 90, No. 7, 807-837.

[27] Zhu L. D. and Xu Y. L. (2005). "Buffeting response of long span cable-supported bridges under skew winds-Part I: theory". *Journal of Sound and Vibration*, 281(3-5), 647-673.

第 9 章 Chapter 9

动静态分布传感技术及结构健康监测理论与设计体系*

吴智深[1,2]，杨才千[1]，李素贞[3]，许斌[4]，张浩[2]，徐赵东[1]，沈圣[1]

（1. 东南大学，城市工程科学国际研究中心，南京 210096；2. 茨城大学，都市系统工学科，日立，日本 31—8511；3. 同济大学，土木工程学院，上海；4. 湖南大学，土木工程学院，长沙）

摘 要：作为提高重大基础工程设施管养水平和防灾减灾能力的一个重要技术和措施，结构健康监测技术近年来引起了国内外的广泛关注和重视，已有不少监测系统被应用于实际工程中。对于大型工程结构，分布式监测技术是建立一个有效的监测系统的首要因素，其次是基于分布式动静态传感监测的解析・评价及预警系统的研究和开发。本文将基于作者研究团队近年来的研究成果并结合大型工程基础设施结构健康监测的特点，简要总结结构健康监测中的动静态分布传感技术、损伤识别及性能评价方法和结构健康监测中的其他相关理论与设计体系。这些分布式传感技术包括光纤传感的 FBG（Fiber Bragg Grating）和基于布里渊散射的分布传感技术以及 HCFRP（Hybrid Carbon Fiber Reinforced Polymer）分布式电传感技术。基于这些分布传感技术，提出了一系列新的结构健康监测理论，其中包括动静态损伤识别理论、由分布应变到结构分布变形转换理论等，并进行了大量的试验验证。随后，简要分析了基于分布式传感技术的结构健康监测系统的设计方法和原则。最后，分别给出了三种传感技术在实际大型工程上的应用。

关键词：分布传感技术；光电传感；结构健康监测；动静态检测；损伤识别及性能解析・评价

Structural identification theories and SHM design methodology based on dynamic and Static distributed sensing techniques

Z. S. Wu[1,2], C. Q. Yang[1], S. Z. Li[3], B. Xu[4], H. Zhang[2], Z. D. Xu[1], S. Shen[1]

(1. International Institute for Urban System Engineering, Southeast University, Nanjing, 210096, China; 2. Department of Urban & Civil Engineering, Ibaraki University, Japan;
3. College of Civil Engineering, Tongji University, Shanghai; China
4. College of Civil Engineering, Hunan University, Changsha; China)

Abstract: Structural health monitoring (SHM) is an important technique to improve the disaster

---

* 第一作者：吴智深（1961-），男，教授，教育部长江学者，主要从事先进纤维复合材料、先进传感技术及结构健康监测和控制、非线性计算力学学等方面的研究和开发，E-mail：zswu@mx. ibaraki. ac. jp

prevention and mitigation of large-scale structures and their daily maintenance, which has been attracting increasing attention from both research and application fields. For the health monitoring of large-scale structures, distributed sensing technique is a key factor. In addition, the dynamic and static analysis and evaluation are also essential for an effective SHM system. This article briefly summarizes the distributed dynamic and static monitoring techniques, structural identification theories, structural performance evaluation and SHM design methodology based on the recent investigations of the authors' research group. The distributed sensing techniques include FBG(Fiber Bragg Grating), Brillouin scattering based optic fiber sensing and HCFRP (Hybrid Carbon Fiber Reinforced Polymer) sensing techniques. Based on the distributed sensing techniques, some dynamic and static structural identification theories are proposed and confirmed. And then, the design principles of SHM are studied. Finally, the applications of the distributed sensing techniques and SHM systems to actual large-scale structures are addressed as well.

Keywords: distributed sensing technique; opto-electronic sensing technology; Structural health monitoring(SHM); dynamic and static monitoring technique; structural identification and performance evaluation

9.1 引言

随着社会的发展，大跨和超大跨空间结构、跨江跨海桥梁（隧道）、高层建筑、城市地铁和轨道交通等新型和复杂建筑结构不断出现。然而，材料老化、环境侵蚀、长期荷载效应、疲劳效应、突变效应以及与地震、台风等自然灾害因素的耦合，将不可避免地导致结构的损伤积累和抗力衰减，从而降低了这些结构的正常使用功能和安全性能，极端情况下甚至引发灾难性的突发事故。例如，1999 年，重庆彩虹大桥突然倒塌，造成上百人员伤亡；2000 年，台湾省连接台北和高雄屏东的高屏大桥突然一个桥墩、二孔桥面约一百公尺断裂坍塌，造成 30 余人受伤，16 部汽车受困；2001 年，宜宾南门大桥桥面一部分突然坍塌；2004 年，法国戴高乐机场倒塌事故，同年阿拉伯联合酋长国迪拜机场一座正在施工的候机楼发生墙体倒塌事故。2007 年，美国明尼苏达首府明尼阿波利斯市内的一座繁忙的跨河立交桥突然倒塌，共有 300 米长的桥体断裂成三段落入密西西比河中，本次事故发生时，正值当地的下班高峰，造成多人伤亡。这些事故的不断发生，造成巨大的经济损失和人员伤亡，带来了恶劣的社会影响。更有甚者，地震、飓风等大型自然灾难的发生对大型工程结构造成的破坏将更为严重、影响规模也更为广泛，大型灾难一旦发生急需对结构进行快速检测、诊断甚至恢复。因此，为了保障结构的安全性、耐久性和正常的使用功能，既建和将建重大基础工程结构急需采用有效的技术手段实时监测和评定其安全状况、提供必要的预警、修复和控制损伤的发展。

9.1.1 结构健康监测的基本概念及分类[1~6]

自上世纪 90 年代中期以来，结构健康监测（SHM）在国内外得到了广泛的研究与应用，目的是对结构进行常规例检、实时监测、风险管理或灾后安全性能评价等。主要用于提高结构体系的安全可靠性、实现长寿命化、减少甚至避免灾难性事件的发生、降低全寿命周期的费用以及考虑其他一些未知的因素等。结构健康监测的核心是在人力参与尽量少的基础上，对结构的正常使用状况进行实时监控、对损伤进行实时监测和识别以及对结构的使用情况进行实时评价，并能给出维护、管理和修复建议。

通常，结构健康监测系统包括：1）传感系统，2）信号发生装置，3）信息传输及处理系统，4）损伤识别及解析·评价系统以及5）系统整合系统，如图1所示。

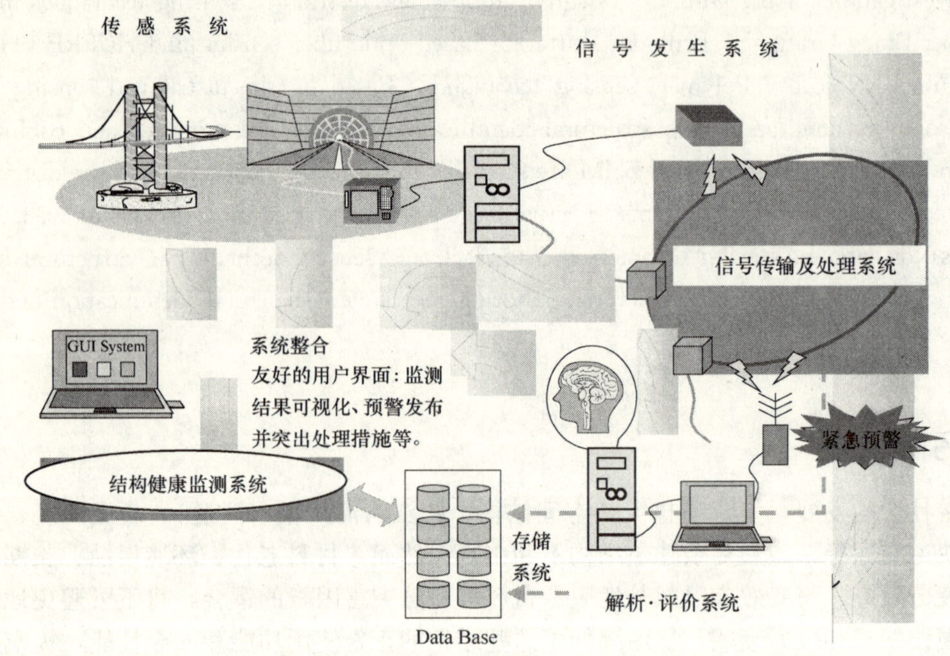

图1 结构健康监测系统（SHM）的组成

传感系统是结构健康监测系统的基础部分，包含各类传感器，如加速度计、应变计、位移计、光纤传感器、温度传感器等。传感器监测的信息经过传输系统存储到存储中心并建立数据库，然后数据经过专家系统进行分析处理、判断结构的健康状况并对损伤进行定位。最后，经过系统和信息整合将监测结果进行可视化显示，并给出实施预报，如果出现损伤则还需给出相应的处理措施。在整个结构健康监测系统中，传感系统和解析·评价系统是核心部分。传感系统提供结构健康监测所需要的最基本、最直观的信息，是整个系统的硬件支撑。而解析·评价系统是整个系统的“大脑”，对所收集的错综复杂信息进行梳理和分析，并结合结构自身特征以及各种损伤识别理论建立对应的健康监测模型，对结构的健康状况进行分析和评价，如果出现损伤，首先对损伤进行定位，然后进行量化，最后给出结构的处理措施。

导入和未导入结构健康监测系统的建筑物维护管理对比如图2所示。未导入结构健康监测系统的结构随着服役年限的增加，结构的可靠性可能会急剧下降，维护成本急剧升高；而对于导入结构健康监测系统的结构，随着服役年限的增加结构的可靠性基本可保持稳定而维护管理成本也能够保持微量增加的状态。

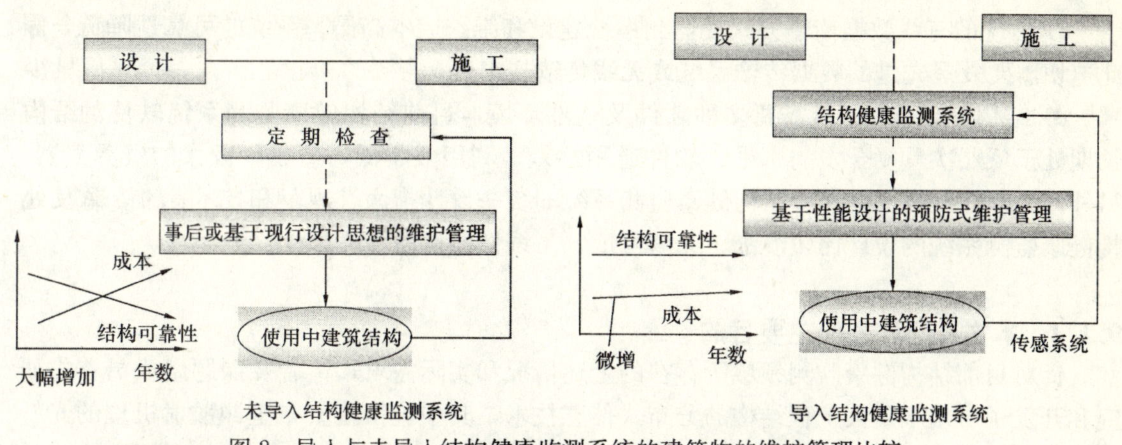

图 2　导入与未导入结构健康监测系统的建筑物的维护管理比较

9.1.2　大型土木工程结构对传感性能的要求

由于大型工程基础设施体积庞大、长距离分布，对传感系统有着一些特殊的性能要求：

a）传感系统要有分布和长标距特性：分布式传感系统可以为大型结构的损伤进行定位和对局部损伤进行定量化监测；

b）耐久性和长期稳定性：由于基础工程设施的设计寿命周期通常大于 50 年，所以传感系统要有比较好的耐久性，同时传感性能也需要有长期的稳定性和可靠性；

c）动、静态测量及精度要求：为了更好反映结构的性能特征、损伤状况，通常需要对结构进行动静态监测，动态监测可以用来在外力不能定量化的情况下反映结构的特征性能（固有频率、振型分布等）的变化，静态监测可以直接反映结构的受力状况（应变大小和变形等）；

d）另外，还要求传感系统具有比较好的经济性能，操作简易。

9.1.3　目前健康监测发展中尚存在的一些问题

用于大型土木工程结构的健康监测系统目前还处于不断完善和发展过程之中，无论是传感系统还是损伤识别及解析、评价系统都还存在诸多亟待完善的部分，主要表现在以下几个方面：

a）多数传感系统不具备分布式传感功能，像基于布里渊散射原理的 BOTDR 光纤传感技术虽然具有分布传感功能，但是应变测量精度（理想状态下的精度为$\pm 50\mu\varepsilon$）和空间分辨率（1m）不高，而且传统的传感元件无法满足长期监测的要求，迫切需要发展先进的、具有良好分布传感特性、耐久性和可靠性的监测传感系统。

b）目前国内外有关结构损伤识别和模型修正的方法均是基于结构整体性态响应的模态分析理论建立的，而模态分析方法已被研究证明其对大型土木工程结构的局部损伤并不敏感，损伤识别的效果较差。如何建立更有效的模态分析理论和方法，并充分利用结构健康监测积累的大量数据，识别结构的状态、评估其安全可靠性能、及时预警，是结构健康监测研究的目标，也是当前迫切需要解决的问题。

c）传统的有线数据传输方式使监测系统建设和维护成本高昂、系统的可靠性降低，因此迫切需要发展先进的数据传输系统或无线传输技术。

d）结构健康监测系统需要多种软件及软件环境，目前结构健康监测系统软件的开发长期处于低层次重复开发的水平、软件通用性差、利用效率低，造成了极大的资源浪费；另外，至今还没有大型工程结构健康监测系统的统一设计指南、规范和技术标准，致使结构健康监测系统的设计比较混乱、无据可依、系统性能良莠不齐。

9.1.4 本文系列研究的主要目的

针对目前结构健康监测领域所存在的上述情况和实际需求，笔者及课题组成员首先研究和开发了可用于重大工程结构的分布式传感技术，其中包括基于布里渊散射机理的光纤传感技术、二次封装长标距分布 FBG（可用于动、静态精确测量）以及 HCFRP 传感技术；这些传感技术的共同特征是具有长的测量标距（可根据具体测量需求进行选定），且可以实现长距离以及大范围的分布监测。基于长标距、分布式传感系统的监测数据，建立一系列新的结构损伤识别方法和理论以及结构健康监测系统的设计方法。

9.2 各分布式传感技术的研究和开发

本节着重介绍三种分布式传感技术的研发工作以及每种传感技术的传感特性和主要应用范围。这三种传感技术分别是基于布里渊散射机理的分布式光纤传技术、长标距分布式 FBG 传感技术以及 HCFRP 分布式传感技术。

9.2.1 基于布里渊散射机理的分布式传感技术的开发[7-16]

9.2.1.1 基于光时域（OTDR）反射技术的监测原理

分布式光纤传感系统是基于光时域反射（OTDR）的技术原理，光纤既可用来感知信号又可用来传输信号，可以探测几十公里内沿着光纤所有位置的被测表面的物理参数，适合于多个行业的线性火灾报警和大型工程的应力、温度和压力等的监测。其原理如下，激光光源发出的光脉冲沿着光纤传输，大部分光往前传输，但一小部分散射光信号会沿光纤反射回来，具体包括瑞利（Rayleigh）散射光、布里渊（Brillouin）散射光和拉曼（Raman）散射光等。这些反射回来的光对多个测量参量敏感，瑞利散射对温度敏感，布里渊散射对应力和温度敏感，而不同位置反射回来的时间不一样，这样就实现了真正的分布式测量。基于光的 OTDR 技术的各物理参数测量如图 3 所示。该图同时也显示了光纤中一个典型的散射光谱。

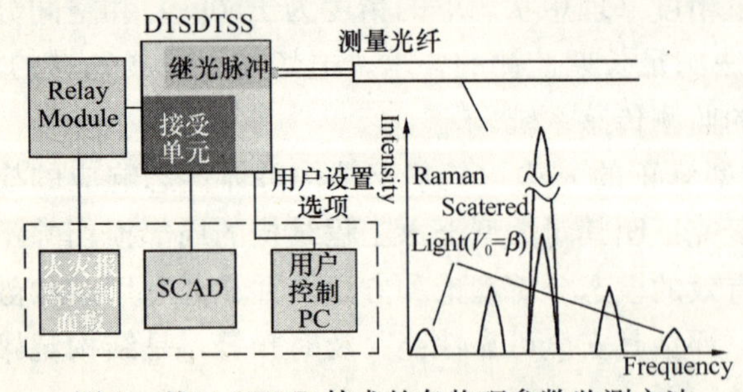

图 3 基于 OTDR 技术的各物理参数监测方法

在这些散射光中，以布里渊散射光作为测量信号的研究和应用最多。根据测量原理和分析技术的不同，可以分为 BOTDR（Brillouin Optical Time Domain Reflectmetry），BOTDA（Brillouin Optical Time Domain Analysis）以及 BOCDA（Brillouin Optical Correlation Domain Analysis）等测量技术。为了提高测量的精度和空间分解能，日本的 NEUBREX 公司在 BOTDA 技术的基础上开发出了 Pulse-PrePump（PPP-）BOTDA 新型的测量技术，测量的空间分辨率提高到了 10cm，应变的测量精度提高到 $\pm 25\mu\varepsilon$。这几种测量技术的共同点是依据布里渊散射原理，且都可以进行长距离分布式监测，不同点是具体测量原理和测量的精度不同。

基于布里渊时域散射机理的分布式监测原理如图 4 所示，脉冲激光以一定频率从光纤一端入射，入射的脉冲光与光纤中的声学声子发生相互作用后产生布里渊散射，其中的背向布里渊散射光沿原路返回到脉冲光的入射端，进入测试仪器的受光部和信号处理单元，可以得到光纤沿线各采样点的散射光功率，按一定间隔变化入射光的频率，实现不同频率下布里渊散射光的功率的测量，可得到光纤沿线各个采样点的散射光谱。理论上，布里渊散射光谱呈洛伦兹曲线，其峰值功率所对应的频率即为布里渊频率。

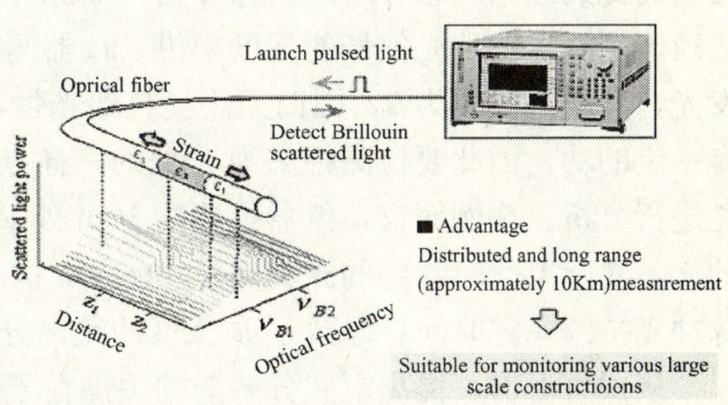

图 4　基于布里渊散射传感技术的测试原理

当光纤某处的应变或温度发生变化时，光纤中的后向布里渊散乱光谱的频率会发生相应的偏移，频率的偏移量与光纤所承受的应变或温度改变量呈良好的线性关系。

因此通过测量光纤中的布里渊散乱光频率偏移即可推算出光纤沿线的应变或温度的变化量。光纤任一点 z 处发生应变或温度改变时，布里渊散乱光的频率偏移量可由下式计算：

$$\nu_B(z)=\nu_B(0)+c_1\Delta\varepsilon(z)+c_2\Delta T(z) \tag{1}$$

式中：$\nu_B(0)$为 z 点处布里渊光的初始频率偏移

$\nu_B(z)$为 z 点处发生应变和温度变化后布里渊光的频率偏移，c_1、c_2 为布里渊光的频率偏移的应变系数和温度系数，由于材质及生产工艺的差别，不同厂家不同型号的单模光纤系数会有所差别。光纤中发生应变或温度变化的位置 Z 可由下式确定：

$$Z=cT/(2n) \tag{2}$$

式中：Z 发生应变或温度变化的位置距起点的距离；

c 为真空中的光速；

T 为发送脉冲光到接收散乱光的时间差；

n 为光纤的屈折率；

空间分解能

对于基于布里渊散乱光技术的分布式光纤传感器而言，空间分解能是最重要的性能指标，任一测点所得到的信息，实际上都是包含该点的一段距离 L 内应变的综合反映。因为 L 段内所有后向散射光在同一时刻 t 到达光纤始端，而处于 L 外的后向散射光在不同于 t 的另一时刻到达光纤始端。因此，L 为理论上可分辨的最小光纤长度，称为空间分解能

(Spatial Resolution)，可由下式计算：

$$L=v\tau/2 \tag{3}$$

式中：L 为空间分解能，v 为光纤中的光速，τ 为发送脉冲光的宽度。

若取 $\tau=10$(ns)，$v=0.2\times109$(m/s)，则 $L=1$(m)，那么在此情况下，在第 X(ns)时刻接收到的，并非只是 $0.1\times X$(m)点处的反射光，而是 $0.1\times X$(m) 到 $0.1\times X-1$(m)这一段光纤所产生的背向散射光。空间分解能的存在，有其必然性，从仪器设备的角度而言无法捕捉到理想状态一瞬间的脉冲光，而脉冲持续的时长，反映到空间距离中必然导致无法精确到一点。从光信号的强度来讲，仪器测量的是散射光的功率，而理论上任一时刻的发光，其功只能视为 0，因而无法接收到光强，发光必须要有一定时长，也就是要入射光有一定的功。因此要检测距离为 Z 处的采样点的应变情况，就需要对 $2*Z/v$ 时刻收到的光进行分析。举例而言，仪器得到的 10m 处采样点的频谱，反映的是在 100ns 时刻所收到的光，实际上此时收到的光其实是 10～9m 这一段光纤的反射光的总和。总而言之，由于脉冲光持续一定时间，导致了 L 长度的距离分解度的存在。

从式可看出，提高空间分解能的关键在于缩短入射脉冲光的脉冲宽，对于 BOTDR 技术，当脉冲宽小于 28ns 时，声子不容易被激发，导致测试精度显著下降，目前 BOTDR 技术的最小空间分解能为 1m。近年开发的 PPP-BOTDA 技术采用两种光源，一种为预泵浦光用于充分激发声子，另一种为探测泵浦光，由于声子被预泵浦光充分激发，探测泵浦光的脉冲宽可以显著降低，从而显著提高空间分解能，目前，PPP-BOTDA 技术的空间分解能可达到 0.1m。

9.2.1.2 基于布里渊散射技术的 BOTDR 和 PPP-BOTDA 监测系统

a) 基于布里渊散射机理的不同监测系统的监测原理

目前，基于布里渊散射的分布式光纤测试技术得到了迅速发展，其测试精度及空间分解能都得到了很大的提高，其中 BOTDR 和(PPP-)BOTDA 为目前最为常用的两种测试技术；日本安藤公司所生产的基于 BOTDR 技术的测试设备 AQ8603 和 Neubrex 公司所生产的基于 BOTDA 技术的测试设备 Neubrescope 为目前较为常用的测试系统，两者的性能参数比较如表 1 所示。相对于 BOTDR 系统，PPP-BOTDA 系统的应变测量精度和距离分辨率更高。

基于布里渊时域散射技术的 BOTDR 与 PPP-BOTDA 监测性能比较　　表 1

设备名	AQ8603				Neubrescope			
采样间距	5cm				5cm			
脉冲光宽度（ns）	10	20	50	100	1	2	5	10
空间分解能(m)	1	2	5	11	0.1	0.2	0.5	1
最大测试距离(km)	10	25	45	55	1	5	10	20
应变测试精度($\mu\varepsilon$)	±40	±40	±30	±30	±25	± 25	± 25	±25

两者的测量原理如图 5 所示，BOTDAR 采用单光源，而 PPP-BORDA 采用双光源并从光纤的两端分别注入。当单模光纤中传播的光功率大于布里渊的极限功率，后向的受激布里渊散射(SBS)就将产生，SBS 跟入射光、Stokes 光和声波等有关。处于光纤两端的可

调谐激光器分别将一脉冲光(泵浦光)与一连续光(探测光)注入传感光纤，当泵浦光与探测光的频差与光纤中某区域的布里渊频移相等时，在该区域就会产生布里渊放大效应(受激布里渊散射)，两光束之间发生能量转移。由于布里渊频移与温度、应变存在线性关系，因此对两激光器的频率进行连续调节的同时，通过检测从光纤一端耦合出来的连续光的功率，就可确定光纤各小段区域上能量转移达到最大时所对应的频率差，从而得到温度、应变信息，实现分布式测量。

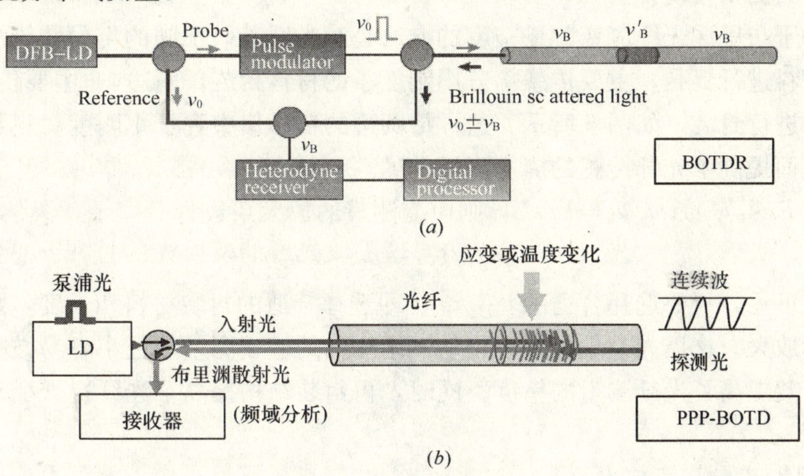

图 5　基于布里渊散射的 BOTDR 及 PPP-BOTDA 传感原理图

BOTDA 系统中信号的检测较容易，在空间分辨率、测量精度等方面更有优势，技术也较为成熟，但双光源的使用以及光源的两端入射使它的应用受到一定的限制。对于时域分析技术，空间分辨性能提高是研究的热点，目前，缩小空间分辨率必须靠缩短脉冲光宽度(Pulse Width)来实现。然而，通常 BOTDA 需要 28ns 来激发光纤中的声波，缩短脉冲光宽度的同时会造成受激布里渊增益(Stimulated Brillouin Gain)的减弱和布理渊频谱形态的劣化。日本 Neubrex 公司的新一代 PPP-BOTDA 产品，改变了传统的脉冲光波形，在测量的短脉冲光之前，利用预泵脉冲光(Pre-Pump Pulse)来激发光纤中的声波，获得了 10cm 的空间分辨率和 $\pm 25\mu\varepsilon$ 的准确度。这一技术的测量原理可由图 5(b)来表示：在测量脉冲光 PD 到达某一光纤位置前，预泵脉冲光 PL 已经激发了声波，因此，宽度仅为 1ns 的 PD 就足以获得该处的洛仑兹型的布里渊增益频谱。

b)基本监测特点与性能评估

对于基于布里渊散射的分布式光纤测试技术，任一测点 Z 处所测得的布里渊频谱实际上是包含该点在内的空间分解能长度范围内频谱的综合反映，如下式所示：

$$\bar{g}(\nu, Z) = \frac{1}{L}\int_{Z-L}^{Z} g(v, v_b(\zeta))\mathrm{d}\zeta \tag{4}$$

由该谱曲线峰值所对应的频率 $\nu_b(\zeta)$ 所推导出的应变也就是空间分解能长度范围内应变的综合反映。当光纤所受均匀分布的应变长度大于空间分解能时，测试设备能够准确测得应变值，然而当光纤所受均匀分布的应变长度小于空间分解能或空间分解能长度范围内应变分布不均匀时，仪器无法测得该范围内准确的应变变化，详细分析参见参考文献[15]。

9.2.1.3 光纤传感增敏、提高距离分辨率及无滑移技术

a）光纤增敏技术

由于基于布里渊散射机理的光纤传感技术的应变监测精度比较低，BOTDR 的应变测量精度为±50$\mu\varepsilon$，而作为最近开发的高测量精度的 PPP-BOTDA 的理想状态下的应变测量精度也只有±25$\mu\varepsilon$。所以，有必要开发一些高新的应用技术来提高基于布里渊散射机理的光传感技术的测量精度。

提高基于布里渊光纤应变测量精度的方法是对光纤采取不同的再封装措施，用两种刚度不同的材料进行封装，主要传感部分用刚度小的材料封装，两边(非主要传感部分)用刚度大的材料进行封装，如图 6 所示。这样把两边的变形集中到了小刚度材料封装的传感光纤部分，从而提高了光纤传感的应变测量精度。

令 $\alpha_E = E_0A_0/E_1A_1$，$\alpha_L = L_0/L$，则应变测量的放大系数为

$$\eta = 1/(\alpha_L + \alpha_E - \alpha_L\alpha_E) \tag{5}$$

由上式可见，只要选择合适的 α_E，α_L，即选择合适的封装材料和长度，则测量到的应变可以进行放大。这极大改善了环境振动下结构物的应变测量值过小容易被噪音淹没的不利情况，有效提高了光纤测量的精度。同时，再封装结构提高了光纤抗严酷环境的性能和安全性。

b）提高距离分辨率技术

为进一步提高基于布里渊散射机理的光传感技术的距离分辨率，笔者所带领的课题组开发了环形安装技术，如图 7 所示。通过环形将普通光纤安装在被监测的结构上，可以有效提高布里渊分布式光传感技术的距离分辨率。

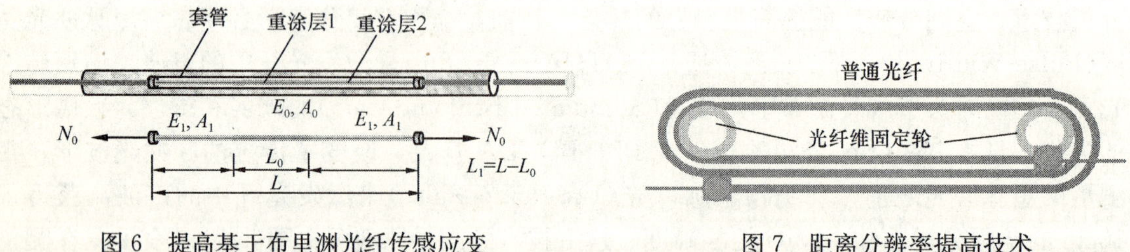

图 6 提高基于布里渊光纤传感应变测量精度的封装示意图

图 7 距离分辨率提高技术

针对混凝土结构的性能特点和测量要求，课题组同时开发了光纤的定点粘贴技术(PFB：Point Fixation Bongding）和全面粘贴技术（OB：Overall Bonding），如图 8 所示。全面粘贴可以给出光纤所对应结构每个点的物理参量；但是对于混凝土这样非均质容易出现裂缝的结构，局部比较大的裂缝出现往往会引起光纤的断裂，从而引起整个监测系统的失效。为了避免该类问题的出现，开发了定点粘贴技术，定点粘贴的另外一个优点是可以将两个粘贴点之间的光纤部分作为长标距传感，从而给出结构的宏观应变。另外，对于混凝土这种非均质材料，如果采用全面粘贴技术，则在测量范围内会出现比较多的不稳定峰值，如下左图所示；而如果采用定点粘贴，则在粘贴范围内光纤的测量值比较平均，如下右图所示。

c）无滑移光纤传感器的开发

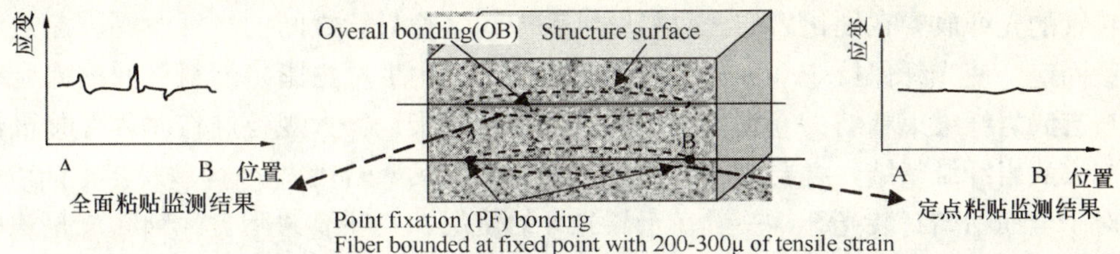

图 8　传感光纤的定点和全面粘结技术及测量结果示意图

根据实际测试需要，提出了 2 种封装方法，一种为全粘接封装，另一种为定点粘接封装，以下将详细介绍两种封装方法的过程。

全粘结封装光纤

如图 9 所示，制作过程如下：①在桌面上垫一层塑料薄膜，在薄膜上刷一层树脂，固定玄武岩纤维束两端保持拉直状态；②在 UV 树脂单模光纤内剥出一段裸光纤，裸光纤长度稍大于监测所需的测试长度，并置于玄武岩纤维束中间，两端拉紧固定，使光纤保持 100～300$\mu\varepsilon$ 左右的预拉应变；③用软橡胶刷刷一层环氧树脂浸透玄武岩纤维，树脂浸透玄武岩纤维长度需稍大于裸光纤长度，以免裸光纤裸露在硬化后的玄武岩纤维外而极易断裂；④在上面再铺一层玄武岩纤维束，悬挂重物保持拉直状态，固定玄武岩纤维束两端，在第二层纤维上再刷一层环氧树脂浸透；⑤上面再铺一层塑料薄膜，用软橡胶刷子在塑料薄膜上将玄武岩纤维束刷平，挤掉环氧树脂内的气泡，尽量使浸润后的玄武岩纤维素密实，并保持宽度厚度均匀一致；⑥环氧树脂凝固硬化后揭去塑料薄膜，放松玄武岩纤维和光纤，剪去两侧多余树脂；⑦大长度制作时，可部分段用树脂含浸，含浸长度根据测试需要，部分段不含浸，不含浸段可剪断与其他光纤或接头熔接。

内部定点粘结光纤

为保证能获得稳定准确的测试结果，可进行一定标据范围内的定点封装，封装后两个粘贴点间的光纤相当于一个长标距光纤传感器，该段光纤均匀变形，可获得一定标距的平均应变。标距大小可根据实际需要确定，从而形成分布式的平均应变测试，如图 10 所示，具体封装过程如下：①在桌面上垫一层塑料薄膜，将一层玄武岩纤维束两端挂重物拉直平铺在塑料薄膜上，固定玄武岩纤维束两端保持拉直状态；②玄武岩纤维束刷树脂含浸，含浸长度可根据测试需要确定，等待凝固硬化；③将 UV 树脂涂敷的光纤套上多个塑料保护管，每段塑料管的长度约为所需定点测试的标距减去 1cm，塑料管间距保持为 1cm；④将

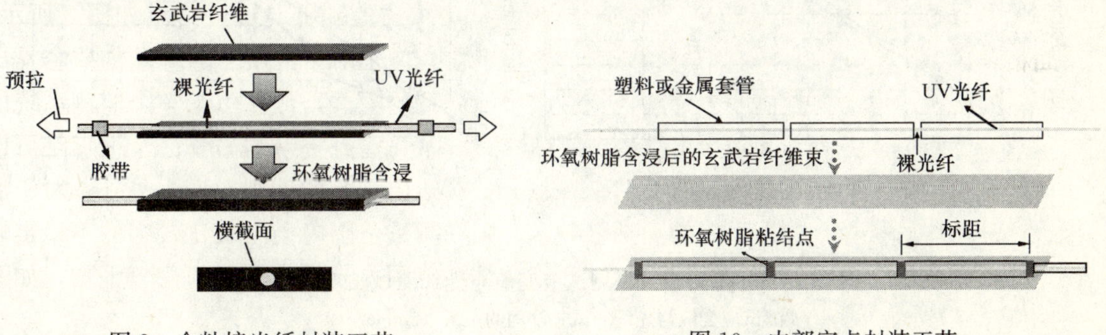

图 9　全粘接光纤封装工艺　　图 10　内部定点封装工艺

带塑料管的光纤放置在硬化后的玄武岩纤维束中间，塑料管之间的光纤段剥去涂敷层，两端拉紧固定，使光纤保持 300$\mu\varepsilon$ 左右的预拉应变；⑤用少量树脂将塑料管间的裸光纤及塑料管与玄武岩纤维束粘结；⑥放松纤维，剪去多余树脂；⑦大规模制作时，部分段塑料管可不与玄武岩纤维粘结，必要时可剪断与其他光纤或接头熔接。

对于全面粘结封装光纤，关键在于裸光纤的使用，经环氧树脂含浸粘结成整体后，可保证光纤无滑移现象，使得测试应变准确性得到保障。对于使用内部定点粘结封装光纤，关键在于套管的使用，使光纤在套管内部处于自由状态，将光纤固定在套管的两端，从而形成传感器内部一定标距范围内光纤的均匀应变，确保光纤测量的应变代表了标距范围内平均应变。

封装光纤的温度补偿

根据基于布里渊散射技术的分布式光纤传感原理，布里渊频率偏移受温度和应变变化的双重影响，为保证封装后的光纤能准确测试结构应变，封装时我们将封装一条伴随光纤用作温度补偿，伴随光纤外套有套管以保证光纤在套管内可自由移动，从而使得伴随光纤仅用作温度测试。加伴随光纤后，封装传感器的横截面如图 11 所示。

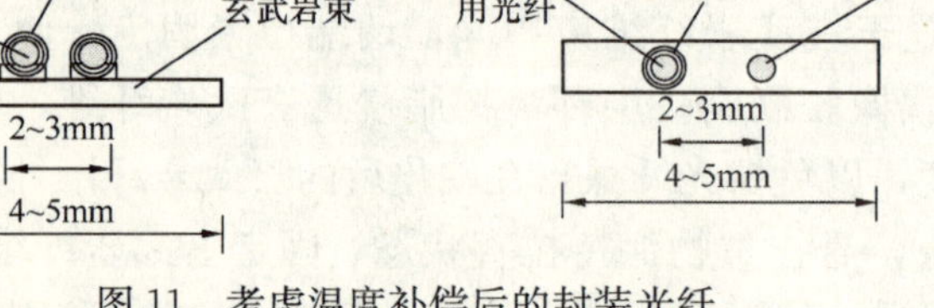

图 11 考虑温度补偿后的封装光纤

9.2.1.4 布里渊分布式传感技术的基础试验研究

a）光纤测试性能的单向拉伸试验研究和验证

BOTDR 传感技术

为验证以上理论推导，不同长度的光纤测试段粘贴在钢板上做单向拉伸试验，采用液压千斤顶加载，将不同粘贴标距的光纤用定点法固定在钢板上，试件应变由粘贴的应变片控制。在荷载等级 800$\mu\varepsilon$ 和 1500$\mu\varepsilon$ 下，测试应变值大小与光纤标距长度间的关系如图 12 所示。

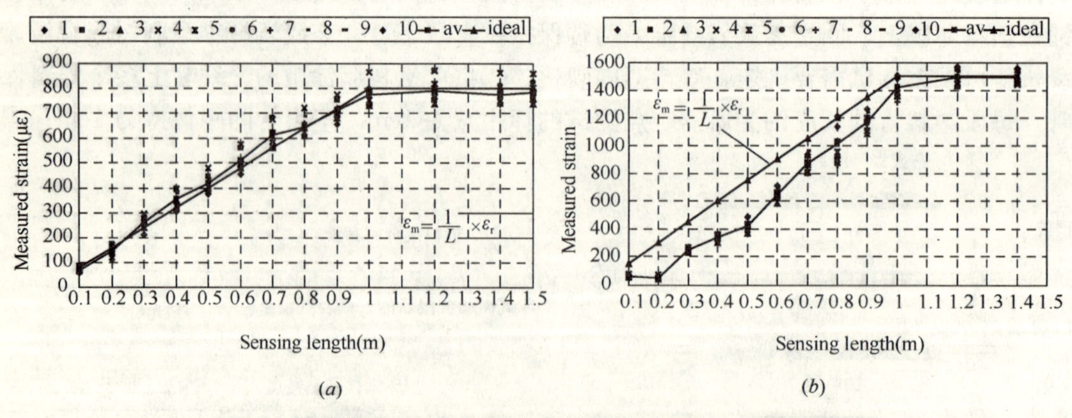

图 12 两种荷载等级下测试应变值与光纤测试长度间的关系

测试应变值与光纤测试长度间的关系（800$\mu\varepsilon$）（*a*）；

测试应变值与光纤测试长度间的关系（1500$\mu\varepsilon$）（*b*）

试验结果表明，光纤的长度与测量的结果有一定的关系。对于 BOTDR 分布式监测技术，当光纤的长度小于 1m 的时候，测量值随着长度的增加而增大；当长度大于 1m 时，测量值与光纤的长度无关，只与光纤所受的实际应变有关。

PPP-BOTDA 分布式光纤传感技术

相比 BOTDR 技术，PPP-BOTDA 的距离分辨率有了明显的提高。在试件应变量级分别为 300、800 及 1500$\mu\varepsilon$ 情况下，不同标距长度的光纤的测试结果如图 13 所示。对于距离分辨率为 10cm 的 PPP-BOTDA 分布式传感技术，可以得到同样的结论。即，对于基于布里渊散射技术的分布式光纤测量技术，当光纤的长度小于所对应的距离分辨率时，所得到的结果小于实际值，且所测得的值所着光纤长度的增加而增加；当光纤的长度大于所对应的距离分辨率时，所测得的值与光纤的长度无关，只与光纤所受的应变幅值大小有关。该试验验证了上节关于测量精度与光纤长度之间理论的正确性。

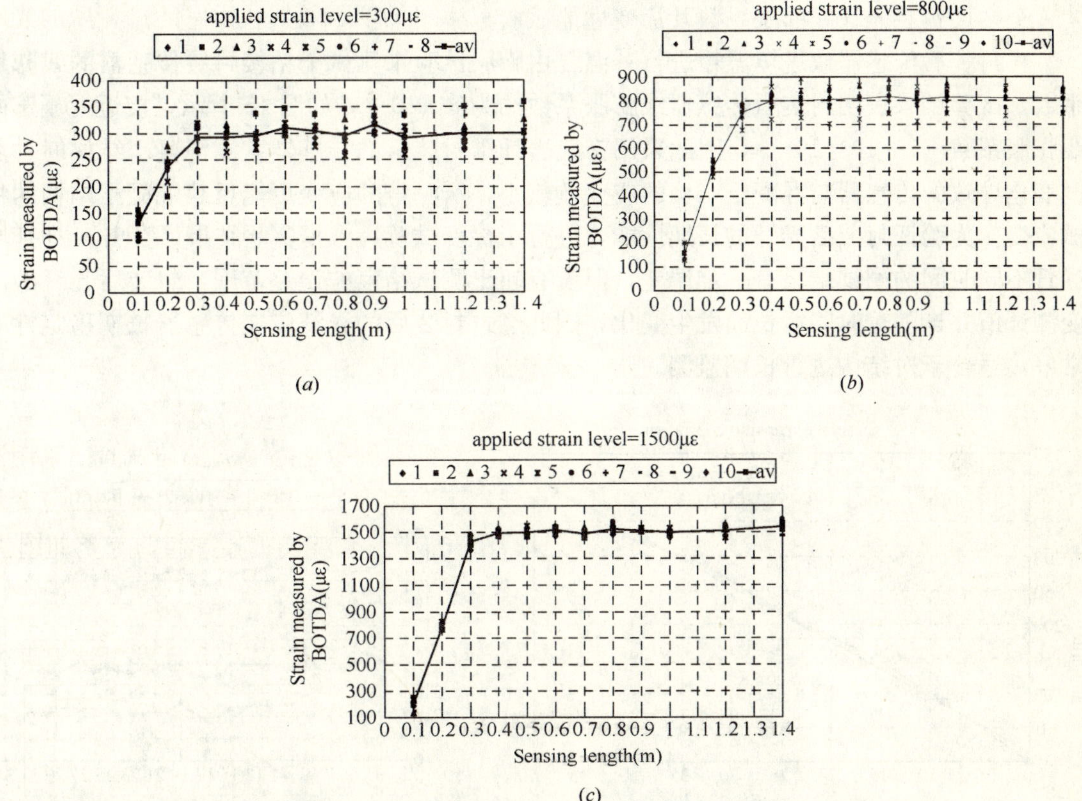

图 13 不同应变段长度下光纤应变测试值

(*a*) 试件应变量级为 300$\mu\varepsilon$；(*b*) 试验应变量级为 800$\mu\varepsilon$；(*c*) 试验应变量级为 1500$\mu\varepsilon$

从图 13 试验结果中，有一点需特别注意，尽管 BOTDA 的空间分解能为 10cm，但 PM 光纤的应变段长度需大于 40cm 才能测得正确的应变值，经初步判断，该现象是由裸光纤与外包材料间的滑移造成的。为了克服监测过程中光纤芯与外层可能存在的滑移，笔者已经结合光纤的增敏技术开发了无滑移增敏光纤。

b) 封装后抗滑移试验研究和验证

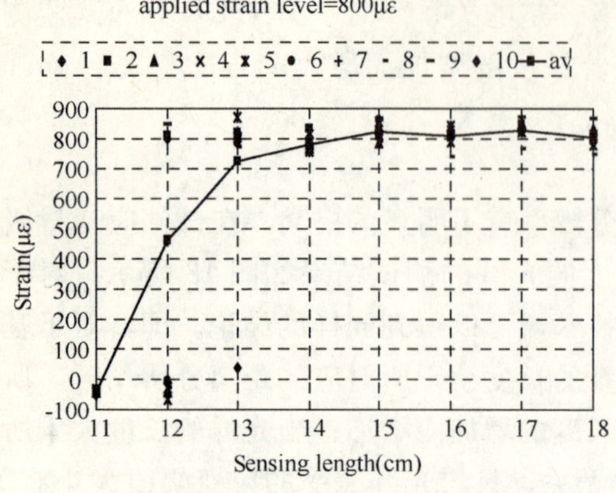

图 14 封装后光纤的抗滑移性能检验

通过单向拉伸试验对现有商品光纤及封装后的光纤进行了抗滑移性能检验，以裸光纤测试结果为基准来检验各种商品光纤及封装后光纤的抗滑移性能，BOTDA 测试技术的空间分解能为 10cm，将各类光纤不同长度的测试段用环氧树脂粘在钢板上测试钢板所受应变，实验结果表明当裸光纤和封装光纤的粘接长度大于 13cm 时即可获得正确的测试值，如图 14 所示，而现有商品光纤由于内部滑移的影响都需要 15cm 以上的粘接长度才能测得正确的应变值。

全粘接封装光纤线膨胀系数及徐变性能试验

将封装后传感器放进恒温箱进行升温自由膨胀试验来测试全粘接封装传感器的线膨胀系数，试验结果表明封装后传感器线膨胀系数约为 8×10^{-6}，如图 15 所示，接近混凝土的线膨胀系数（$7\sim10\times10^{-6}$），因此采用玄武岩纤维封装后的传感器监测混凝土结构时，温度变化导致的传感器与混凝土结构的应变差较小，所测得的应变更能反映混凝土结构的机械应变。为验证封装后光纤的长期性能，进行了徐变试验，根据实验结果可看出（图 16），尽管不同时间的测试结果有一定误差，但没有超出测试结束的误差范围，封装后光纤的应变测试值并未随着时间增长而发生变化，因此经过封装后的光纤应变传感器长期稳定性非常好，适合于对结构进行长期监测。

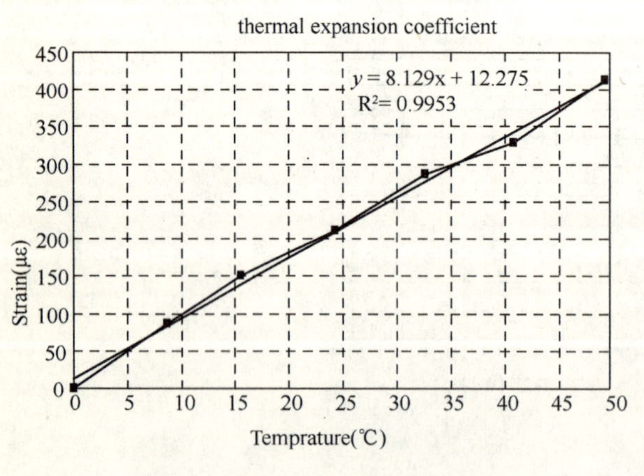

图 15 封装后光纤的线膨胀系数

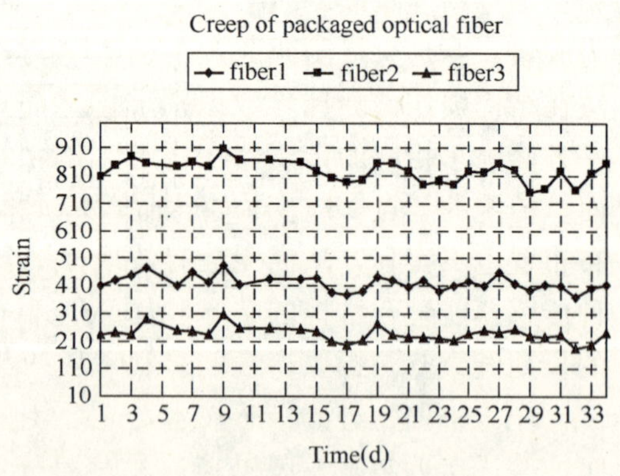

图 16 封装后光纤的徐变性能

c）光纤传感增敏的试验研究和验证

试验在单向拉的受力模式下进行，增敏封装后的光纤分为增敏段和连接段。增敏段裸纤外套塑料套管进行保护，连接段采用 CFRP 环绕包裹并用环氧树脂充分浸润，形成对裸纤的良好保护。光纤端部用环氧树脂固定在加载装置上。光纤标距全长 48cm，增敏段 16cm，连接段 32cm。根据式增敏系数定义，可得增敏系数为 3。

图 17(a) 是各级加载下 3 次测量应变值与电测百分表测量平均应变值的比较。由该可

见，根据实测应变均值与电测百分表测量平均应变值拟合得到的增敏系数为 3.251，与理论增敏系数的误差为 9%，图 17（b）表示了增敏光纤在各级加载下测量值误差。与裸纤所存在的±40με 的测量误差相比，增敏光纤测测量误差下降了 50%以上，完全证明了增敏的有效性，试验的具体情况和研究结果请参照参考文献[34]。

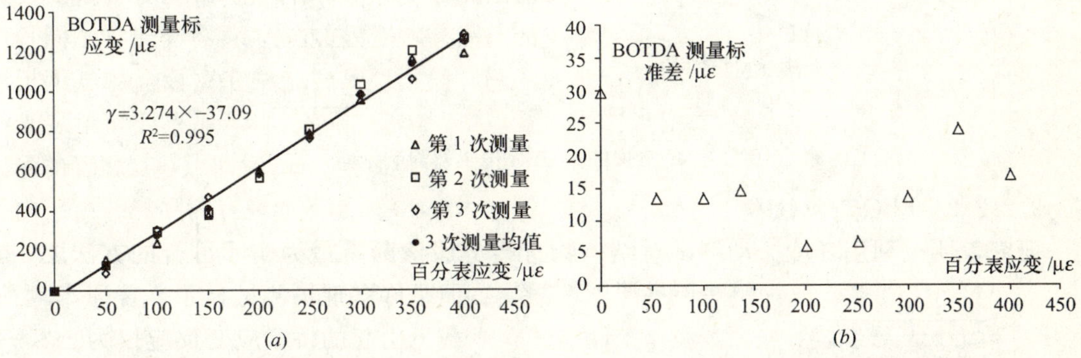

图 17　（a）实测增敏系数 η 示意图；（b）增敏光纤测量标准差分布示意图

9.2.2　长标距 FBG 传感器的开发及研究［17～22］

尽管基于布里渊散射机理的光纤传感技术可以提供长距离、大范围的分布式监测，但是应变的测量精度有限，最好只能达到±25με，且不能进行动态监测。为了能够进行精确和动态监测，开发了长标距 FBG 传感技术，并提出了一系列基于长标距 FBG 动、静监测的结构损伤识别和健康监测理论。

9.2.2.1　FBG 长标距化封装

结构的健康监测包括整体与局部监测。以加速度计为代表的整体监测的传感器已经较为成熟，但是大量的研究及工程实际表明，仅仅基于模态（频率、振型等）的整体监测方法在结构损伤诊断和参数识别上仍然面临较大的困难。而基于应力（应变）的局部监测，一方面由于相应的传感器在稳定性和耐久性上远远不能满足结构长期监测的需要，另一方面，传统的“点”应变片常因局部应力集中和裂缝出现而失效，并且在大型土木结构上分布布置也不实际，从而无法有效地捕捉到事先不可预知的结构破坏。很自然的，如果有一种传感器能融合结构的整体和局部信息，且满足稳定性和耐久性要求，将为大型土木结构的长期健康监测提供崭新的有利手段。

作为一种先进的传感技术，光纤传感器因其质量轻体积小、抗电磁干扰性强、抗化学腐蚀等优点近年来发展迅速。尤其是光纤布拉格光栅，传感机理成熟，高精度，高灵敏度，易构建多点或分布式传感网络，已成为最有前景的光纤传感器之一。但是，纤细的裸光纤光栅部分特别脆弱，抗剪能力差，直接将其作为传感器无法胜任土木工程粗放式施工和恶劣的服役环境。因此，对其进行二次开发，即封装或增减敏处理，是将光纤光栅在土木工程等领域推广应用的重要环节。综合以上认识，我们开发了一种新型的长标距 FBG 应变传感器，提出了分布式平均应变动静态测量技术。长标距 FBG 封装后的形态如图 18 所示；各长标距 FBG 传感器可以进行串联从而实现准分布式监测，并可形成二维甚至三维监测网络，如图 19 所示，具体的封装工艺步骤参照参考文献[18～21]。

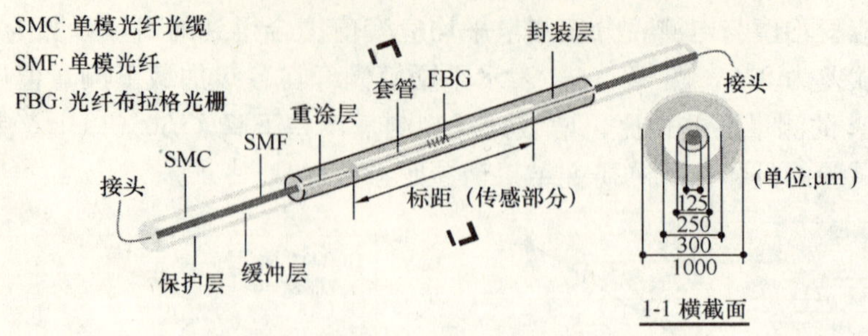

图 18 单个长标距 FBG 传感器封装示意

9.2.2.2 FBG 增敏技术

研究表明，利用环境振动测试对结构物进行健康诊断是最为实际可行的方法之一。但是许多现场测试显示，普通车辆荷载下引起的桥梁应变反应仅为±20～30$\mu\varepsilon$ 左右，甚至更小，而一般 FBG 的稳定精度大概±4$\mu\varepsilon$，加上环境测试噪音大，不定因素多，这对有效地应用应变测量结果带来了很大的困难。基于这样的认识，在原有传感器开发的基础上，提出了一种用于微小应变测试的高精度长标距 FBG 应变传感器的设计构想。FBG 的增敏原理与上节所述的基于布里渊散射光纤传感技术的增敏原理相似，其基本原理是在光栅两边用刚度比较大的材料进行封装。目的是使标距范围内变形集中到光栅范围之内，使 FBG 部分的应变远大于其它部分或者是极端的情况长标距范围内的变形全部由 FBG 部分承担，测量的较大应变值再通过简单的换算反推较小的平均应变量。但是，在同样的标距下，同样的封装材料，FBG 的增敏效果要比基于布里渊散射机理的光纤传感的增敏效果好。主要是因为基于布里渊散射的光纤传感技术的空间分辨率比较大，如 PPP-BOTDA 的空间分辨率最好但是也达到了 10cm 左右，而 BOTDA 的空间分辨率则达到了 100cm；FBG 的栅距长度通常为毫米量级。

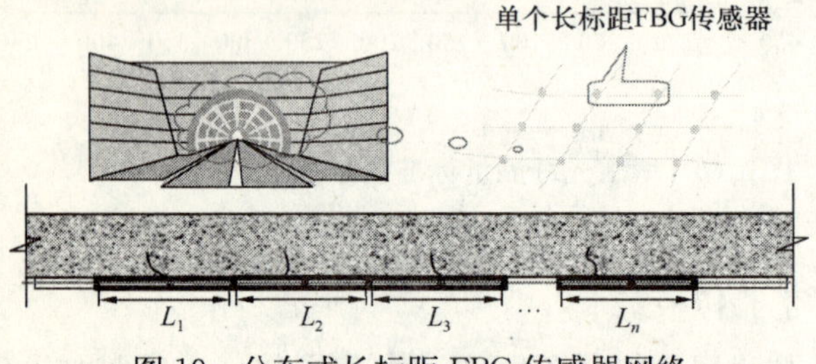

图 19 分布式长标距 FBG 传感器网络

9.2.2.3 长标距 FBG 传感器的拉伸试验验证及其应用

将三个应变片（SG5-1，SG5-2，SG60）和两个封装 FBG 传感器（FBG10 和 FBG25，其测量标距分别为 100mm 和 250mm）分别粘贴在一块矩形钢板上，如图 20 所示，钢板的尺寸为 50mm×3mm 且设计了两个直径为 7mm 的圆孔。以应变片 SG5-1 的测量值为基准，SG5-2，SG60，FBG10 和 FBG25 的相对测量值分别为 1.1450，1.0146，1.0088，1.0070。测量所得到的应力-应变曲线如图 20 所示。这些结果表明，通过封装可以在不影响测量精度的前提下增加传感器的测量标距和使用性能。

两个应变片（SG60 和 SG5，测量标距分别为 60mm 和 5mm）和一个封装 FBG 传感器（FBG60，测量标距为 60mm）粘贴在混凝土圆柱上，其直径为 150mm 高度为 300mm，如图 21(*a*)所示。每个传感器所测得的应变随时间的变化如图 21(*b*)所示，从该图可以发现，FBG60 和 SG60 所测得的应变吻合很好，但与作为点传感器的 SG5 所测得的应变存在一定的

差异。这些研究结果表明，封装的长标距 FBG 传感器可以用来测平均拉伸和压缩应变。

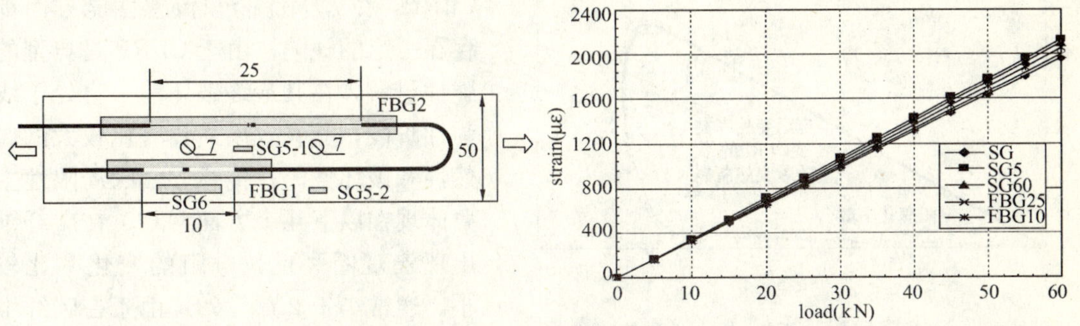

图 20　长标距 FBG 单向拉伸试验及试验结果

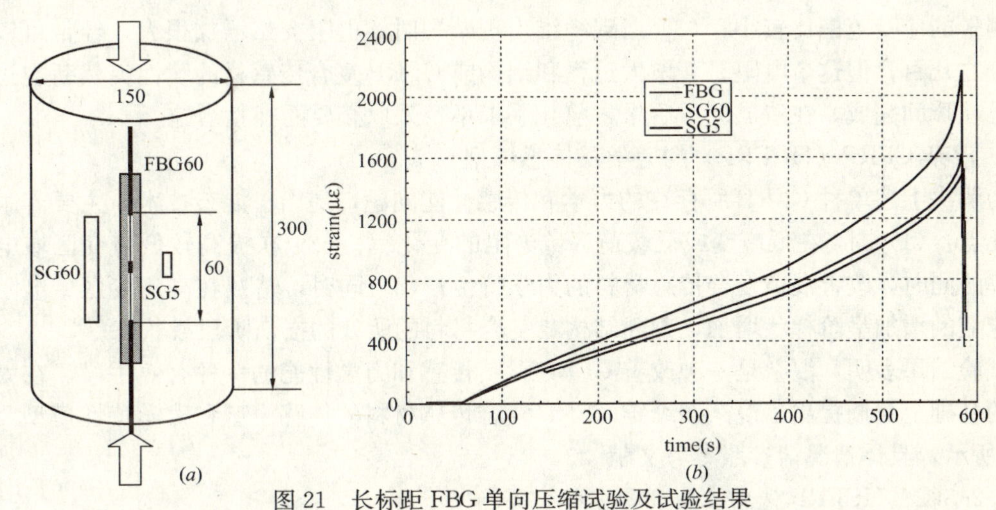

图 21　长标距 FBG 单向压缩试验及试验结果

9.2.3　基于碳纤维压阻效应的分布式传感技术的研究和开发[23-33]

研究表明碳纤维具有良好的导电特性，在小应变范围内即碳纤维出现微观和宏观断裂之前电阻随着应力的增加大致呈线性增加，称之为压阻效应。而正是由于这些特性，碳纤维可以用做传感材料甚至可以制成碳纤维智能复合材料，同时使开发具有自健康诊断和损伤监测功能的 CFRP 或 CFRP 加固结构（如智能复合结构）成为了可能。

9.2.3.1　基本监测设备、混杂碳纤维复合材料（HCFRP）力学及传感模型

a）基本监测设备

由于 CFRP 材料中的碳纤维具有良好的导电性和压阻效应，故碳纤维可以用做传感材料并可以通过测量电阻变化来判断结构的应力和损伤状态。电阻测量系统简单，只需要一个直流电发生器和电压测量仪即可。而且，还可以把 HCFRP 视为应变片，利用测应变的仪器和电路进行电阻测量；然后通过压阻效应以及电阻变化率与 HCFRP 中各纤维应变和损伤状态的关系推导出结构的应变和损伤状况。

b）单种碳纤维 CFRP 的力学及传感模型

基于大量试验研究，提出了只含单种碳纤维的 CFRP 材料的力学和传感模型，如

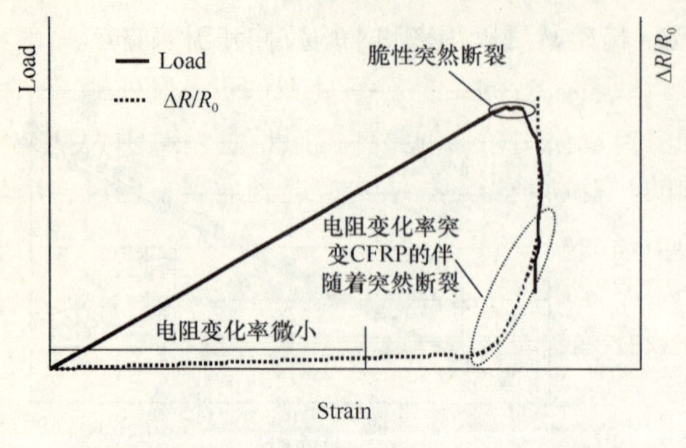

图 22 单种碳纤维的 CFRP 的力学和传感模型

图 22所示。对于只含单种碳纤维的CFRP，在力学性能和传感性能方面都存在一定的缺陷。由于 CFRP 是典型的脆性材料，在其最终破坏前一直处于线弹性阶段，其失效模式具有比较大的偶然性和破坏性。在传感方面的局限性主要体现在以下几个方面：（1）在碳纤维出现宏观断裂前，电阻的变化率比较小，通常小于 2%；（2）由于在碳纤维出现宏观断裂前的一个很大应变范围内，电阻变化率不大而且受偶然因素影响初始电阻的误差比较大，所以电阻测量法所能够准确测量的应变范围比较小；（3）当碳纤维出现断裂时，电阻突然增加很大，经常可以达到百分之几百，但这个电阻显著增大过程和结构或 CFRP 复合传感材的脆性突然断裂过程往往是在瞬间完成，在造成严重后果之前几乎来不及采取必要的补救措施。

c）混杂 CFRP（HCFRP）的力学及传感模型

为解决上述单种 CFRP 所存在的力学和传感性能问题，提出了混杂技术。主要着眼于以下几点：（1）提高在低应变或低载荷区域电阻的变化率；（2）在提高较低应变区域电阻变化率的同时，改善混杂复合传感材料的力学特性；（3）同时，造成在高应变区域电阻变化随应变的增加呈阶梯状增加，判断结构或复合材料的所处的损伤阶段或状态。

试验结果表明，混杂是一种改善 CFRP 材料传感和力学性能的一种有效手段，在大量试验的基础上我们提出了混杂碳纤维 CFRP 复合传感材料的传感模型和力学特性模型，如图 23 所示，具体情况请参照参考文献[23，33]。

9.2.3.2 HCFRP 传感性能理论分析

详细分析了 HCFRP 的灵敏度与其各组成成分性能之间的关系，以及电阻阶跃与 HCFRP 中不同种类碳纤维断裂之间的关系，详细情况请参照参考文献[48]。

9.2.3.3 基于 HCFRP 的关键传感技术开发

a）传感增敏技术

通常 HCFRP 的灵敏度系数比较低，只有 2 左右。为了提高其灵敏度系数，将预含浸的碳纤维束进行预拉伸处理，如图 24 所示。增敏处理工艺、效果及其它具体情况请参照文献[30]。

b）分布式传感技术

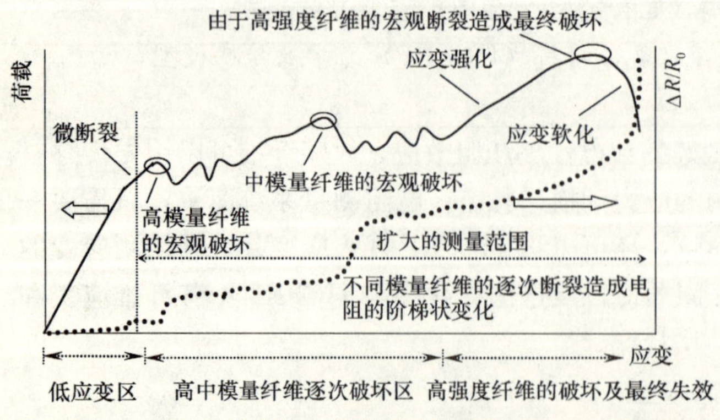

图 23 混杂碳纤维复合材料的力学特性和电阻变化模型

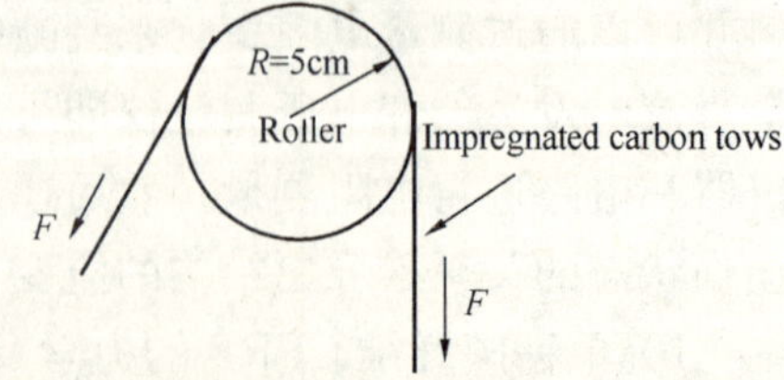

图 24 HCFRP 传感增敏处理示意图

通过在 HCFRP 上安装电极可以实现分布式监测，如图 25 所示。分别测量不同电极之间的电阻可以实现整体测量和分布测量，或者二者的结合。测量 HCFRP 最外端两个电极之间的电阻可以实现整体测量，整体测量可以给出结构的整体健康信息。对于大型结构整体测量和健康评估是非常有必要和有用的，特别是当这些结构遭遇到地震、台风、爆炸等突然灾害时。

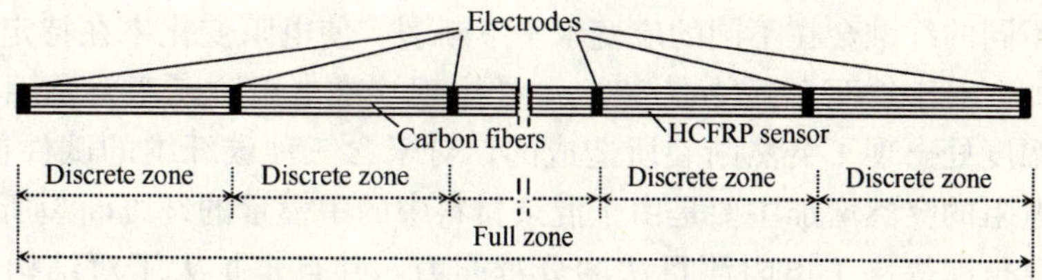

图 25　HCFRP 的分布式传感示意图

通过整体测量，如果电阻没有明显的变化说明结构仍处于安全状态。如果发现电阻有明显的变化，则通过监测每两个电极之间的电阻变化可以对损伤进行定位。同时，通过每两个电极之间的电阻测量也可以给出结构的应变分布。

9.2.3.4　基于 HCFRP 的传感技术试验研究和验证

a）单种碳纤维 CFRP 的传感性能研究

碳纤维的电阻变化与应变的变化有着密切的关系，在小应变范围电阻变化与应变的变化呈正比例变化，且具有比较稳定的线性关系。这些关系表明，无论是碳纤维可用做传感材料。当碳纤维出现部分破坏时，卸载后存在部分残留电阻变化率，可以利用这种残留的电阻变化率来记忆材料曾经历的最大载荷、应变或损伤状况。这些结果表明，利用电阻测量法对 CFRP 材料的应变状态和损伤状况进行实时监测是一种潜在经济有效的测量手段，具体研究情况参照[23,24]。

为详细探讨碳纤维电阻变化率与应变之间的线形关系以及这种关系的稳定性，以高强度的碳纤维为材料做了两个试件。用重物方法以 2 天一次的频率加载。总共进行了 20 次加载循环，表 2 列出了这两个试件电阻变化率与应变拟合曲线的线形系数和线形相关系数，线形相关系数大于 0.985，灵敏度系数介于 2.81～3.33 之间。该结果表明碳纤维可以作为应变传感器使用。

试件 1 和 2 电阻变化率——应变拟合曲线的线形系数和线形相关系数　表 2

Cycle number	试件 1		试件 2	
	a (GF)	R^2	a (GF)	R^2
1	3.2127	0.9867	2.8496	0.9845
2	3.3348	0.9935	3.3095	0.9972
3	2.9545	0.9904	2.8717	0.9926
4	2.8098	0.9969	3.3309	0.9922
5	2.8777	0.9965	3.2281	0.9904
6	3.3348	0.9935	3.3354	0.9987
7	2.9752	0.9923	3.3602	0.9992
8	2.8284	0.9971	3.3434	0.9951
9	3.3307	0.9901	3.3373	0.9916
10	2.8098	0.9969	3.1519	0.9966

b）混杂 HCFRP 的力学及传感性能试验研究

含三种碳纤维(高、中模量和高强度，比例为 4∶1∶1)的混杂 HCFRP 片材的传感性能和力学行为的关系如图 26 所示。由于这种试件中含有两种模量较高的纤维，所以反映在载荷—应变曲线上是该曲线具有两个甚至多个类似金属的塑性变形阶段，随后也出现两个甚至多个应变强化阶段。从图 26 可以看出，由于混杂复合材料中的碳纤维具有不同的弹性模量，所以不同的纤维会在不同的应变水平下断裂，使电阻变化率在特定的应变幅度处出现突然的增加，对于这两个试件在 3300$\mu\varepsilon$ 处出现一次电阻阶跃主要是因为高模量的纤维在这个应变幅度处出现了突然断裂所造成的，对于含三种碳纤维的试件在 6700$\mu\varepsilon$ 左右处也出现一次电阻的突然增加主要是由于混杂材料中的中模量的纤维断裂所致。这样的电阻变化比较明显也即提高了电阻测量法的分辨能力，并且也扩大了对结构的应变监测范围；同时混杂也避免了 CFRP 作为结构材料的突然断裂。对于混杂碳纤维复合材料的电阻变化率呈明显阶梯状变化，一般情况下每一种纤维的断裂会使电阻产生一次突跃，纤维的种类越多阶梯的数量也越多，每经过一个台阶（即碳纤维每断裂一次）电阻变化率随着应变的增加也将以更大的速率增加。

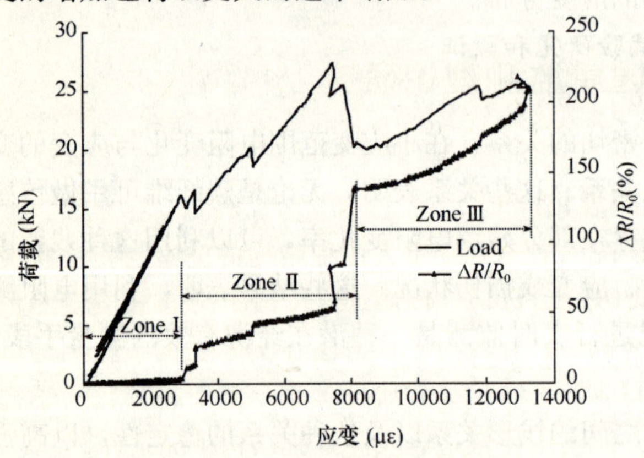

图 26　含高模、中模和高强度碳纤维的 HCFRP 的力学及传感性能研究

现以含三种碳纤维的混杂材料的电阻变化—应变关系为例说明这种阶梯状增加的机理和应用。可以明显将电阻变化率—应变曲线划分为三个区域，如图 26 所示。在 Zone Ⅰ，我们发现在这个低应变区域由于碳纤维的压阻效应电阻变化率随着应变的增加呈线性增加。在稍高应变区（Zone Ⅱ），随着应变的增加其中的高模量纤维开始破坏随后是中模量纤维的破坏，这种破坏造成电阻变化呈现阶梯状阶跃式增加，阶跃的幅度主要取决于该高模量纤维的含量多少，阶梯的数量主要取决于材料中高模量纤维的种类多少；在这个阶段，随着载荷的继续增加一般会出现应变强化，在这个强化阶段电阻会继续呈现缓慢近线性的增加。第三个阶段（Zone Ⅲ）对应着材料的最终破坏即高强度纤维的断裂，这个阶段也属于应变强化阶段，材料仍可以承受一定的载荷，同时电阻也会进一步的增加，在这个应变强化阶段，电阻变化率呈抛物线形快速增加直到最终断裂（主要是高强度纤维的逐渐断裂造成的）。

c）增敏试验验证

图 27 为增敏和未增敏 HCFRP 传感器的电阻变化率与应变的关系，从该图可以看出通过增敏措施可以大大提高 HCFRP 的灵敏度。通常的 HCFRP 的灵敏度介于 1.5～3.0 之间，而增敏 HCFRP 传感器的灵敏度达到 1000 以上。

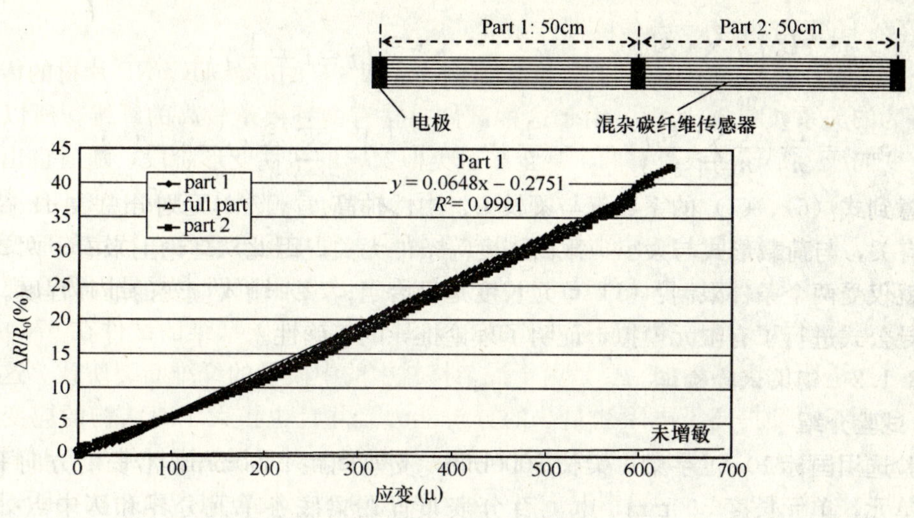

图 27　增敏与未增敏 HCFRP 性能比较

9.3 基于分布传感技术的结构健康监测理论研究

9.3.1 基于分布应变监测的结构变形监测理论[34]

通过结构的应变分布评估结构的变形情况，对于判别重大工程结构的健康与安全状况而言具有非常重要的意义。本方法是基于 PPP-BOTDA、FBG 或 HCFRP 分布式应变测量技术，提出了一种监测结构变形分布的新型方法，并探讨了该方法在实际结构中应用的可行性。基于该方法所得到的结构变形分布与应变分布呈显式线性关系，且所有参数与结构受载形式及大小、截面刚度条件均无关，方便实用比较适合实际工程结构工况条件。

9.3.1.1　简支梁条件下结构变形—应变显式线性关系推导

由于实际结构中荷载条件的任意性，因此必须设法将其模拟出来。根据材料力学，改变荷载形式及大小等效于改变梁上弯矩分布，即曲率分布，而实际梁的曲率分布在数值上等于共轭梁的荷载分布。因此，利用应变分布计算荷载曲率分布，既可等效共轭梁的荷载分布，又可模拟实际梁任意荷载条件。

简支梁模型如图 28 所示，由于简支边界条件的共轭条件仍是简支，因此图 28 所示简支梁即为其本身的共轭梁。设梁全长为 L，梁截面抗弯刚度为 EI，沿长度方向将梁均分为 n 个单元，则单元长度 $l=L/n$。图中虚线表示任意荷载形式下的实际梁弯矩图，实线表示共轭梁等效均布荷载。

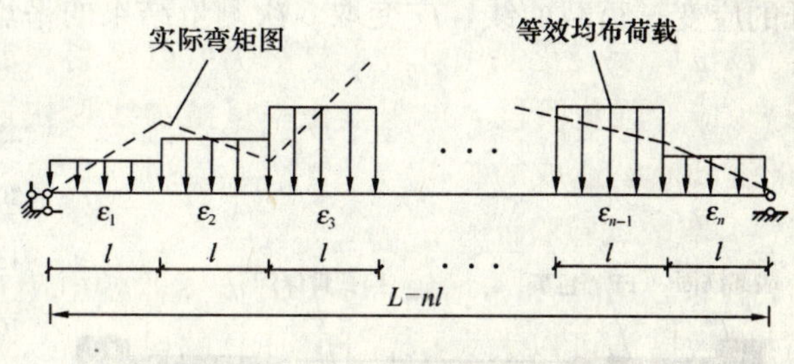

图 28　简支梁与其共轭梁示意图

具体推导过程详见文献[34]，单元 j 处和结构最大变形分别为

$$v_j=\frac{L^2}{n^2}\left(\frac{j}{n}\sum_{i=1}^{n}\frac{\bar{\varepsilon}_i}{y_i}\left(n-i+\frac{1}{2}\right)-\sum_{i=1}^{j}\frac{\bar{\varepsilon}_i}{y_i}\left(j-i+\frac{1}{2}\right)\right) \tag{6}$$

$$v_{\max}=\frac{L^2}{n^2}\left(\frac{1}{n}\sum_{i=1}^{n}\frac{\bar{\varepsilon}_i}{y_i}\left(n-i+\frac{1}{2}\right)\cdot\left(j+\frac{1}{2}\right)-\sum_{i=1}^{j}\frac{\bar{\varepsilon}_i}{y_i}(j-i+1)\right) \tag{7}$$

注意到式（6）、（7）的系数仅与梁长 L、中性轴高度 y_i、单元划分总数 n、待求单元位置 j 有关，与荷载形式与大小，截面刚度等条件无关。因此从理论上来看，两式的计算准确度就仅受两个条件限制：（1）单元长度是否恰当；（2）输入应变的准确程度。同时对上述推导公式进行了有限元模拟，证明了理论推导的正确性。

9.3.1.2　钢梁试验验证

a）试验介绍

试验选用国标 10a 工字梁，梁长 3000mm，支座间距 2500mm，沿长度方向平均划分为 5 个单元，单元长度 500mm，电测百分表布置在梁底各单元分界和跨中点处（P1～P5），布设细节如图 29 所示。梁底布设应变增敏光纤，对比光纤采用普通裸纤，二者均为定点布设，增敏光纤结构设计如图 30 所示，传感器全长 480mm，增敏段长度为 240mm，增敏系数 η 为 2，传感器间粘贴长度为 10mm。

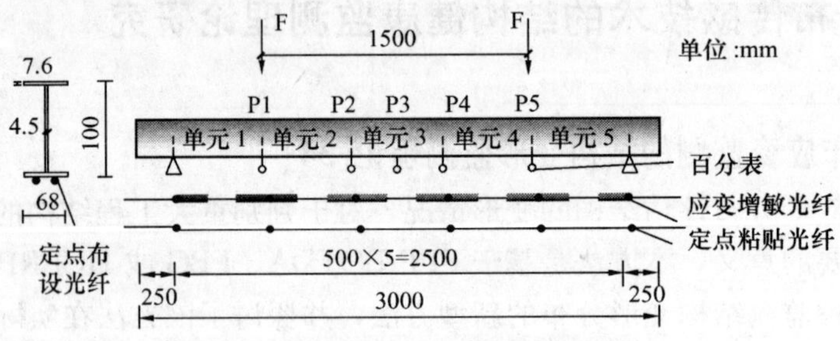

图 29　钢梁光纤布设及加载示意图

试验采用四点加载，加载点间距 1500mm，分 8 级加载，每级 3kN，共加至 24kN，每级加载重复测量 5 次。

b）试验结果与分析

由于结构、荷载的对称性和材料的均匀性，沿梁轴线测量得到的实测应变和变形基本呈对称分布，故本文仅选取钢梁左侧半梁变形进行分析。图 31 是单元 1～3 增敏光纤与裸纤的应变—荷载曲线，应变取 5 次测量结果的平均值，单元 4、5 实测应变与单元 2、1 基

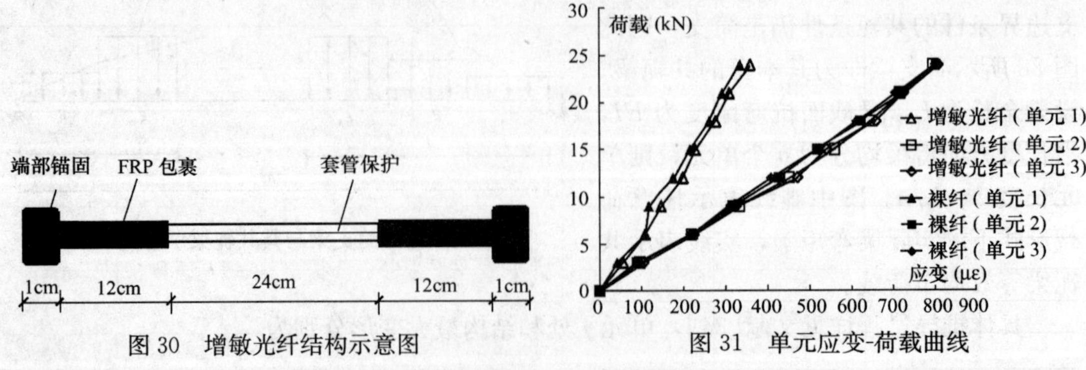

图 30　增敏光纤结构示意图　　图 31　单元应变-荷载曲线

本一致。由测量结果对比可以看出，钢梁各截面荷载-位移曲线基本呈良好的线性关系。0～6kN 二者测量结果基本一致，荷载大于 6kN 后，增敏光纤的测量应变增长幅度较裸纤有所增大，12～15kN 时二者差距达到最大值，此时二者差距达到 50～60$\mu\varepsilon$，约占测量值的 10%，随后差距又逐渐减小，24kN 时二者测量值又基本一致。注意到单元 2、3 的应变值约为单元 1 的两倍，与图 29 梁单元划分及加载位置基本对应。根据式（6）、式（7），代入 $L=2500$mm，$n=5$，$y=50$mm，得到单元分界点 P1、P2 和梁最大变形点 P3 的变形计算公式如下：

P1：$$v=2000\varepsilon_1+3500\varepsilon_2+2500\varepsilon_3+1500\varepsilon_4+500\varepsilon_5 \quad (8)$$

P2：$$v=1500\varepsilon_1+4500\varepsilon_2+5000\varepsilon_3+3000\varepsilon_4+1000\varepsilon_5 \quad (9)$$

P3：$$v=1250\varepsilon_1+3750\varepsilon_2+6250\varepsilon_3+3750\varepsilon_4+1250\varepsilon_5 \quad (10)$$

式中 ε_1～ε_3 为增敏光纤和裸纤 5 次测量的平均值。

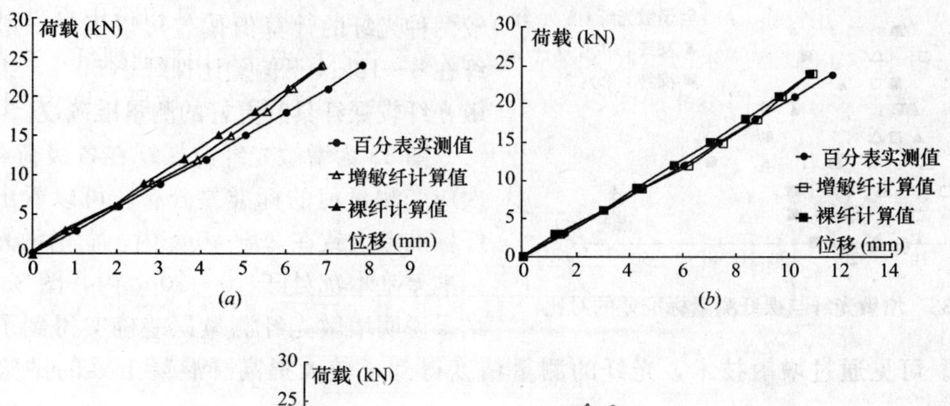

图 32　基于 PPP-BOTDA 分布应变所监测的各点变形与传统测量结果的比较

(*a*) 点 P1；(*b*) 点 P2；(*c*) 点 P3

两种光纤变形监测值与 LVDT 实测值偏差百分比　**表 3**

荷载/kN	增敏光纤			裸纤		
	P1	P2	P3	P1	P2	P3
3	17%	13%	13%	27%	23%	22%
6	7%	1%	2%	7%	1%	3%
9	6%	0%	1%	13%	6%	7%
12	5%	3%	3%	13%	6%	6%
15	7%	2%	0%	12%	5%	6%
18	8%	1%	1%	13%	4%	6%
21	12%	5%	7%	13%	6%	8%
24	12%	7%	9%	13%	7%	9%

基于裸纤测量数据得到的P1～P3的变形计算值与实测值对比如图32(a)～(c)所示，计算值与实测值偏差百分比如表3所示。表中所列偏差最大值仅为12%～13%，说明通过分布式应变测量，式(11)、式(12)能够比较准确地计算钢结构的变形分布。当荷载为3kN时，各点基于PPP-BOTDA分布应变的应变监测值与LVDT实测值偏差较大，这是由于单元真实应变很小(不超过100$\mu\varepsilon$)，应变测量准确度受测量误差控制所致。荷载增大时，随着测量误差在结果中所占比例迅速减小，变形计算值更加接近真实值。比较同种光纤条件下各点的测量误差可以发现，在相同荷载下P2、P3的计算偏差要比P1小50%。可能原因是式(13)、式(14)中纯弯单元应变的权重较大。在相同荷载作用下，纯弯单元应变的理论值应为剪弯单元平均应变的两倍(此说法不妥，纯弯段的应变为定值，而剪弯段的应变为分布应变)，因此相同的测量误差在纯弯单元应变中所占比例就低于其在剪弯单元应变中比例，从而提高了变形计算准确度。比较两种光纤的计算值偏差可以发现，增敏光纤在6～18kN下偏差比裸纤小50%，说明增敏光纤较裸纤具有更好的测量准确度。

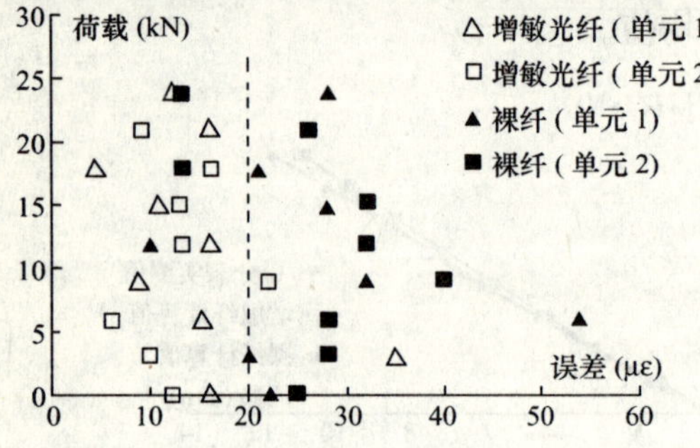

图33 增敏光纤与裸纤测量标准差的对比

图33为增敏光纤与裸纤在各级荷载下5次应变测量值的标准差分布。可以看出，裸纤标准差大致在30～40$\mu\varepsilon$内，而增敏光纤的标准差基本都处于10～20$\mu\varepsilon$内，图33所示结果说明增敏光纤测量误差确实得到了成比例降低。可见通过增敏技术，光纤的测量精度得到了较大提高。混凝土梁的试验验证参考[34]。

9.3.2 基于长标距应变分布的动力学模态分析理论[35-36]

以典型的欧拉梁为例，如图34所示，每个节点包括两个自由度（一个平移和一个转角）。假设一标距为L_m的应变传感器安装在梁的底部，其连接的起点和终点的转角位移对应结构第i和j个自由度。

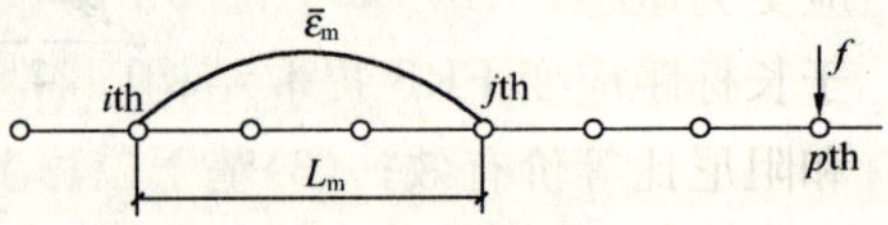

图34 欧拉梁模型

则在时刻t或频率ω，测量的长标距应变为：

$$\bar{\varepsilon}_m(t)=\frac{h_m}{L_m}\cdot[v_i(t)-v_j(t)]=\eta_m\cdot[v_i(t)-v_j(t)]$$

$$\bar{\varepsilon}_m(\omega)=\frac{h_m}{L_m}\cdot[v_i(\omega)-v_j(\omega)]=\eta_m\cdot[v_i(\omega)-v_j(\omega)] \tag{11}$$

h_m为传感器到中性轴的距离，v为各自由度上的位移。该长标距应变和作用在第p个自由度上的激励之间的频响函数（Frequency Response Function，FRF）可以写成：

$$H_{mp}^{\bar{\varepsilon}}(\omega)=\frac{\bar{\varepsilon}_m(\omega)}{P_p(\omega)}=\eta_m\cdot\frac{[v_i(\omega)-v_j(\omega)]}{P_p(\omega)}=\eta_m\cdot(H_{ip}^d(\omega)-H_{jp}^d(\omega)) \tag{12}$$

其中$H_{ip}^d(\omega)$、$H_{jp}^d(\omega)$是对应第i、j个自由度的位移频响函数。从传统的模态分析理论可知，一个ND自由度线性时不变结构体系，对应第l个自由度的位移FRF为：

$$H_{lp}^{d}(\omega)=\sum_{r=1}^{N}\frac{\varphi_{lr}\varphi_{pr}}{M_r(\omega_r^2-\omega^2+2j\zeta_r\omega_r\omega)}=\sum_{r=1}^{N}\frac{{}_rA_{lp}^{d}}{\omega_r^2-\omega^2+2j\zeta_r\omega_r\omega} \tag{13}$$

这里，φ_{lr} 为第 r 阶模态、第 l 个自由度的振型系数，ω_r 为第 r 阶模态频率；M_r 及 ξ_r 为第 r 阶模态质量及模态阻尼比；位移模态常数则定义为

$$_{r}A_{lp}^{d}=\frac{\varphi_{lr}\varphi_{pr}}{M_r} \tag{14}$$

将式（12）代入式（11）可得：

$$\begin{aligned} H_{mp}^{\bar{\varepsilon}}(\omega)&=\sum_{r=1}^{ND}\frac{\eta_m(\varphi_{ir}-\varphi_{jr})\varphi_{pr}}{M_r(\omega_r^2-\omega^2+2j\xi_r\omega_r\omega)}\\ &=\sum_{r=1}^{ND}\frac{\eta_m({}_rA_{ip}^{d}-{}_rA_{jp}^{d})}{\omega_r^2-\omega^2+2j\xi_r\omega_r\omega}=\sum_{r=1}^{ND}\frac{{}_rA_{mp}^{\bar{\varepsilon}}}{\omega_r^2-\omega^2+2j\xi_r\omega_r\omega} \end{aligned} \tag{15}$$

长标距应变模态常数因而写成：

$$_{r}A_{mp}^{\bar{\varepsilon}}=\eta_m\ ({}_rA_{ip}^{d}-{}_rA_{jp}^{d})\ =\frac{\eta_m\ (\varphi_{ir}-\varphi_{jr})\ \varphi_{pr}}{M_r}=\frac{\varphi_{pr}}{M_r}\delta_{mr} \tag{16}$$

这里，定义第 r 阶模态长标距应变（Modal Macro-Strain，MMS）为：

$$\delta_{mr}=\eta_m(\varphi_{ir}-\varphi_{jr}) \tag{17}$$

比较式（13）和式（15）

$$\frac{{}_rH_{lp}^{d}(\omega)}{{}_rH_{mp}^{\varepsilon}(\omega)}=\frac{{}_rA_{lp}^{d}}{{}_rA_{mp}^{\bar{\varepsilon}}}=\frac{\varphi_{lr}}{\delta_{mr}}=\frac{\varphi_{lr}}{\eta_m(\varphi_{ir}-\varphi_{jr})} \tag{18}$$

可以发现：位移 FRF 和长标距应变 FRF 的比值与荷载及时频都无关，仅与自由度的空间位置有关。这个结论提供了三点重要信息：（1）从时频域上看，长标距应变 FRF 是个更类似位移 FRF，而不同于速度或加速度 FRF 的物理量，因此对低频响应更为敏感；（2）基于位移 FRF 和基于长标距应变 FRF 提取结构固有频率和阻尼比等价有效；（3）基于位移 FRF 和基于长标距应变 FRF 提取的特征向量有所不同，它们之间的相互关系和位移与长标距应变测量之间的映射关系完全相同，换言之，MMS 是个直接的应变模态测量。以下将结合模态测试详细说明这些重要特征。

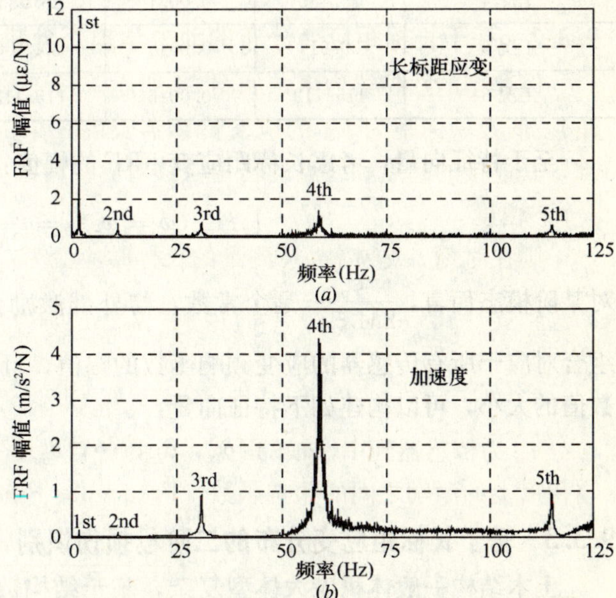

图 35　实测长标距应变 FRF 和加速度 FRF 比较
（*a*）长标距应变 FRF 幅值谱；（*b*）加速度 FRF 幅值谱

同一模态测试下的长标距应变 FRF 和加速度 FRF 幅值谱如图 35 所示。显然，长标距应变 FRF 在低频呈现了更加显著的峰值，但是随着频率增大，峰值迅速减小。这一现象可从

下式长标距应变 FRF 和加速度 FRF 的相互关系解释：

$$H_{mp}^{\bar{\varepsilon}}(\omega)=\eta_m\cdot(H_{ip}^{d}(\omega)-H_{jp}^{d}(\omega))=-\frac{\eta_m}{\omega^2}\cdot(H_{ip}^{a}(\omega)-H_{jp}^{a}(\omega)) \tag{19}$$

当 ω 很小，尤其 $\omega\leqslant1$ 时，长标距应变频谱在共振频率处具有更敏感的峰值显示。这是一个重要特征，表明长标距应变测量在柔度大的结构监测中具有很大潜力，例如长大跨桥，其固有频率通常数值小且十分密集，导致利用加速度测量在准确识别模态特征时面临较大困难，而动态位移测量由于相应的测试手段不够成熟或实际应用条件的限制仍然无法有效使用。

比较式（12）和式（14）可以发现，位移和长标距应变两个 FRF 表达式的分母相同，说明两者在识别结构固有频率和阻尼比时等价有效，也就是说，传统的基于位移或加速度测量识别结构频率和阻尼比的方法都适用于利用长标距应变测量的情况。同一模态测试下，利用长标距 FBG 传感器、加速度计及应变片测量识别的结构固有频率和阻尼比的比较结果见表 4。可以看到，基于不同传感器识别的参数结果基本一致，尤其是频率识别结果几乎完全相同，具有很高的精度。

基于不同传感器识别的固有频率和阻尼比 **表 4**

传感器	长标距 FBG 传感器		加速度计		应变片	
	频率（Hz）	阻尼比	频率（Hz）	阻尼比	频率（Hz）	阻尼比
工况 1	0.732	0.0307	0.732	0.0385	0.732	0.0260
工况 2	6.256	0.0090	6.256	0.0078	6.256	0.0079
工况 3	11.017	0.0047	11.017	0.0050	11.017	0.0048
工况 4	25.085	0.0031	25.085	0.0033	25.085	0.0036
工况 5	30.914	0.0044	30.914	0.0047	30.914	0.0049
工况 6	115.112	0.0016	115.112	0.0016	115.112	0.0017

至于特征向量，考虑长标距应变 FRF 的幅值：

$$|{}_{r}H_{mp}^{\bar{\varepsilon}}(\omega=\omega_r)|=\frac{\varphi_{pr}}{2M_r\xi_r\omega_r^2}\cdot\delta_{mr} \tag{20}$$

对某阶模态而言，$\frac{\varphi_{pr}}{2M_r\xi_r\omega_r^2}$ 是个常数，与外部激励无关。假设共安装了 N 个应变传感器，组合对应于所有传感器的应变频响函数的幅值，且只考虑它们之间的相互关系而忽略绝对数值的大小，可以构建如下特征向量：

$$\{\delta_{1r},\ \delta_{2r},\ \cdots,\ \delta_{mr},\ \cdots,\ \delta_{Nr}\}^{T} \tag{21}$$

9.3.3 基于长标距应变分布的二阶段损伤识别[37-39]

土木结构一般体积庞大体型复杂，基于结构数学模型的损伤诊断因待识别参数太多，逆算法或优化算法往往收敛困难，而且算法的有效性非常容易受实际测量噪音和随机误差的影响。考虑结构损伤的定位和定量是两个不同层面的损伤识别，我们认为将其分离，逐步实现是解决目前大型结构物损伤识别中未知参数多、优化解和逆解析不稳定的较有前景的有效手段。基于长标距应变分布动态测试的二阶段损伤识别方案如图 36 所示，首先利用模态向量直接进行无模型的损伤定位，然后结合数学模型如有限元模型进行损伤定量。多数结合数学模型进行损伤定量方法的实质都是求解由模态参数和物理参数的变化向量构建的灵敏度方程组。经过第一阶段的损伤定位后，大部分完好无损的区域物理参数无变化，损伤定量的目标锁定在损伤定位后的局部区域，从而使得未知数数目锐减，这对方程组的优化求解带来了极大便利。

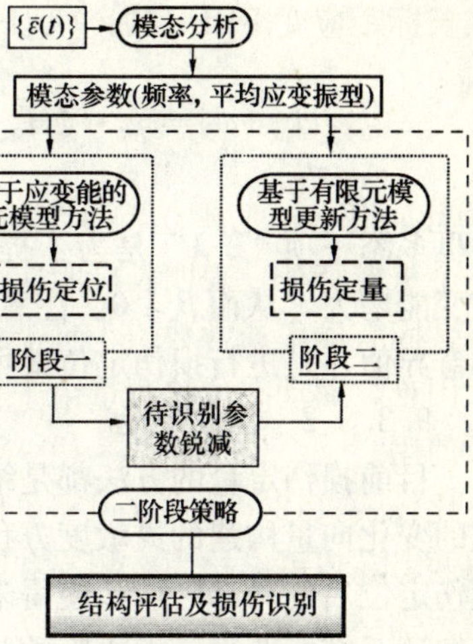

图 36　二阶段损伤识别方案

9.3.3.1　损伤定位

模态向量涵盖了传感器所处位置及周围的空间信息，比较适合损伤定位。注意与位移振型相似，MMSV 在构建的过程中，只考虑了向量各分量的相互关系，而忽略了各分量绝对数值的大小。因此，对结构的不同状态（如损伤前和损伤后）进行比较时，不同工况下获得的 MMSV 需要按照相同的准则标准化，或者换言之，定义的损伤指纹必须在 MMSV 乘以任一常数的情况下都保持原值不变。

Stubbs et al.（1992）曾经利用位移振型构建了基于模态应变能的损伤指纹，文献（Doebling et al.，1996；Sohn et al.，2003）对各种已有的无模型损伤定位方法进行实验比较研究，结果表明，该损伤指纹是个比较有效的损伤定位指标。根据这些成果可知，对于一个长度 L、自由度 ND 的梁，存在以下近似关系：

$$\frac{EI_{\mathrm{k}}}{EI_{\mathrm{k}}^{*}} \approx \frac{\sum_{i=1}^{ND} \int_{x_{\mathrm{k}}}^{x_{\mathrm{k}}+L_{\mathrm{k}}} \left(\frac{d^2\chi_{\mathrm{k}i}^{*}}{\mathrm{d}x^2}\right)^2 \mathrm{d}x \Big/ \int_0^L \left(\frac{d^2\chi_{\mathrm{k}i}^{*}}{\mathrm{d}x^2}\right)^2 \mathrm{d}x}{\sum_{i=1}^{ND} \int_{x_{\mathrm{k}}}^{x_{\mathrm{k}}+L_{\mathrm{k}}} \left(\frac{d^2\chi_{\mathrm{k}i}}{\mathrm{d}x^2}\right)^2 \mathrm{d}x \Big/ \int_0^L \left(\frac{d^2\chi_{\mathrm{k}i}}{\mathrm{d}x^2}\right)^2 \mathrm{d}x} \tag{22}$$

上标 * 代表损伤结构的参数，EI 为梁单元的抗弯刚度，x 为沿梁长度方向的坐标，$\chi_{\mathrm{k}i}$ 特指第 k 个梁单元，第 i 阶模态的挠度振型，注意这里单元的挠度振型和上述节点的位移振型 φ 具有不同含义。等式（22）的本质含义就是建立梁单元模态应变能和抗弯刚度的关系，由于 MMSV 相当于直接的应变模态测量，考虑到：

$$\frac{d^2\chi_{\mathrm{k}i}}{\mathrm{d}x^2} = \frac{\delta_{\mathrm{k}i}}{h_{\mathrm{k}}} \tag{23}$$

这里定义损伤指纹向量为：

$$\{\beta_1, \beta_2, \cdots, \beta_{\mathrm{k}}, \cdots, \beta_{\mathrm{N}}\}^{\mathrm{T}} \tag{24}$$

包括 N 个分量，对应于长标距应变传感器的总数，其中，

$$\beta_k = \frac{\sum_{i=1}^{ND}\left[(\delta_{ki}^*)^2 L_k / \sum_{k=1}^{N} (\delta_{ki}^*)^2 L_k\right]}{\sum_{i=1}^{ND}\left[(\delta_{ki})^2 L_k / \sum_{k=1}^{N} (\delta_{ki})^2 L_k\right]} - 1 \tag{25}$$

这里多添一项“-1”是为了提供更明显的损伤显示，因为一般情况下，损伤会导致模态应变能增大，从而$\beta_k \geqslant 0$。注意这只是个近似的结论并非绝对成立，组合所有向量分量识别奇异值才是进行损伤定位最可行的方法。

9.3.3.2 损伤定量

目前损伤定量的方法都是结合数学模型开展的，其实质都是求解由模态参数和物理参数的变化向量构建的灵敏度方程组。经过第一阶段的损伤定位后，损伤定量的目标锁定在损伤定位后的局部区域，使得未知数数目锐减。由于结构的自振频率测量精度高，高速光纤解调系统能获得较大幅度的频谱，目前我们采用自振频率进行结构的损伤定量。特征值和结构刚度及质量之间的灵敏度方程组为（Friswell and Mottershead，1995）：

$$[S]_{M\times Nu}\cdot\{\Delta p\}_{Nu} = \{\Delta\lambda\}_M \tag{26}$$

M为测得的特征值阶数，Nu为待识别的未知量数目，$\{\Delta p\}$代表结构损伤亦即待识别参数向量，$\{\Delta\lambda\}$是测得的特征值变化向量，$[S]$是灵敏度矩阵，其中元素可以表示为：

$$S_{ij} = \{\varphi_0^i\}^{\mathrm{T}}\left[\frac{\partial[K_0]}{\partial p_j} - \lambda_0^i\frac{\partial[M_0]}{\partial p_j}\right]\{\varphi_0^i\} \tag{27}$$

这里下标“0”代表结构损伤前的状态，K、M是刚度和质量矩阵，$\{\varphi_0^i\}$是无损伤结构的第i阶振型，p_j为待识别的第j个参数。从以上等式可知，灵敏度矩阵仅与结构损伤前的初始状态有关，只要结构的基准模型已知，根据式（22）即可进行参数识别，即基于数学模型的损伤定量诊断。具体流程如下：建立无损伤结构的有限元模型，比较实测数据与数值计算结果进行有限元模型更新，确立无损结构的基准模型，由此获得K_0、M_0及$\{\varphi_0^i\}$，据式（27）构建灵敏度矩阵S；从式（19）可以看到，经损伤定位后，待识别结构参数$\{\Delta p\}$的数目锐减，比较损伤前后结构的特征值求出$\{\Delta\lambda\}$，从而获得灵敏度方程组，求解即可定量识别损伤。试验验证请参照文献[37～39]。

9.3.4 基于相对长标距应变模态向量统计数据的无模型损伤识别

9.3.4.1 基本原理

尽管二阶段损伤识别方法在解决土木结构待识别参数多反演困难的问题上有了显著的改善，但在第二阶段损伤定量时，仍然依赖结构的数学模型，导致模型更新或优化这一复杂且具有显著不适定性的过程无法逾越。考虑到应变的分布特征，理想情况下MMSV各分量的相对关系不变，我们提出了另一种对MMSV进行标准化的方法，据此定义损伤指纹，并建立损伤指纹和损伤程度之间的定量关系。

取MMSV中的一个分量，假设第b个MMS（δ_{br}）为基准，其他分量与该分量的比值构成另一向量：

$$\{\alpha_{1r},\ \alpha_{2r},\ \cdots,\ \alpha_{mr},\ \cdots,\ \alpha_{Nr}\}_{(N-1)}^{\mathrm{T}} = \left\{\frac{\delta_{1r}}{\delta_{br}},\ \frac{\delta_{2r}}{\delta_{br}},\ \cdots,\ \frac{\delta_{mr}}{\delta_{br}},\ \cdots,\ \frac{\delta_{Nr}}{\delta_{br}}\right\}_{(N-1)}^{\mathrm{T}} \tag{28}$$

定义为目标特征向量，其中 α_{mr} 为第 m 个传感器的目标特征。通过比较损伤前后结构的目标特征，可以定义损伤指纹向量为：

$$\{\beta_{1r},\ \beta_{2r},\ \cdots,\ \beta_{mr},\ \cdots,\ \beta_{Nr}\}^{T} \tag{29}$$

其中，

$$\beta_{mr}=\frac{\alpha_{mr}^{*}-\alpha_{mr}}{\alpha_{mr}}\times 100\% \tag{30}$$

上标 * 代表损伤结构的参数。注意该方法理论上对采用结构的任意阶模态信息都适用。由于一般情况下，低阶测量精度高，因此采用第一阶测量结果比较合适，同时可参考采用二阶或二阶以上信息进行损伤诊断的结果。值得注意的是，如果事先能对损伤的集中程度进行估计，例如图 37（*b*）中“$n=4$”表示传感器的标距一分为四，预估计损伤集中在传感器标距四分之一长度的范围内，则可以进行更加精确的损伤定量。

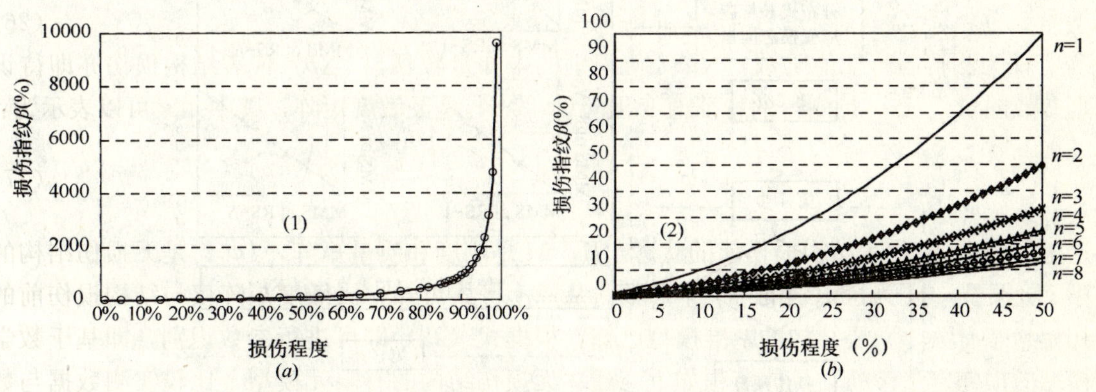

图 37　损伤指纹与损伤程度

9.3.4.2　具体操作流程

图 38 给出了基于分布式长标距应变动态测试的结构损伤识别方法实现的流程图，总体上分为四个步骤，具体说明如下：

第 1 步：选择参考传感器。分布式传感网络的 N 个传感器分为两大类，一类是参考传感器，另一类是普通传感器，参考传感器的选择是本方法的关键。应该确保参考传感器安装在损伤发生概率最小的位置，如所受内力不大，远离结构薄弱部位且服役环境好的构件或区段。原则上至少选择一个参考传感器，最好考虑若干个，这样可以对比采用不同参考传感器所得的结果，也可防止某个参考传感器因所在位置发生损伤而无法提供“参考”的不利情况。

第 2 步：采集并初步处理数据，获得长标距应变模态向量（MMSV）这个重要参数。假设被监测结构上安装的分布式 FBG 传感网络包括 N 个传感器，则基于上述硬件系统，N 条长标距应变时程的原始数据被记录。根据测量数据处理的一般原则，初步判断数据的有效性，分析异常数据产生的原因。结合对动态激励形式的确定，按照试验模态分析求取特征向量的方法，构建分布式长标距应变模态向量。

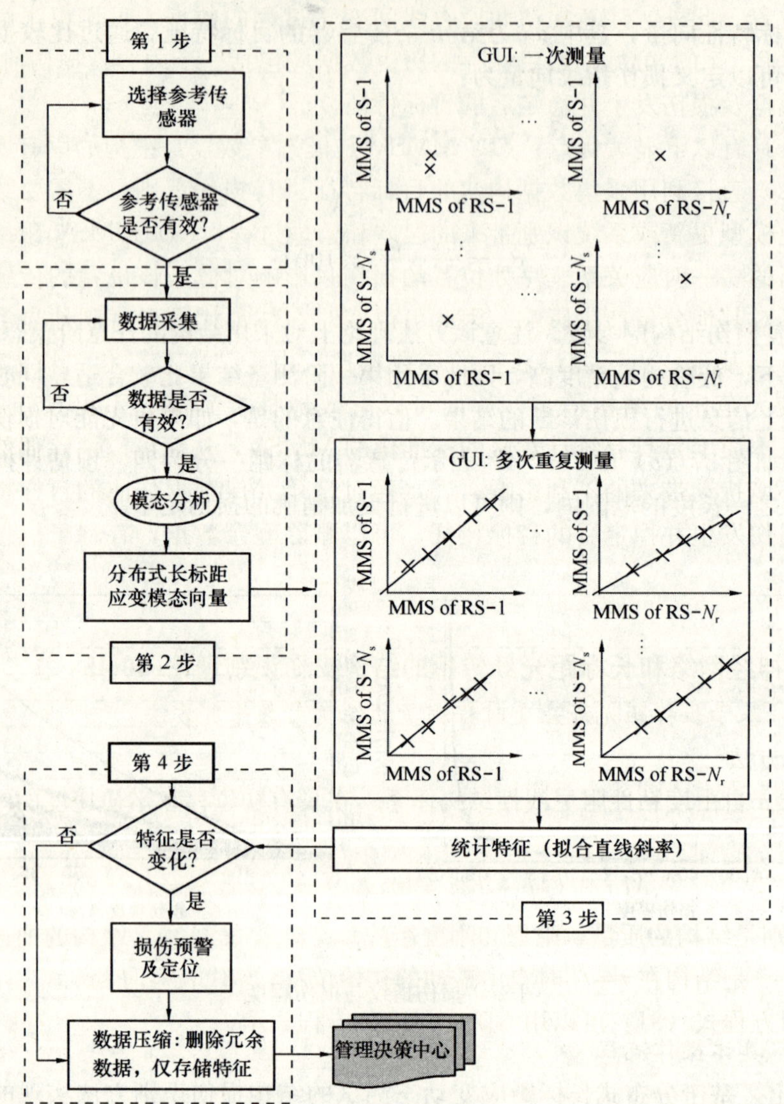

图 38 基于相对长标距应变模态向量统计数据的无模型损伤识别流程

第 3 步：参数的实时图形再现及目标特征的确定。假设考虑 N_r 个参考传感器，则普通传感器有($N-N_r$)个。设计一图形用户界面(GUI)，该界面可以直接或者经切换显示(($N-N_r$) * N_r)个坐标图，其中坐标图以某个参考传感器的模态向量分量为 x 轴，某个普通传感器的模态向量分量为 y 轴。每进行一次测量，所有坐标图上都显示一个对应点。以其中一个坐标图为例，在一个时间段内重复多次测量后，多个点被记录。若无损伤发生，这些点理论上应该线性相关。因此实际考虑测量噪音的影响时，可以对这些测量点进行线性拟合。所得直线的斜率记为该时间段的目标特征。

第 4 步：基于目标特征的结构损伤识别及有效数据的高效率存储。考虑一定的时间长度，如果测得的数据始终不偏离该目标特征，可以及时删除多次测量积累的数据，仅保留和存储目标特征。如果一定时间长度内，数据偏离目标特征，并显示了明显的规律性，即拟合后的直线斜率明显变化，可以及时进行损伤预警，对应的坐标图同时显示了损伤所在的位置。此外，注意保留两段直线间变化过程的数据，可用以了解损伤发生的瞬间和发展过程。由上可见，最后保存的历史数据应包括三个内容：损伤前的目标特征，损伤过程的数据及损伤发生并稳定后的目标特征。

这一方法具有以下显著优点：(1) MMSV 是模态参数，是结构的固有属性，与外部荷载无关；(2) 直接利用采集数据构建的目标特征进行损伤识别，不需要结构的数学模型，可以避免模型更新或系统识别带来的误差和繁琐过程；(3) MMSV 跟结构的局部损伤有直接清晰的一一对应关系。一处位置的损伤只影响对应于该位置的传感器的模态向量分量，而不对其他分量产生影响，因此可以实时在线地用图形直观地显示损伤的发生和位置；(4) 与基于“点”测试获得的参数相比，由于“分布”和“长标距”，MMSV 包含了覆盖较大范围的结构信息，从而可以有效地捕捉到事先不可预知的结构损伤；(5) 目标特征本质上是具有统计意义的标准化的 MMSV，从而可以消除偶然误差，减少测量噪音和环境扰动带来的不利影响；(6) 通过记录并保存损伤前的目标特征，损伤过程的数据及损伤发生并稳定后的目标特征，可以删除冗余数据，有利于信息的高效率储存和管理。

9.3.5 基于神经网络和长标距光纤传感的结构损伤识别理论[40-45]

9.3.5.1 基于长标距光纤应变时程测量的结构参数识别方法

(1) 一般方法

考虑一个 n 自由度粘性阻尼线性结构体系。在具有初位移和零初速度的结构自由振动可以描述为：

$$M\ddot{x}+C\dot{x}+Kx=0, x_{t=0}=x_0, \dot{x}_{t=0}=0 \tag{31}$$

M，C，K 分别是结构的质量，阻尼和刚度矩阵；$\ddot{x}$，$\dot{x}$，x 分别是结构的加速度，速度和位移；$x_{t=0}=x_0$ 是结构在 $t=t_0$ 时自由振动的初始位移；0 代表零向量。

状态空间方程式 (31) 可以用一阶向量微分方程表示：

$$\dot{Z}_t=AZ_t \tag{32}$$

状态向量 Z_t 可以定义为：

$$Z_t=\begin{Bmatrix}\dot{x}_t\\ x_t\end{Bmatrix} \tag{33}$$

系统矩阵 A 是：

$$A=\begin{bmatrix}-M^{-1}C & -M^{-1}K\\ I & 0\end{bmatrix} \tag{34}$$

状态空间方程式 (32) 的连续时间解为：

$$Z_t=e^{A(t-t_0)}Z_{t_0} \tag{35}$$

$Z_{t_0}=Z_0$ 是 t_0 时刻的初始状态向量。如果方程的时间间隔 $t-t_0$ 用 kT 表示，T 表示时间间隔，状态方程的离散解可以表示为：

$$Z_k=e^{AT}Z_{k-1}, \quad (k=1, \cdots, K) \tag{36}$$

Z_k，Z_{k-1} 分别是 kT 和 $(k-1)T$ 实时状态变量。

图 39 详细的描述了基于结构应变时程和人工神经网络的目标结构参数识别方法的具

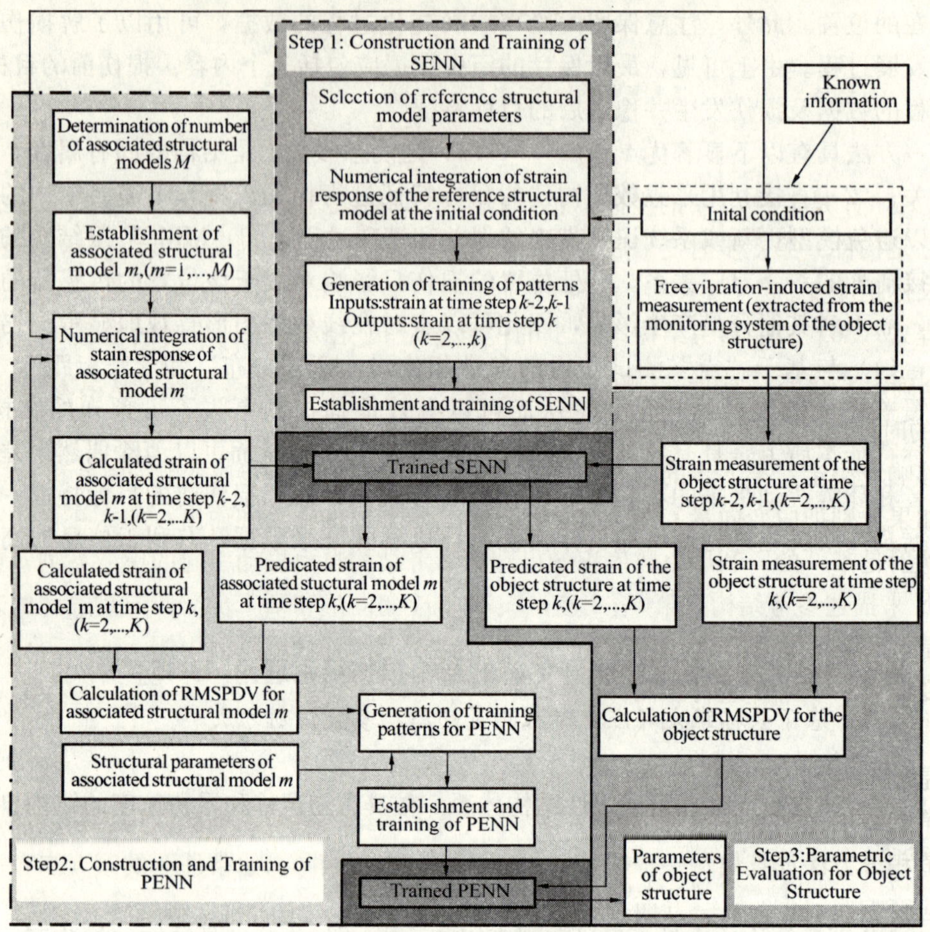

图 39　基于动态宏观应变和神经网络的参数识别流程图

体过程。第一步，构建一个参考结构，利用其在确定初始条件下的自由振动引起的长标距应变响应的时间序列建立和训练一个神经网络 SENN。该网络本质上为参考结构的一个非参数化模型。

比如，为了对图 40(*a*)所示桁架结构，可以建立每一个单元节点位移和构件应变之间的函数关系。如图 40(*b*)所示两节点 i 和 j 的桁架结构构件 n 第 k 步时的长标距应变可以由如下方程得出。

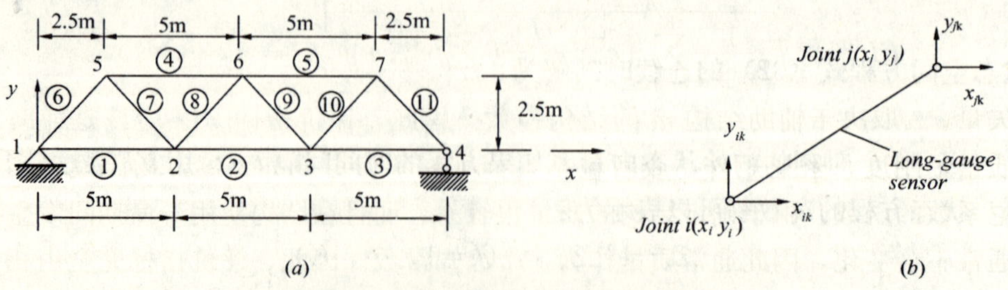

图 40　桁架结构长标距应变测量

(*a*) 桁架结构模型；(*b*) 利用长标距传感器进行应变测量

$$l_{nk}=\sqrt{[x_i+x_{ik}-x_j-x_{jk}]^2+[y_i+y_{ik}-y_j-y_{jk}]^2} \tag{37}$$

$$l_{n0}=\sqrt{[x_i-x_j]^2+[y_i-y_j]^2} \tag{38}$$

$$\varepsilon_{nk}(t)=\frac{l_{nk}-l_{n0}}{l_{n0}} \tag{39}$$

x_i，y_i，x_j，y_j 是节点 i，j 的坐标；x_{ik}，x_{jk}，y_{ik}，y_{jk} 是节点 i，j 第 k 步时 x，y 方向的位移；l_{nk} 是构件 n 第 k 步时变形后长度；l_{n0} 是构件 n 没有变形时长度，ε_{nk} 表示单元 n 第 k 步时的长标距应变。

方程式（36）表明参考结构第 k 步的位移响应 x_k 由第 $k-1$ 步时的位移 x_{k-1} 和速度 $\dot{x}_{k-1}$ 完全确定。此外，速度 $\dot{x}_{k-1}$ 响应可以由 $x_{k-1}-x_{k-2}$ 的位移改变量除以从第 $k-2$ 步到第 $k-1$ 步的时间间隔得到。容易理解，在某一确定时间步长内的应变响应由在相应时间步长的位移响应决定，第 k 步的应变响应完全由 $k-1$ 步和 $k-2$ 步应变响应 ε_{k-1} 和 ε_{k-2} 确定。因此，如果分别把应变向量 ε_k 和 ε_{k-1}，ε_{k-2} 选作 SENN 的输入和输出，那么在输入和输入之间的映射是唯一的。利用由数值积分得到的参考结构自由振动引起的应变系列，可以训练 SENN 来描述参考结构在第 $k-2$ 步，$k-1$ 步应变向量和第 k 步应变向量之间的映射关系。于是，SENN 可以作为参考结构的非参数化模型，并且可以运用如下的方程对参考结构的应变向量一步一步进行预测。

$$\varepsilon_k^f=SENN_\varepsilon(\varepsilon_{k-2},\ \varepsilon_{k-1}),\ (k=2,\ \cdots,\ K) \tag{40}$$

ε_k^f 是用训练的 SENN 预测的第 k 步应变。

在第二步中，构筑一定数量的参考结构具有不同结构参数的辅助结构。一方面，可以采用数值积分方法计算出每一个辅助结构在与第一步的相同初始条件下的第 k 步自由振动引起的应变响应 $\varepsilon_{m,k}$，另一方面，将积分所得到的辅助结构的应变响应作为 SENN 的输入，根据方程式（40）可以预测应变响应 $\varepsilon_{m,k}^f$。由于辅助结构的参数与参考结构不同，预测的响应与由数值积分计算结果将存在误差。第 k 步的误差向量可以由下式计算出：

$$E_{m,k}=\{e_{m,k}^{(1)}\cdots e_{m,k}^{(j)}\cdots e_{m,k}^{N_m}\}^T=\varepsilon_{m,k}^f-\varepsilon_k,$$
$$(m=1,\ 2,\ \cdots,\ M,\ j=1,\ \cdots,\ N_m,\ k=2,\ \cdots,\ K) \tag{41}$$

N_m 表示测量轴向应变的结构构件总的数目。方程（41）的上标 T 表示这是向量的转置。

与辅助结构 m 相对应，定义了差向量 $E_{m,k}$ 的函数 $EI_m\in R^{N_m}$ 作为评价指标。对应每个辅助结构，定义了预测误差向量均方根（RMSPDV）作为评价指标如下：

$$EI_m=\{EI_m^{(1)}\cdots EI_m^j\cdots EI_m^{N_m}\}^T,\ (j=1,\ \cdots,\ N_m) \tag{42a}$$

$$EI_m^{(j)}=\sqrt{\frac{1}{K-1}\sum_{k=1}^{K}\left(e_{m,k}^{(j)}-\frac{1}{k}\left(\sum_{k=1}^{K}e_{m,k}^{(j)}\right)\right)^2},(j=1,\cdots,N_m) \tag{42b}$$

差矢量 $e_{m,k}^j$ 取决于辅助结构 m 的结构参数。因此，评价指标 EI_m 应该是辅助结构 m 的结构质量，刚度，阻尼矩阵，或者包括几何和材料参数例如密度，杨氏模量，截面尺寸，阻尼系数的函数。由于确定结构的质量很容易，而且在结构使用寿命期间发生损伤结构质量通常不会变化，因此通常质量作为一个已知常数。因此，评价指标完全由相关结构的刚度矩阵 K_m 和阻尼矩阵 C_m 确定并且可以由一般函数关系表示：

$$EI_m=f(K_m,C_m) \tag{43}$$

因此如果已知方程(43)的反函数，根据评价指标 EI_m 就能确定结构参数。为此，建立和训练第二个神经网络，参数识别用神经网络（PENN）来描述评价指标和结构参数之间的关系：

$$(K_m, C_m)=f^{-1}(EI_m)=PENN(EI_m) \tag{44}$$

用辅助结构的结构参数和相应的 RMSPDV 组成的数据样本来训练 PENN。PENN 训练后，把它应用到第三步，利用目标结构的长标距应变时程作为 SENN 的输入，计算得出 RMSPDV，并作为 PENN 的输入，就可以得到目标结构的结构参数。

9.3.5.2 基于结构动力响应测量时程的结构参数识别方法

以上提出的运用结构应变测量时程的结构参数识别方法具有一般意义。运用目标结构的位移、速度和加速度等不同类别的动力响应时程，可以建立相应的实现方法。

对于结构的强迫振动的情况，运用结构的位移和速度时程信息，也可以直接对结构参数进行识别。图 41 所示为利用神经网络进行参数识别的基本思想，该法直接运用结构在动力激励下响应进行识别。为了对目标结构进行识别，首先创建了一个与目标结构具有相同结构尺寸和拓扑结构的参考结构，并且建立一个神经网络（ENN）来描述其动力响应，即建立参考结构的一个非参数模型。结构参数的识别将由第二个神经网络来实现。

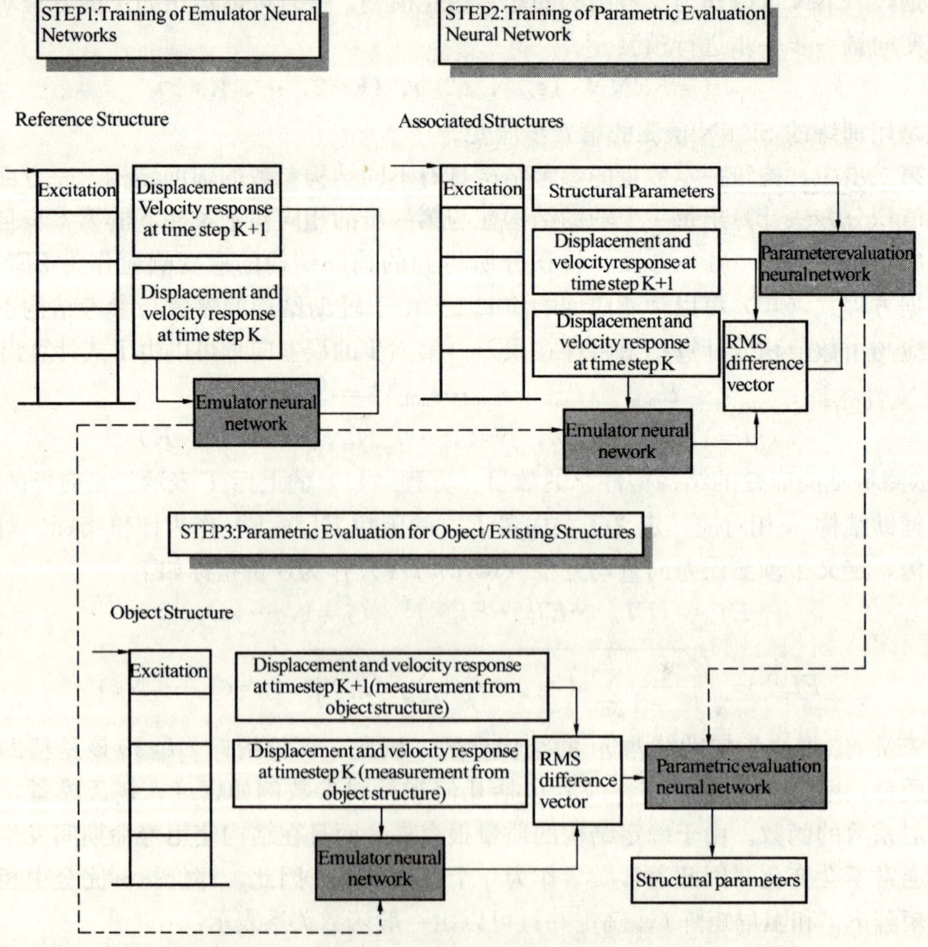

图 41 利用神经网络进行参数识别的步骤

与上面介绍的基于应变测量时程的参数识别方法类似，运用结构位移和速度响应的结构参数识别方法也分三步进行，如下图所示。在第一步中，构建参考结构并创建和训练 ENN 网络。利用特定动力激励下的动力响应对其进行训练，将训练好后的 ENN 模型能够作为参考结构的非参数模型。步骤 2，构建辅助结构，并计算其动力响应，并通过训练好的 ENN 进行预测。由于结构参数的变化，预测所得到的动力响应与精确值存在差异，它们之间的差异与辅助结构的结构参数有关。为了描述差异，引入误差向量的均方根。建立和训练参数评价神经网络 PENN 来描述评价指标与结构参数的关系。步骤 3，利用安装在结构上的传感器来测量目标结构的速度和位移，基于 ENN 得到评价指标，然后根据 PENN 可以识别出结构参数。

9.3.6　基于加速度响应的能量损伤识别方法[46、47]

9.3.6.1　理论推导

N 自由度体系的运动微分方程为：

$$M\ddot{x}(t)+C\dot{x}(t)+Kx(t)=F(t) \tag{45}$$

式中，M、C 和 K 分别为系统的质量矩阵、阻尼矩阵和刚度矩阵，$x(\mathrm{t})=[x_1(t),\ x_2(t),\ \cdots,\ x_{\mathrm{N}}(t)]^{\mathrm{T}}_{\mathrm{N}\times 1}$ 和 $F(t)=[f_1(t),\ f_2(t),\ \cdots,\ f_{\mathrm{N}}(t)]^{\mathrm{T}}_{\mathrm{N}\times 1}$ 分别为系统的位移向量和激励力向量。

不失一般性地，由瑞利法知该系统阻尼可表示为

$$C=\alpha M+\beta K \tag{46}$$

其中 α 和 β 是两个比例常数，其与系统前两阶振型的阻尼比有关，将式(45)变换到频域内为

$$(K-\omega^2M+j\omega C)X(\omega)=F(\omega) \tag{47}$$

设系统的模态坐标为 Q，则利用该模态坐标代替式(45)中的物理坐标，左乘 ϕ^{T}，并考虑到正交性条件：

$$\phi_{\mathrm{s}}^{\mathrm{T}}C\phi_{\mathrm{r}}=\begin{cases}0 & (r\neq s)\\ C_{\mathrm{r}} & (r=s)\end{cases} \tag{48}$$

可得

$$(K_{\mathrm{r}}-\omega^2M_{\mathrm{r}}+j\omega C_{\mathrm{r}})Q=F_{\mathrm{r}} \tag{49a}$$

对于第 r 阶模态则有

$$(K_{\mathrm{r}}-\omega^2M_{\mathrm{r}}+j\omega C_{\mathrm{r}})q_{\mathrm{r}}=F_{\mathrm{r}} \tag{49b}$$

由式(49b)可得该系统的第 r 阶模态坐标为

$$q_{\mathrm{r}}=\frac{F_{\mathrm{r}}}{K_{\mathrm{r}}-\omega^2M_{\mathrm{r}}+j\omega C_{\mathrm{r}}} \tag{50}$$

其中，$K_{\mathrm{r}}=\phi_{\mathrm{r}}^{\mathrm{T}}K\phi_{\mathrm{r}}$，$M_{\mathrm{r}}=\phi_{\mathrm{r}}^{\mathrm{T}}M\phi_{\mathrm{r}}$，$C_{\mathrm{r}}=\phi_{\mathrm{r}}^{\mathrm{T}}C\phi_{\mathrm{r}}$，$F_{\mathrm{r}}=\phi_{\mathrm{r}}^{\mathrm{T}}F(\omega)=\sum_{j=1}^{n}\varphi_{j\mathrm{r}}f_j(\omega)$ 分别为系统的第 r 阶模态刚度、模态质量、模态阻尼及模态激励力；ϕ_{r} 为 r 阶模态向量；$\varphi_{j\mathrm{r}}$ 为 j 节点在 r 阶振型下的振型系数。

结构上任意点 l 的响应为

$$x_l(\omega)=\sum_{r=1}^{N}\varphi_{lr}q_r \tag{51}$$

将作用于 p 点的激励力向量 $F=[0,\cdots f_p(\omega),\cdots 0]^T$ 代入式 $F_r=\phi_r^T F(\omega)=\sum_{j=1}^{n}\varphi_{jr}f_j(\omega)$中，得

$$F_r=\phi_r^T F(\omega)=\sum_{j=1}^{n}\varphi_{jr}f_j(\omega)=\varphi_{pr}f_p(\omega) \tag{52}$$

从而有

$$x_l(\omega)=\sum_{r=1}^{N}\frac{\varphi_{lr}\varphi_{pr}f_p(\omega)}{K_r-\omega^2 M_r+j\omega C_r} \tag{53}$$

测量点 l 与激励点 p 之间的频响函数为

$$H_{lp}(\omega)=\frac{x_l(\omega)}{f_p(\omega)}=\sum_{r=1}^{N}\frac{\varphi_{lr}\varphi_{pr}}{K_r-\omega^2 M_r+j\omega C_r} \tag{54}$$

根据位移导纳、速度导纳及加速度导纳三者之间的关系[46]

$$H_a=j\omega H_v=-\omega^2 H_x \tag{55}$$

其中，H_a、H_v 和 H_x 分别为加速度响应、速度响应和位移响应。由式(54)和式(55)可得系统频域下的加速度响应为

$$H_{a,lp}(\omega)=-\omega^2\cdot\sum_{r=1}^{N}\frac{\varphi_{lr}\varphi_{pr}}{K_r-\omega^2 M_r+j\omega C_r} \tag{56}$$

则相应的频域响应第 r 阶模态为

$$H_{a,lp}^{r}(\omega_r)=\frac{-\omega_r^2\varphi_{lr}\varphi_{pr}}{K_r-\omega^2 M_r+j\omega C_r} \tag{57}$$

由式（56）和式（57）可以看出，系统振型与频域响应之间存在一种比例关系，也就是说 r 阶模态下每个测点的振型与其频域响应之间存在一种比例关系。一般情况下，振型对于损伤的识别是一个非常重要的参数。因此，如果找到基于加速度反应的能量函数，此函数与振型关系密切，就有可能明显提高损伤识别的成功率。在桥梁工程中，一般利用加速度计来测量振动数据和做损伤识别，其中频域响应方程可根据所测得的加速度响应得到。作者经过一系列理论推导[47]得到了基于加速度反应的能量函数。

$$E=\frac{1}{2\pi}\int_{-\infty}^{\infty}E(\omega)\mathrm{d}\omega=\frac{1}{2\pi}\int_{-\infty}^{\infty}\left|\tilde{\ddot{x}}(\omega)\right|^2\mathrm{d}\omega=\frac{1}{2\pi}\int_{-\infty}^{\infty}\left|FFT(\ddot{x}(t))\right|^2\mathrm{d}\omega \tag{58}$$

式中，$\ddot{x}(t)$为加速度反应，$\tilde{\ddot{x}}(\omega)=|FFT(\ddot{x}(t))|$ 为$\ddot{x}(t)$的傅立叶变换值的绝对值，考虑到 $f=2\pi\omega$，上面的等式能被改写为

$$E=\int_{-\infty}^{\infty}E(f)\mathrm{d}f \tag{59}$$

式中，$E(\omega)$ 是加速度反应的功率谱密度函数。

9.3.6.2 能量损伤识别方法

式（58）和式（59）给出基于加速度的能量识别指标的表达式，其可用于确定结构损伤的位置和数量。结构在损伤与未损两种情况下存在能量差异 ΔE，如下所示

$$\{\Delta E\} = |\{E_d\} - \{E_0\}| \tag{60}$$

式中，E_d 为由式（58）计算的损伤结构的加速度能量和 E_0 为由式（58）计算的未损伤结构的加速度能量。结构损伤位置可通过该加速度能量差异 ΔE 的最大值来预测。

另外一种能量指标是基于振型曲率差异提出的，亦即能量曲率差，其可表示如下

$$\left.\begin{aligned}\{\Delta E''\} &= |\{E''_d\} - \{E''\}| \\ E''_i &= (E_{i+1} - 2E_i + E_{i-1})/h^2\end{aligned}\right\} \tag{61}$$

式中，$\Delta E''$为能量曲率差，E''_d 和 E''分别为结构损伤与未损伤时的能量曲率，E''_i 为结构第 i 结点处的能量曲率，E_i 为结构第 i 结点处的能量，h 为第 $i+1$ 个测点到第 $i-1$ 个测点的距离。与上述方法相同，结构损伤位置可以通过结构损伤和未损伤时加速度能量曲率之差的最大值来确定。

应当注意的是式(57)中忽略了激励力，且式(58)的能量表达对不同激励均具有较好的适应性。而该方法在复杂激励力作用下的适用性还有待进一步研究。试验验证参照文献[47]。

9.4　结构健康监测系统设计体系

9.4.1　结构健康监测系统的设计流程和原则

结构健康监测技术越来越受到业界和研究者的关注，逐渐成为重大工程结构健康与安全的保障，也是研究重大基础工程结构的损伤累积破坏和动力灾变演化规律的重要手段。一个完整、实用的结构健康监测系统的设计如图 42 所示，一般要遵循以下几个主要过程和原则。

1）监测内容的完整性和合理性：监测内容（物理参数）的选择关系传感器的选择和监测系统特性，概括而言监测内容可分为三类——荷载、响应和长期性能监测（腐蚀、蠕变等）；荷载主要包括工作荷载（车辆、自重等）和环境荷载（温度、湿度、环境震动等），结构的响应主要包括局部性态和整体性态响应；通常为了获得全面的监测数据，需要进行动、静态监测相结合。

2）传感器的选择与布设原则：传感器的选择首先要根据所要监测的物理参数进行选择，同时兼顾精度、数据传输和经济性等，通常传感器可分为“点”传感器和“分布”式传感器；对于点传感器由于价格高，数据传输比较复杂，合理布设显得尤为重要，通常布设在结构的特征性截面上以便获得结构整体的响应；对于大型工程结构，为了获得更好的监测效果和可信度，通常需要布设一定的分布式传感器进行分布式监测，目前可用于分布监测的传感技术相对比较少。

3）解析・评价软件系统设计：进行设时首先要明确健康监测系统的等级（实时在线监测、定期在线监测还是定期检测），要考虑海量数据的存储（通常可设定阈值，小于该值可不存储）、历史数据的对比分析以及根据所监测结构和物理参数特性建立结构模型；同时，所设计的软件系统要有比较好的通用性和易操作性。

此外，课题组一直倡导对重大工程结构进行分布式监测，基于所研发分布式传感的结构健康系统设计策略如图 43 所示。

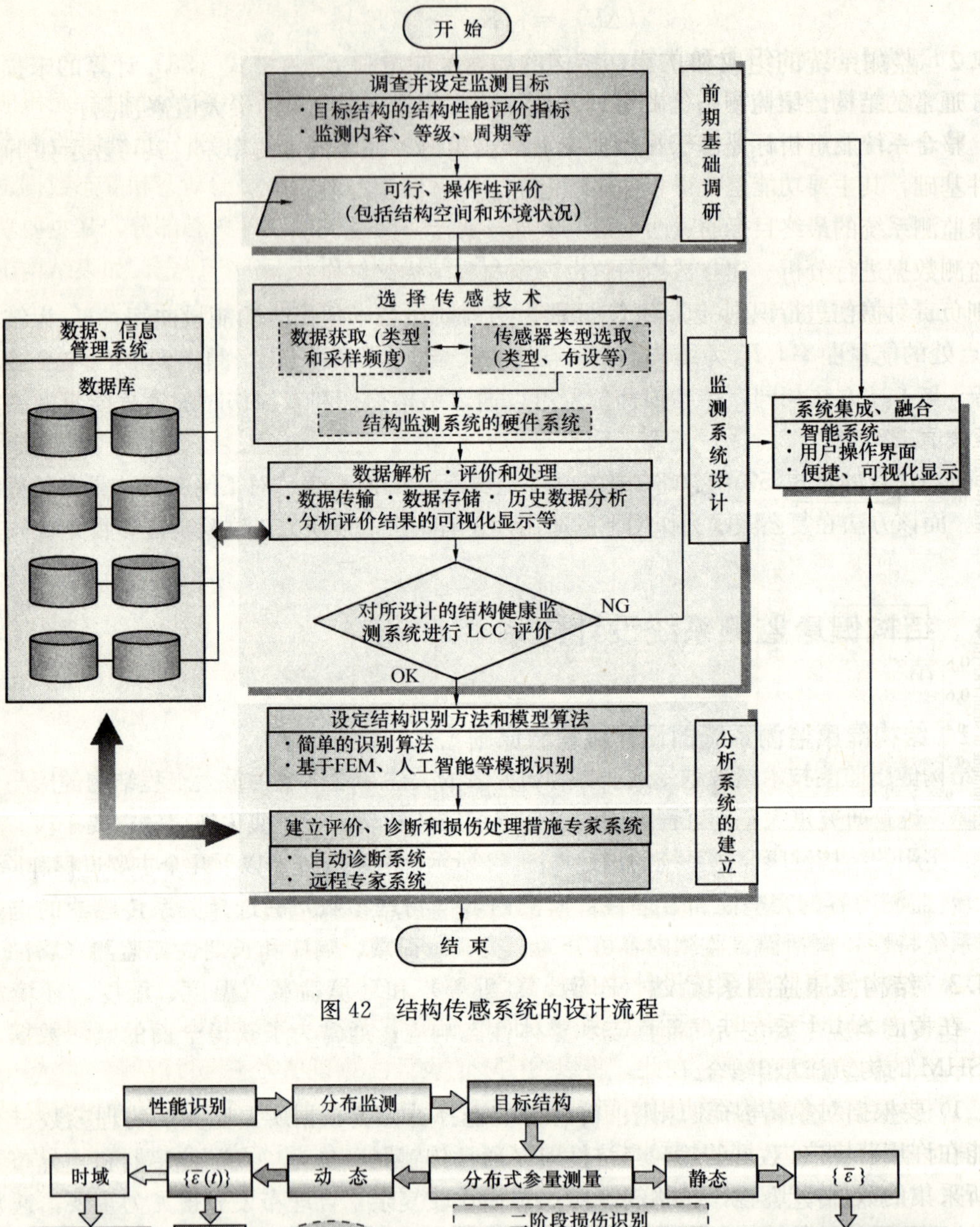

开 始
调查并设定监测目标
・目标结构的结构性能评价指标
・监测内容、等级、周期等
可行、操作性评价
(包括结构空间和环境状况)
前期基础调研
选 择 传 感 技 术
数据获取(类型和采样频度)
传感器类型选取(类型、布设等)
结构监测系统的硬件系统
数据、信息管理系统
数据库
数据解析 ・评价和处理
・数据传输 ・数据存储 ・历史数据分析
・分析评价结果的可视化显示等
监测系统设计
系统集成、融合
・智能系统
・用户操作界面
・便捷、可视化显示
对所设计的结构健康监测系统进行LCC评价
NG
OK
设定结构识别方法和模型算法
・简单的识别算法
・基于FEM、人工智能等模拟识别
建立评价、诊断和损伤处理措施专家系统
・自动诊断系统
・远程专家系统
分析系统的建立
结 束

图42 结构传感系统的设计流程

性能识别
分布监测
目标结构
时域
{ε(t)}
动 态
分布式参量测量
静态
{ε}
非线性、动态激励
频域
模型分析
二阶段损伤识别
无模型损伤定位
工程结构(钢结构、RC结构等)
模型参数
减少未知参数数目
基于神经网络的性能识别
基于分布传感的无模型损伤识别
损 伤 定 量
裂缝、腐蚀监测
变形监测等
无模型损伤识别
基于模型参数识别
结构整体性能评价
健康监测、损伤识别和诊断

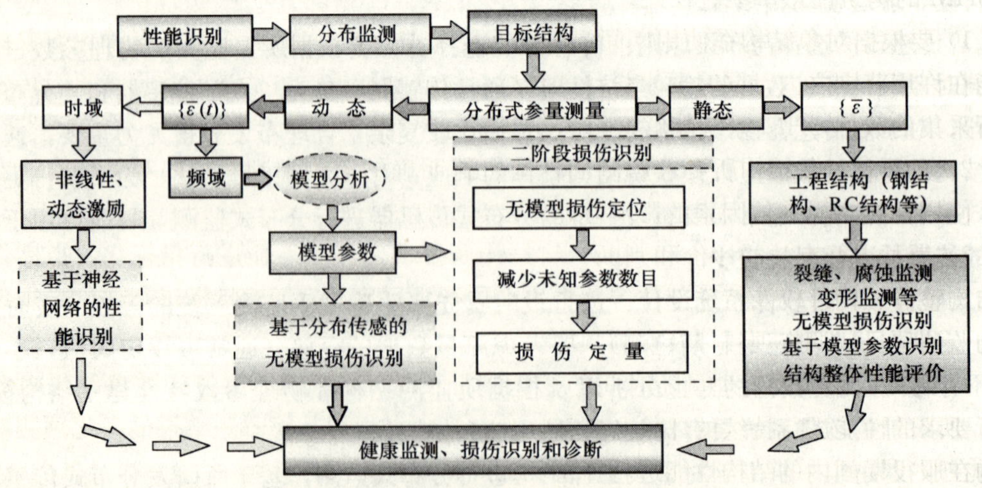

图43 基于分布传感的SHM设计策略

9.4.2　监测系统的组成和功能

通常的结构健康监测系统由传感系统、信号发生装置、数据传输及处理系统和损伤识别、整合系统及解析、评价软件系统等组成，如图 44 所示。前三个部分是健康监测系统的硬件基础，其主要功能是获得和存储进行桥梁结构健康诊断所需要的动态和静态数据。结构健康监测系统的最终目标通过两个部分实现，属于监测系统的软件核心部分，其主要功能是对监测数据进行分析、建立适当的解析模型及对结构的健康状况进行评定，如果结构出现损伤则负责对损伤进行识别(包括定位和定量)并给出结构的功能恢复措施或建议。

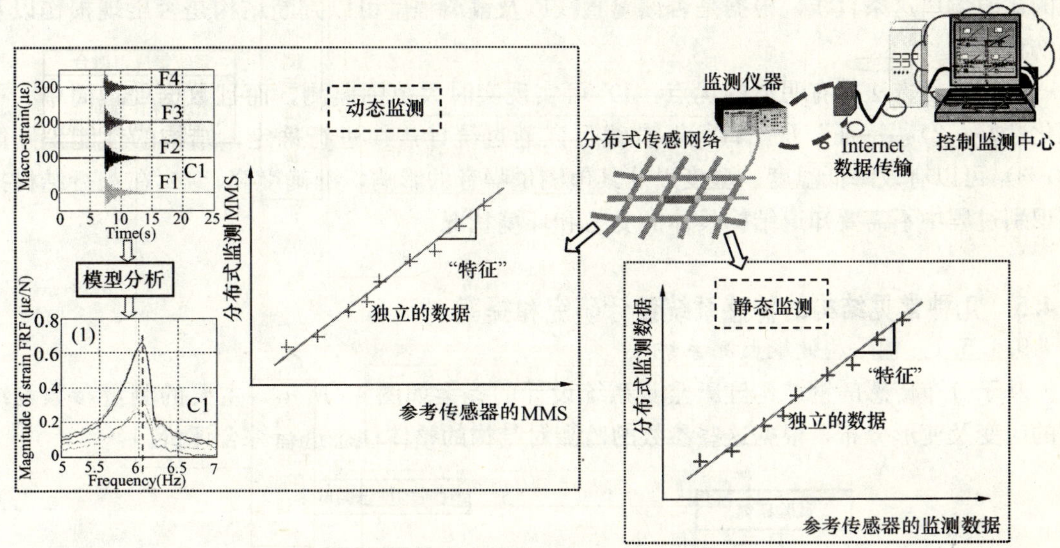

图 44　基于分布传感的动、静态监测策略

9.4.3　结构健康监测系统设计中的注意事项

在考虑 SHM 系统时，要留意两个方面的注意事项，一是 SHM 的设计注意事项，二是 SHM 的构筑注意事项。在此，将以 SHM 的设计注意事项为主进行说明。

1）要根据对象结构在使用期间内的性能要求进行 SHM 设计，目的是确保对象结构的性能在许用范围之内，同时要求设计的传感系统必须能够进行适当的数据采集、传输，基于所采集的数据 SHM 系统能够进行劣化预测、评价和判断；

2）必须对对象结构所要求监测的性能进行明确的监测和判断，如对象结构的安全性能、使用性能、受其他因素的影响性能以及耐久性能等；

3）为了进行有效的评价和判断，将结构所要求监测的性能进行指标化，指标化时分为结构整体、构件以及特殊部位等几个水平进行指标化，为了对指标进行评价必须设定必要的监测项目、调查项目以及实施频度等；

4）对对象结构的物理量随时间的变化进行长时间的监测，并基于这些结果对对象结构所要求的性能进行持续的评价和判断，而且还可以根据基于监测结果的劣化预测对对象结构在服役期间内的结构性能进行评价和判断；

在对结构性能进行评价和判断的过程中，对结构所要求的性能进行明确化，而且还必须明确对象结构的设计寿命周期。

9.4.4 基于分布传感的静、动态监测策略

基于分布式传感技术的动、静态监测策略如图 44 所示。该监测策略不是基于个别或数个测量的数据，而是依据大量的实时多次监测数据对结构的健康和损伤状况进行判断，所以可靠性和稳定性比较高。

结构没有损伤时，分布监测值与参考值的比值在一直线上，如上图所示，这一规律可以视为结构的“特征”，这一特定的直线称为结构的“特征线”。如果结构出现损伤，则二者的比值偏离这条直线，根据是否偏离直线以及偏离程度可以判断结构是否出现损伤以及损伤程度。

这种识别方法具有明显的优点：1）可实现实时大范围监测，而且数据处理简单、可视化性好；2）“特征”及“特征”直线可以通过统计规律进行确定，结构的性能识别简单；3）可以有效降低温度、湿度以及其他测量噪音的影响，准确度高；4）在进行结构损伤识别过程中不需要知道结构具体的受力和环境状况。

9.4.5 几种常见结构的传感系统设计研究和提案

9.4.5.1 钢结构健康监测系统

基于分布传感的钢结构健康监测系统设计的提案如图 45 所示，主要的测量参数为结构的应变及变形分布，根据这些参数的监测对结构的整体性态进行综合评价。

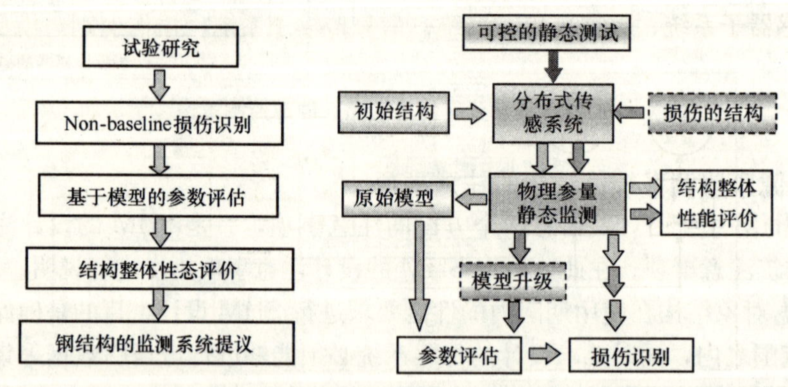

图 45 钢结构的健康监测系统

9.4.5.2 RC 混凝土结构

基于分布式传感技术的钢筋混凝土结构健康监测系统的一种方案如图 46 所示。与钢结构有所不同的是，钢筋混凝土结构还需要对混凝土裂纹的出现和发展（裂缝宽度）进行识别和监测。基于布里渊散射机理和 FBG 光纤传感技术都可以对裂缝进行大范围的分布式监测，与传统的“点”式裂缝计相比，该方法具有无可比拟的优势。

根据所监测宏观应变分布及公式（6）和式（7），可以得到结构的变形分布和最大变形。这种方法可以替代传统的位移传感器监测结构的变形分布和最大变形，克服了传统监测技术在结构变形监测中的诸多不便和障碍，开发了一种新型便捷的大型工程结构变形监

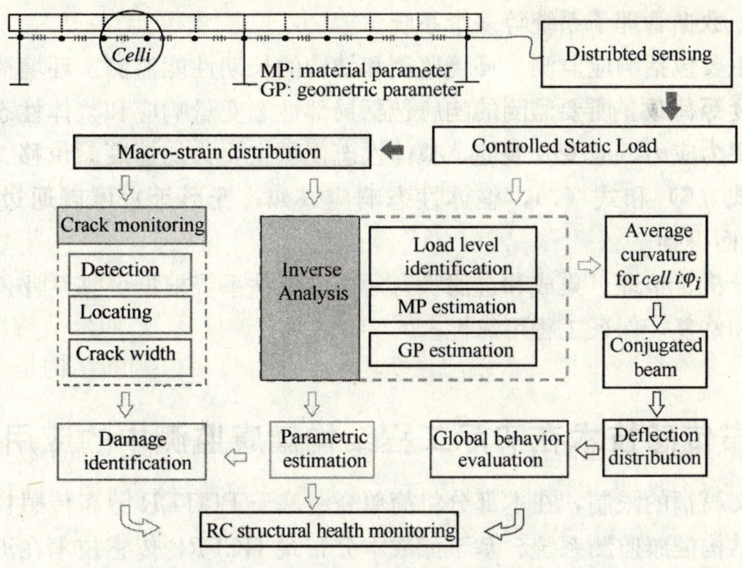

图 46　钢结构的健康监测系统

测技术。另外，在钢筋混凝土结构上，利用所开发的长标距分布式传感器可以对较大范围（传感标距）内的平均应变（Marco-strain）进行测量。这样可以完全避免"点"式传感器（加速度计、应变片等）测量准确性低和易失效等缺陷。

9.4.5.3　地铁隧道监测系统

针对地铁隧道的工作环境和结构特点，所提案的结构健康监测系统如图 47 所示。该系统包括传感器子系统、数据采集子系统、信号传输子系统、损伤识别与模型修正以及安

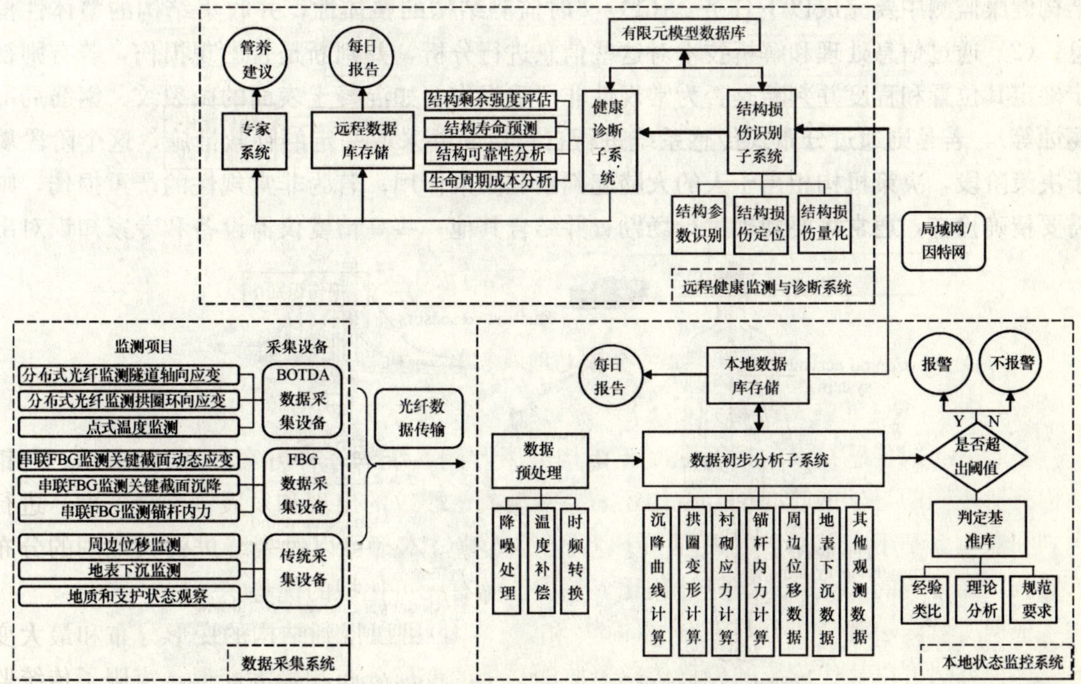

图 47　拟建立的健康监测和灾变预警系统组成图

全评定子系统、数据管理子系统等多个系统。

监测内容主要包括响应监测、环境监测和结构的长期性能监测。环境荷载主要有风荷载、温度、湿度等荷载。所要监测的响应包括局部性态变量响应和整体性态变量响应。局部性态响应主要为应力、应变及裂缝，整体性态变量主要为加速度、位移、温度及结构腐蚀度等。根据式（6）和式（7），整体性态响应（如，变形等）可以通过局部性态响应（应变分布等）而得到。

数据经过分析、整理、集成和过滤噪音后，由评估系统根据可靠性评价准则提出预警信号及时维护和修复，保证工程可靠性。

9.5 各分布传感技术在实际工程结构健康监测中的应用和探讨

由于受论文篇幅的限制，在本部分将简单介绍基于BOTDR分布传感技术的日本某高速公路桥梁的结构健康监测系统、基于低成本分布式HCFRP传感技术在沪宁某高速公路桥梁极限破坏实验研究中的应用以及长标距FBG传感技术在实际桥梁动静态监测中的应用。

在设计结构健康监测体系时，首先要考虑一套合适的传感系统。对于大型土木工程结构，传感系统既要能够反映结构的整体性能信息也要能够反映结构的局部信息（即要能够对损伤进行分布监测），因而分布式传感系统对于大型土木工程结构健康监测系统的建立非常重要。分布式传感系统对于结构健康管理的重要性和作用就如同神经网络对于人体健康管理的作用，如图48所示。如同人类的分布式神经系统作用一样，分布式传感系统在结构健康监测中要完成以下任务：（1）实时监测结构的整体性，并收集结构的整体性信息；（2）通过信息处理和解析技术对这些信息进行分析，并判断是否出现损伤，若有则初步确定其位置和程度并判断是否为常规性非严重损伤（如混凝土表面的微裂纹、钢筋局部腐蚀等），若是则通过分布式传感系统进行自行诊断并采取一定的补救措施，这个阶段属于决策阶段，决策机构相当于人的大脑起判断和决策作用；若为非常规性的严重损伤，则需要精确检查，通常需要到现进行场勘查并结合其他一些高精度仪器设备和专家知识对出

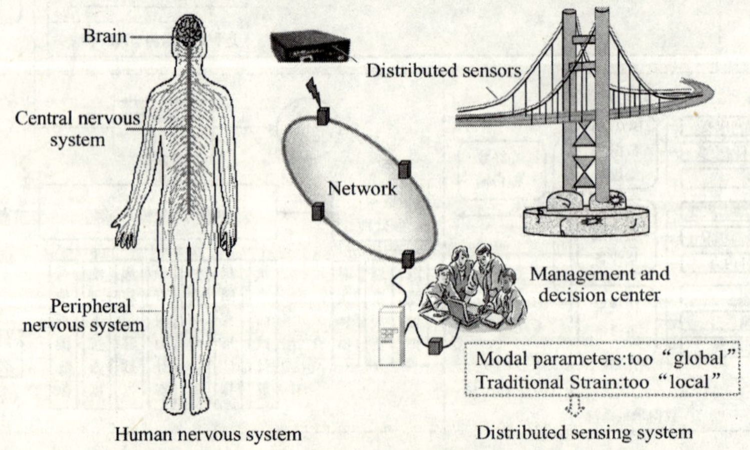

图48 分布式传感系统在结构健康监测中的作用

现了问题的部位进行具体详细的检查和评定。

其次，根据结构的具体特征以及结构损伤识别方法和理论，建立结构的损伤识别、解析评价和预警系统。除此之外，设计一个完善的结构健康监测系统还要考虑数据的传输以及存储等问题。

9.5.1　基于BOTDR传感系统的高速公路实桥监测系统[48]

应用BOTDR光传感技术对日本某一高速公路桥进行结构健康监测和预警，由于该桥在运营中出现了比较严重的病害，利用作者与日本有关公司最近开发的P-PUT（PBO-Prestress Upgrading Technique）技术对该高速公路桥进行了FRP加固补强。为了安全起见，于2003年安装了光纤传感系统对桥梁的运行和结构健康状况进行实时监测。监测的物理量包括梁的上翼缘压缩应变、下翼缘的拉伸应变以及PBO-FRP加固层的拉伸应变。安装此光纤传感系统的具体目的包括以下三点：（1）监测混凝土基体与张拉PBO纤维片材加固层的预应力保持性能；（2）实时监测PBO层与混凝土的剥离；（3）监控交通状况以及整个桥梁的结构健康状态。

9.5.1.1　光纤的铺设情况

光纤在桥梁上的铺设如图49所示。该桥由桁G1到G12共12个桁构成，在其中的5桁即G2，4，6，7和12的加固层PBO的表面布置了光纤，在桁G12的上缘、下缘以及底面的混凝土表面也铺设了分布式传感光纤。为了不影响张贴PBO布，在混凝土表面铺设的光纤时，先在相应的位置开设一条小槽，然后把光纤用环氧树脂粘贴在槽内，最后用putty填封。

总体来说，这些分布光纤可以分为两个系统，系统1是用全面粘贴法铺设在桁G12，7，6，4和2底面PBO布表面，在每个桁上布置的光纤测量长度为19m，两个桁之间的自

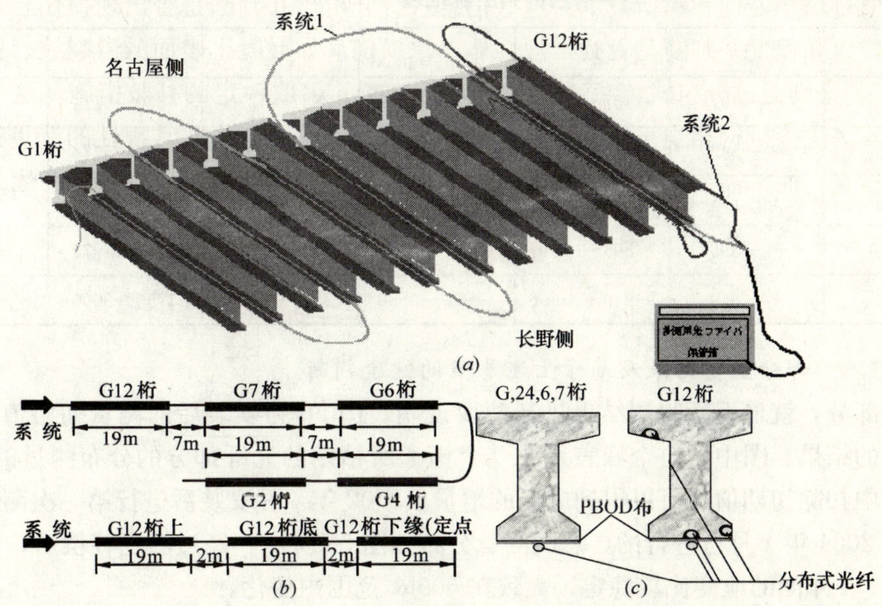

图49　光纤在桁上的总体布置图

(a) 各系统的布置图；(b) 各系统的长度；(c) 光纤布置位置的截面图

由部分光纤的长度为7m，如图49(*b*)所示；系统1用于分布监测这5个桁的PBO加固层的应变分布。系统2分布桁G12的上下缘及底部，测量部分光纤的长度为19m，两测量部分之间的自由部分长度为2m，如图49(*b*)所示，每部分的测量长度用于监测混凝土的拉伸或压缩应变。

9.5.1.2 判断基准和所采取的对应策略

各损伤阶段的判断基准和所采取的相应措施如表5所示。

各阶段的判断基准和采取的相应措施 **表5**

Level	应变	预想的状态	对应措施
D*	10000$\mu\varepsilon$～	出现剥离	查明原因·精密检查
C*	7500～9999$\mu\varepsilon$	可能剥离	进行精密检查
B*	5000～7499$\mu\varepsilon$	微观剥离	增加监测频率
A*	H～4999$\mu\varepsilon$	无异常	
H	大于预应力10%的压缩应变	可能出现异常*	增加监测频率
I	大于预应力15%的压缩应变	出现异常*	查明原因·精密检查

注：对于表中的Level A～D，取（上述的应变/$\sqrt{\text{number of PBO layer}}$）。

根据上述的基准，结合初期预应力程度、温度、布的层数等条件对桥梁的健康状况进行综合判断和评定。根据表5所述的基准，制定了每个桁的具体破坏状态基准，具体如表6所述。在这里所设定的标准温度为5℃，如温度偏离设定温度则需要进行温度补正，对于该测量系统通常温度每升高或降低1℃，则引起应变增加或降低2$\mu\varepsilon$。

各桁的判断标准值（单位$\mu\varepsilon$） **表6**

	I	H	A	B	C	D
G2 3层	～−407	−407～−271	−271～2887	2887～4330	4330～5774	5774～
G4 3层	～−543	−543～−362	−362～2887	2887～4330	4330～5774	5774～
G6 4层	～−781	−781～−521	−521～2550	2550～3750	3750～5000	5000～
G7 5层	～−895	−895～−597	−597～2336	2336～3354	3554～4772	4772～
G12 4层	～−849	−849～−566	−566～2550	2550～3750	3750～5000	5000～

9.5.1.3 部分监量结果及基于上述基准的健康判断

在本部分，就最近所监测结果进行简要说明。图50为安装后的测量初始值和最近5次所测量的结果。图中的5个峰表示在5个桁上所粘贴的光纤部分的分布测量的应变值。从这些值中扣除初期值变可以得到应变的增量。2002年5月安装后进行第一次测量，最近的测量是2004年4月份进行的，总的测量分布如图50所示。经过温度补偿后，测量结果大体一致。1年间的应变比较稳定，大致在600$\mu\varepsilon$范围内变化。

将图50展开就可以得到各桁每部分的测量结果，如将第一个峰展开，可以得到桁12的应变分布图，如图51所示。将测量结果和表5的各桁的健康判断基准相对照，可以发现各桁还处于 Level A，即没有异常出现且各桁处于健康状态。这些监测结果一方面证明了 BOTDR 可以用来对实际重大结构如桥梁等进行分布式的监测，另一方面也证明了 P-PUT 工法是一种可靠的混凝土结构加固改造技术。

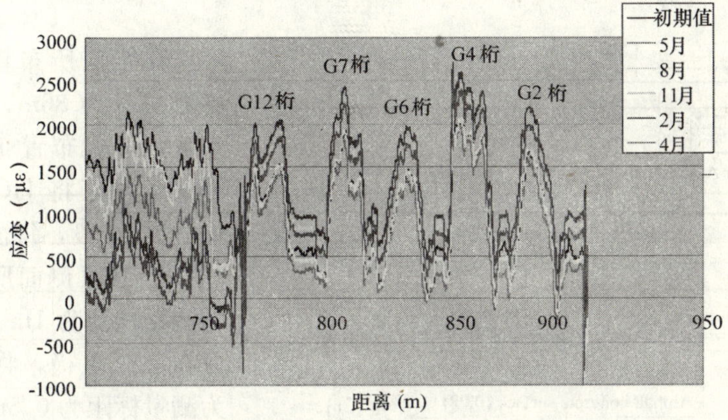

图 50 各桁的 PBO 表面分布监测结果

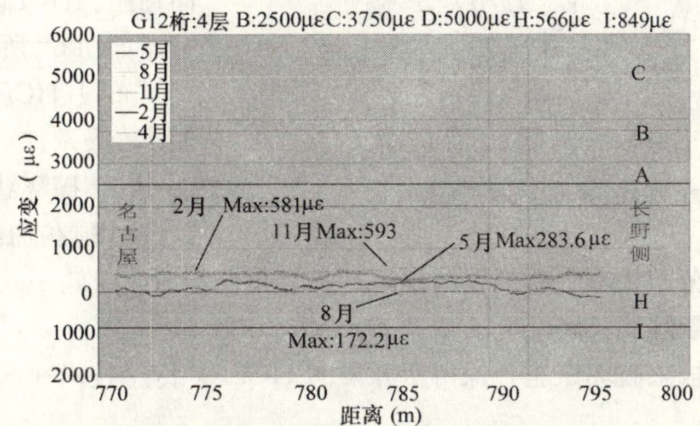

图 51 桁 12 的 PBO 表面分布监测结果

9.5.2 基于 HCFRP 传感技术的实桥极限破坏健康监测[49,50]

新兴塘大桥为沪宁高速公路跨越新兴塘河流的一座大桥，位于锡山市东亭乡新屯村附近。该桥正面照片如图 52 所示。在沪宁高速公路江苏段扩建中，综合考虑有关桥梁的病害、重要性以及航道拓宽等要求，需对部分桥梁进行拆除重建。于是，在拆除前江苏沪宁高速扩建指挥部提出进行极限承载力的实桥试验研究，其主要试验目的之一是通过试验，研究桥梁的破坏规律，探讨如何对结构进行有效地健康监测和预警。

图 52 新兴塘大桥正面照片

9.5.2.1 HCFRP传感系统的布设

作为研究的一部分，将HCFRP分布传感系统布设在该桥的底部和侧面在试验过程中对桥梁的结构健康状况进行监测。混杂碳纤维传感器（HCFRP）主要布置在箱梁的底部和侧面。

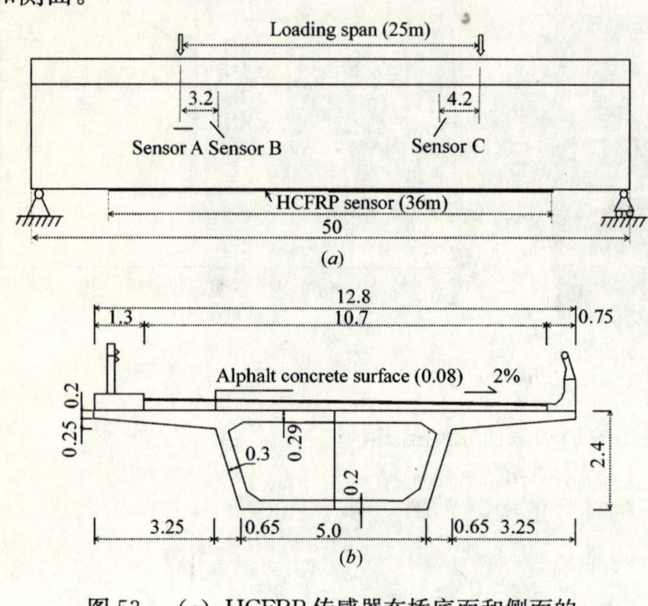

图53 （a）HCFRP传感器在桥底面和侧面的布置；（b）桥的截面尺寸图

在梁底所布置的HCFRP传感器总长为36m，从中跨向两边布置，每边布置18m，为了进行分布监测，将HCFRP传感器平均分为24段，每段的测量标距为1.5m。在梁侧面所布置的传感器的测量标距为1m，每根传感器上分布三个电极，将传感器平均分为测量标距为0.5m的两部分，该类传感器大部分与梁底面呈45°方向粘贴。HCFRP传感器在桥上的分布如图53所示。

布设HCFRP传感器的目的如下：

（1）探测结构出现的裂纹以及监测结构在加载过程中的健康状况（其中包括应变分布状况、新裂纹的出现和扩展情况等）；

（2）监测已有裂纹的扩展变化。

（3）通过实桥极限破坏试验，探讨所开发的HCFRP传感技术在实际土木工程结构中应用的可行性。

9.5.2.2 两类传感器的传感性能试验室标定

HCFRP传感器制成后，在试验室进行了部分单向拉伸试验以检验传感器的测量性能并进行初步标定。试验结果表明，电阻变化率与应变存在着很好的线性关系，电阻变化率—应变曲线可以用直线来拟合，且线性相关系数均大于0.994。在碳纤维出现宏观断裂之前，混杂碳纤维传感器可以用来测量结构的应变或应力。在碳纤维出现宏观破坏之后，利用电阻的突然跳跃来监测混凝土结构的不同破坏阶段。具体研究结果参照参考文献[49]。

9.5.2.3 加载系统及结构的力学性能

综合考虑各种因素，采用钢筋和水袋结合半中跨加载方式，如图54所示。采用力—位移双控方式进行加载，即在试验初期，采用力控制分级加载方式进行加载，接近极限状态（转为水袋控制加载）后，通过位移反馈来控制加载，并制定了严格的加载程序。

加载过程中荷载-位移曲线如图55所示。在“A”点，混凝土梁上开始出现新裂缝；在“B”点，钢筋开始出现屈服；“C”和“D”点分别定义为“屈服点”和“极限点”。根据不同阶段的力学特性，荷载—位移曲线可以分为几个区域。OA对应于弹性区域，该区域内结构的变形与荷载成线性关系。AB对应塑性发展阶段，在该阶段混凝土裂纹逐渐出现，结构的刚度逐渐下降。BC 对应钢筋的屈服和塑性铰的出现。当结构进入结构屈服阶段 CD 后，尽管承载能力不会明显增加，但结构的变形急剧增加。过了点“D”，结构则进入最终破坏阶段 DE。

图 54　桥梁的加载方式

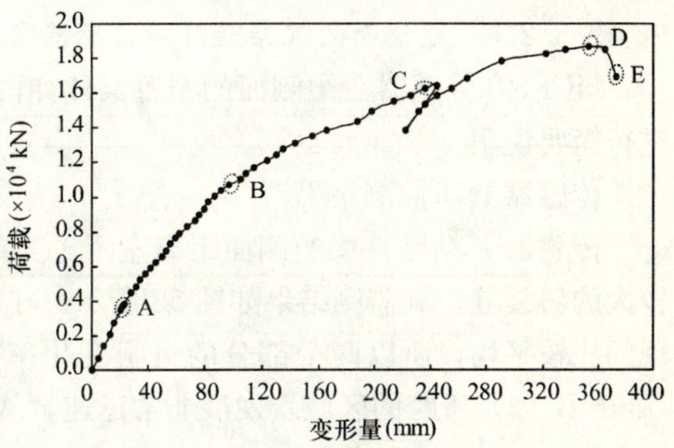

图 55　加载过程中 PC 桥的力学特性

9.5.2.4　梁底的分布监测结果

以梁左边所布置的 12 段混杂碳纤维传感器为例，对测量结果进行简要说明。从跨中向左边桥墩侧，混杂碳纤维传感器段的标号依次记为 01，02，03，04，05，06，07，08，09，10，11 和 12。图 56 为梁底各传感器所测量的结果，由于 01 段在测量中有几个数据有问题，暂时没有把该段的电阻变化率—载荷图表示出来。

实验结果表明，对于梁底部分，靠近桥墩的地方电阻变化率非常小，说明在加载过程中，这一部分的应力/应变增加不明显。而在往中跨的方向上，传感器的电阻变化逐渐变得明显，说明在靠近中跨的方向上，应力/应变随载荷的增加比较明显。而且可以发现 02、03 和 04 的测量结果大致相同，说明在这些传感器所布置的位置即从中跨向左 6m 左右的区域内应变状态大致相同。在加载到 550t 时，梁的底部开始出现一些细小的裂纹，如图 57 所示，观测结果表明裂纹主要集中在 02，03 和 04 段所在的位置，而靠近桥墩部分底部没有出现裂纹。裂纹的出现造成这些段在电阻变化率上一些突然的增加，如图 56 所示。但当载荷加载到 1061t 的时候，06 和 07 段的电阻出现突然增加，此现象表明由于载荷的增加导致混杂碳纤维传感器中的高强度纤维出现了部分断裂。该现象表明，桥梁结构即将

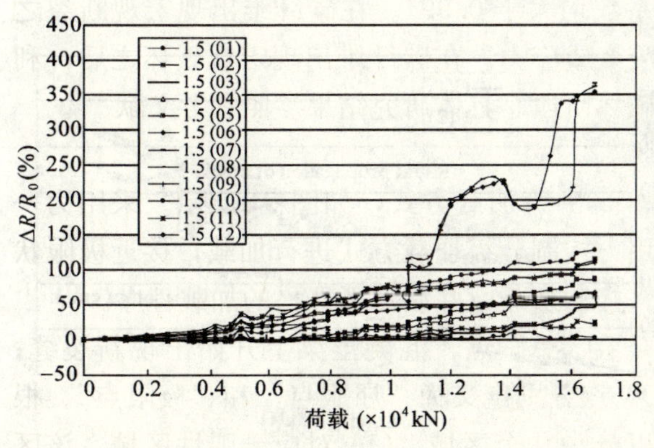

图 56　梁底部 HCERP 传感器 $\Delta R/R_0$ 与载荷的关系

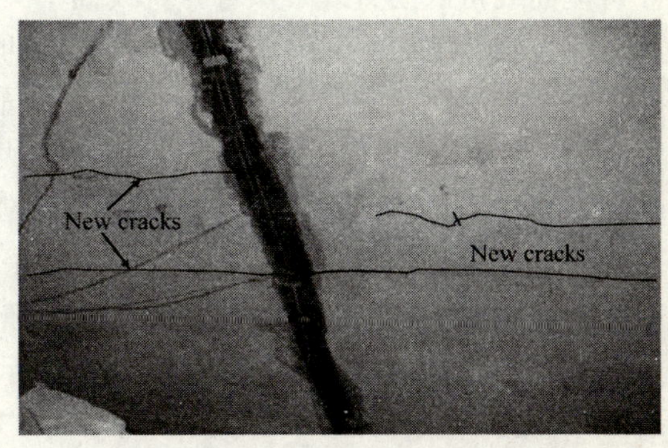

图 57　1.5（10）附近裂纹出现和扩张情况

进入到比较危险状态，内部的钢筋已经开始出现屈服并出现部分断裂。观察结果表明，当载荷达到 1061t 后，06 和 07 段所对应的侧面裂纹出现比较大的裂纹并且发展速率比较快。

9.5.2.5 梁侧面的混杂碳纤维传感器所监测的结果

HCFRP 传感器在梁侧面的布置具体如图 53 所示，现仅以 Sensor B 和 C 的监测结构进行简要说明。

传感器 B 所监测结果

传感器 B 粘贴在梁的侧面距离左测 1/4 跨右边 3.4m 处，试验前在此处已经出现了比较大的斜裂缝，其监测结果如图 58 所示。在 1470t 之前，裂纹在 B（1）和 B（2）部分出现的比较平均，所以两个部分的电阻变化率比较相似。当载荷达到 1470t 时原有的裂纹（处于 B（2）所跨的区域）发展非常迅速，从而导致 B（2）部分被拉断，从这些结果可以推断出，在 1470t 时 B（2）区的局部应变已经达到了 15000$\mu\varepsilon$ 以上，因为该混杂碳纤维传感器中的高强度的碳纤维的断裂应变为 15000$\mu\varepsilon$。在 B（2）出现断裂后，B（1）部分的电阻变化率出现了减小现象。可能是由于 B（2）处的裂纹的迅速发展导致了 B（1）处的裂纹出现了微闭合。B 传感器处的裂纹发展情况如图 59 所示。

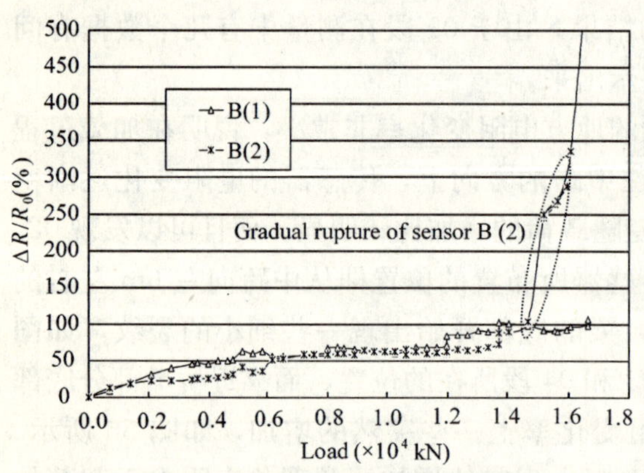

图 58 传感器 B 的监测结果

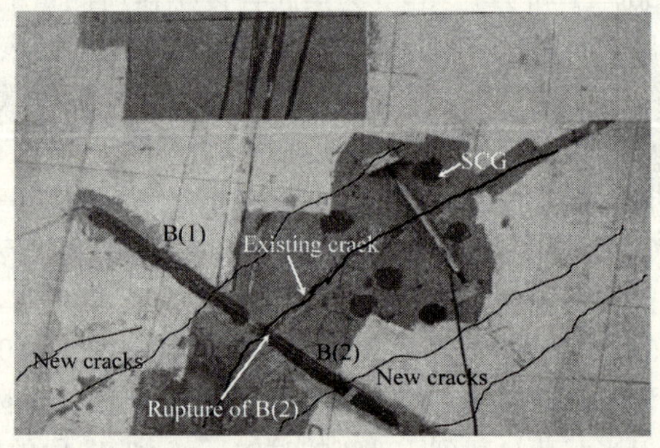

图 59 传感器 B 处的裂纹的出现及发展情况

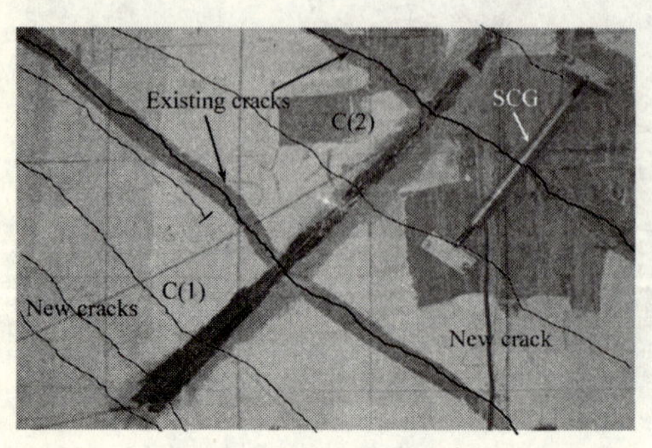

图 60 传感器 C 处的裂纹的出现及发展情况

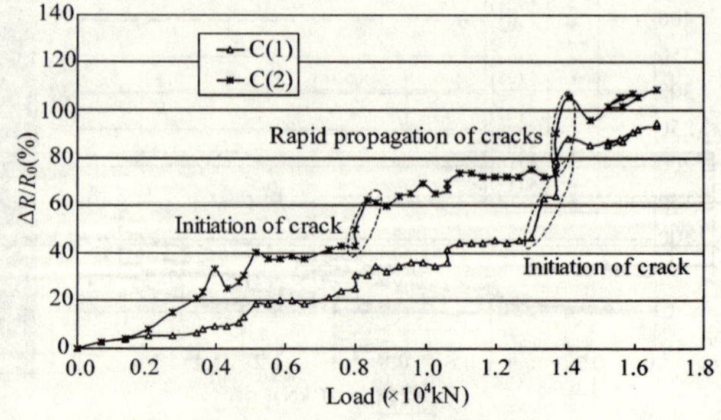

图 61 传感器 C 的监测结果

传感器 C 所监测结果

传感器 C 在右 1/4 跨左侧 4.2m 处与底面大致成 45°布置，如图 53 所示。C(2)跨过一条原有的裂纹，如图 60 所示。传感器 C 的电阻变化率和载荷的关系如图 61 所示，由于 C(2)跨过一条原有裂纹且该裂纹发展比较快，所以 C(2)部分的电阻变化率比 C(1)部分大些。在载荷达到 1200t 之前，电阻随载荷大致呈线性增长。在 1371t 处电阻出现突然增加，可能是由于裂纹的出现或迅速增加所导致。

9.5.2.6　裂纹扩展宽度监测

HCFRP 传感器所监测的裂缝宽度可用下述公式得到

$$l=\frac{\mathrm{d}R/R}{K}L \tag{62}$$

其中，l 为所要监测的裂纹宽度，K 为 HCFRP 传感器的灵敏度系数，L 为 HCFRP 传感器的长度。为了验证 HCFRP 传感器在监测裂缝方面的正确性，在 Sensor B 及 C 附近同时安装了正弦裂缝计(SCGs)，如图 59 和图 60 所示。从这些图可以发现，随着荷载的增加，裂缝的宽度总体呈逐渐增加的趋势，但也有部分加载点呈减小的趋势。这是由于在某些加载阶段，随着荷载的增加一些局部裂缝会出现闭合现象。图 62 和图 63 表明，HCFRP 传感器和正弦式传感器在裂缝监测方面性能相似，都可以用于对裂缝进行有效监测。

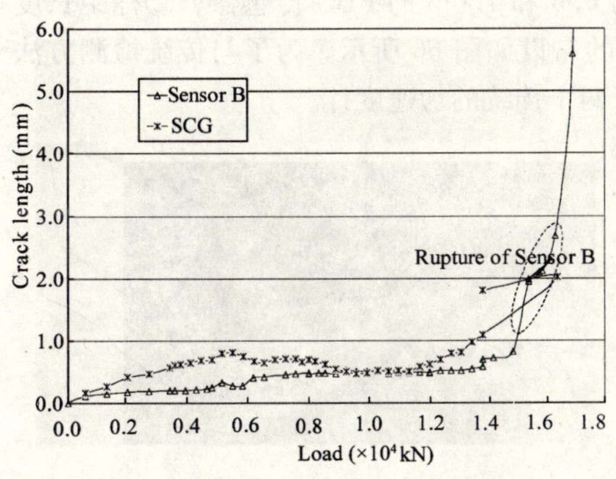

图 62　Sensor B 和 SCGs 在裂缝宽度测量方面的比较

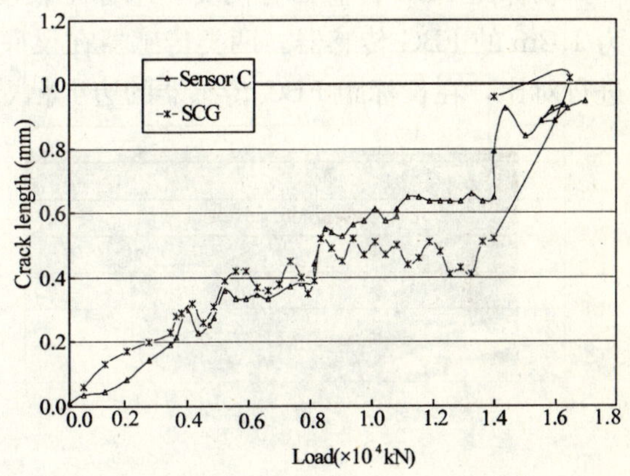

图 63　Sensor C 和 SCGs 在裂缝宽度测量方面的比较

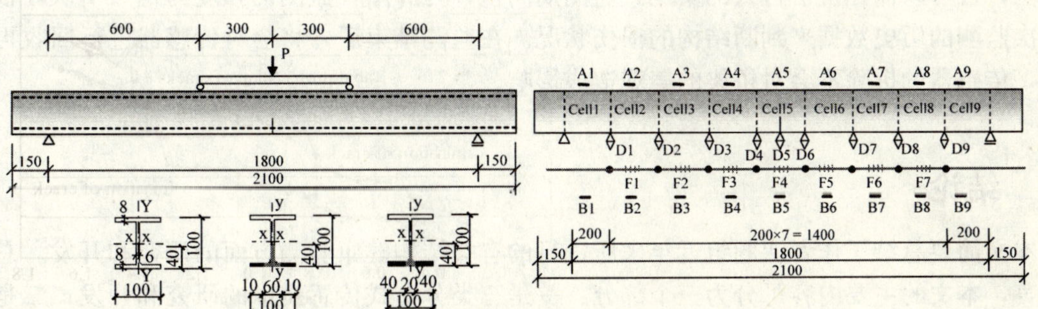

图 64　实验梁及传感器布置

9.5.3 基于长标距 FBG 传感技术的实际结构的健康监测

课题组将所开发的长标距 FBG 动静态传感技术在日本一实际桥梁上进行应用研究，监测大桥的动静态响应，并与传统的方法进行对比。目前，该传感系统正在安装。该桥为日本茨城县横跨涸沼川的川根大桥，如图 65 所示，建于 1963 年 10 月共有 6 跨，每跨的跨度为 20m。长标距 FBG 传感器布置在边跨处，布置长度为 10m，如图 65 所示。

图 65　位于日本茨城县的川根大桥

长标距 FBG 传感器分两类：①标距长度为 0.5 和 1.0m 的 FBG 传感器；②标距长度为 1.2m 的 FBG 传感器。两类传感器在梁底部的布设如图 66 所示。为了与传统检测方法进行对比，在长标距 FBG 传感器的边上粘贴了两个传统的加速度计。

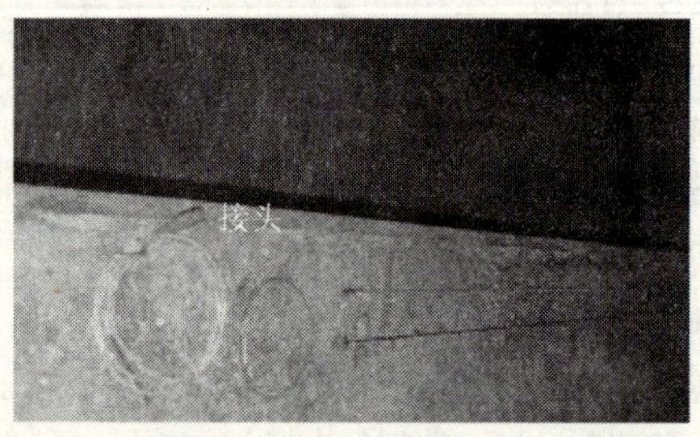

图 66　各传感器在桥上的布置

由于光纤传感器比较小巧，在桥上的粘贴比较牢固且不会对桥本身和正常运营带来任何影响，且可以保证桥下行人的安全。监测期间预计为 2 年，监测的频度为每月 1 次，根据每次监测的历史数据来判断结构的损伤状况。在监测结束后，将会对传感器进行回收再利用，传感器的拆除不会对桥梁的美观造成影响。

9.6 结论

本文简要总结了作者课题组近年来所进行的有关结构健康监测方面的研究和开发。总体而言，本文的主要内容可分为三个部分：一是三类分布式传感技术的研究和开发；二是基于分布式传感监测的损伤识别理论和解析·评价方法的研究；三是分布传感技术和损伤识别理论及解析·评价方法在实际工程结构上的应用和探讨。由于受文章篇幅的限制，结构健康监测系统设计体系没有作为本文的重点主要进行介绍和探讨。

基于布里渊散射机理的光传感技术可以对结构进行全分布式静态监测，提高其应变测量精度和距离分辨率是推广该分布传感技术的关键；长标距FBG传感技术的传感精度高，动态性能好，同时可以进行动静态监测，通过多个传感器的串联可实现准分布式监测；HCFRP分布传感技术的成本低廉、设备简单，监测精度较高，主要用于静态监测。

在研究过程中提出了多个基于分布传感监测的结构损伤识别理论和分析评价方法，大量试验研究证明这些理论和方法可以降低环境噪声的影响，对结构的健康状况进行高精度的分析和评价，并能够对结构损伤进行精确的定位和量化监测。

参考文献

[1] 吴智深等。混凝土结构健康监测技术(混凝土技术系列丛书)，日本土木工程学会，2007年。(日文)

[2] Wu Z S, Xu B and Harada T. Review on structural health monitoring for infrastructures. Journal of Applied Mechanics JSCE, 2003, 6: 1043－54.

[3] Ou JP. Some recent advances of intelligent health monitoring systems for civil infrastructures in mainland China. Proc. Of the First Inter. Conference on Structural Health Monitoring and Intelligent Infrastructure, 13-15 Nov., 2003, Tokyo, Japan, pp. 131-144.

[4] Procs. The 1st, 2nd and 3rd Inter. Conferences on Structural health Monitoring of Intelligent Infrastructure, 2003, Tokyo, Japan, 2005, Shenzhen, China and 2007, Vancouver, Canada.

[5] 李惠，欧进萍。斜拉桥结构健康监测系统的设计与实现，土木工程学报，2006, 39(4): 39-53.

[6] *Proc. of Inter. Workshop on Structural Health Monitoring*. Editor: F.K Chang. (ed.). 1997, 1999, 2001, 2003, 2005. Stanford, USA.

[7] Hao Zhang, Zhishen Wu. Performance evaluation of BOTDR based distributed optic fiber sensors for crack monitoring. Structural health monitoring, 2008, 7: 143-156.

[8] Wu ZS, Iwashita K, Xue S and Zhang H. Experimental Study on Crack Monitoring of PC Structures with Pulse-prepump BOTDA-based Distributed Fiber Optic Sensors. Journal of concrete research and technology, Japan Concrete Institute, 2007.5, 18(2): 49-57. (In Japanese)

[9] Wu ZS and Zhang H. A standard test method for evaluating crack monitoring performance of distributed fiber optic sensors. Proc. of The 3rd Inter. Conference on SHM of Intelligent Infrastructure, 2007, Vancouver, Canada.

[10] Zhang H, Wu ZS. Performance evaluation of BOTDA based distributed optic fiber sensors for crack monitoring. Proceedings of the Japan concrete institute, Technical paper, 2007, pp739-744.

[11] Zhang H, Wu ZS. Performance evaluation of PPP-BOTDA based distributed optic fiber sensors. Structure and infrastructure engineering, 2008. (In press)

[12] Wu ZS, Xu B and Takahashi T, Chen H. Performance of a BOTDR optical fibre sensing technique for crack detection in concrete structures. Journal of Structure and Infrastructure Engineering, 2006, pp. 1-12.

[13] Wu ZS Zhang H and Yang CQ. Development and performance evaluation of no-slippage optical fiber

as Brillouin scattering based distributed sensors. Structural Health Monitoring.（Submitted for possible publication）

［14］ Wu ZS，Xu B，Hayashi K，Machida A. Distributed fiber optic sensing for a full-scale PC girder strengthened with prestressed PBO sheets. Engineering Structures，2006，28(7)：1049-1059.

［15］ Wu ZS. Recent developments in fiber optic based distributed monitoring technologies，The Fourth China-Japan-US Symposium on Structural Control and Monitoring，16-17，October，2006，Hangzhou，China.（Keynote paper）

［16］ 呉智深，許斌，原田隆郎．都市インフラに関する構造ヘルスモニタリングの現状と展望－展望論文．応用力学論文集，日本土木学会，2003，6：1043-1054.

［17］ Wu ZS and Li S. Structural damage detection based on smart and distributed sensing technologies. Proc. of the 2nd International Conference on Structural Health Monitoring of Intelligent Infrastructure（Keynote），Shenzhen，China，2005：107-120.

［18］ Li S and Wu AS. Development of Distributed Long-gage Fiber Optic Sensing System for Structural Health Monitoring. Structural Health Monitoring［J］，2007，6：133-143.

［19］ Li S，Wu ZS and Takumi T. A health monitoring strategy for RC flexural structures based on distributed long-gage fiber optic sensors. Journal of Applied Mechanics［J］，2007，10：983-994.

［20］ Li S，Wu ZS and Zhou LL. Health monitoring of flexural steel structures based on distributed fiber optic sensors. Structure and Infrastructure Engineering［J］. 2008（In press）

［21］ Li SZ and Wu ZS. Characterization of long-gage fiber optic sensors for structural identification. Proc. of SPIE's Smart Structures & Materials and Nondestructive Evaluation for Health Monitoring & Diagnostics Symposium，2005，USA.

［22］ Li S and Wu ZS. Structural damage detection based on smart and distributed sensing technologies". Proc. of 2st Inter. Conference on Structural Health Monitoring and Intelligent Infrastructure，2005，16-18，Nov.，Shenzhen，P. R. China.

［23］ Wu，Z. S. and Yang，C. Q. 2004. "Structural sensing and diagnosis with hybrid carbon fibers". 2nd Inter. Workshop on SHM of Innovative Civil Engineering Structures，September，Manitoba，Canada 3-22.（Keynote paper）

［24］ Yang CQ，Wu ZS. Theoretical/experimental Investigation on electrical resistance change of continuous carbon tows. Proc. Of 1st Inter. Conference on SHM and Intelligent Infrastructure，Tokyo，Japan，pp. 1295-1304，Nov. 13-15，2003.

［25］ Yang CQ，Wu ZS. Distributed sensing of RC beams with HCFRP sensors. Smart Structures and Materials 2005：Smart Structures and Integrated Systems，Proceeding of SPIE，2005，5765：376-385.

［26］ Yang CQ，Wu ZS. Corrosion monitoring of PC beams with a novel type of HCFRP sensors. The 2nd Inter. Conference on SHM of Intelligent Infrastructure，Nov. 16-18，2005，pp. 563-568，Shenzhen，P. R. of China.

［27］ Yang CQ，Wu ZS. Self-diagnosis of hybrid CFRP sheets-strengthened structures. Smart Materials and Structures，2005，14：39-51.

［28］ Yang CQ，Wu ZS. In-situ Corrosion Monitoring of PC Structure with Hybrid Carbon Fiber Reinforced Polymer（HCFRP）Sensors. Smart Materials and Structures，2007，16：1050-1060.

［29］ Wu ZS，Yang CQ. Broad-based Structural Sensing with Hybrid Carbon Fibers. Smart Structures and Systems，2007，3(2)：133-146.

［30］ Yang CQ，Wu ZS. Self-structural health monitoring of RC structures with HCFRP sensors. Journal of Intelligent Material Systems and Structures，2006，17(10)：895-906.

[31] Yang CQ，Wu ZS，Takahashi T. Self-Diagnosis of Hybrid CFRP Rods and As-Strengthened Concrete Beams. Journal of Intelligent Material Systems and Structures，2006，17(7)：609-618.

[32] Yang CQ，Wu ZS. Distributed sensing of RC beams with HCFRP sensors. SPIE Proceeding：Smart Structures and Materials 2005：Smart Structures and Integrated Systems ，2005，5765：376-385.

[33] Wu ZS，Yang CQ. Electrical and mechanical characterization of hybrid CFRP sheets. Journal of Composite Materials，2006，40(3)：227-244.

[34] Li S，Wu ZS. A model-free method for damage locating and quantifying in beam-like structure based on dynamic distributed strain measurements. Computer-Aided Civil and Infrastructure Engineering，2008，23：404-413.

[35] Li S，Wu ZS. A non-baseline algorithm for damage locating in flexural structures using dynamic distributed macro-strain responses. Earthquake Engineering and Structural Dynamics [J]. 2007，36：1109-1125.

[35] Wu ZS，Li S. Two-level damage detection strategy based on modal parameters from distributed dynamic macro-strain measurements. Journal of Intelligent Material Systems and Structures [J]，2007，18(7)：667-676.

[37] Doebling SW，Farrar CR. Computation of structural flexibility for bridge health monitoring using ambient modal data. Proc. Of 11th ASCE Engineering Mechanics Conference，1996，pp. 1114-1117.

[38] Li S，Wu ZS . Modal analysis on macro-strain measurements from distributed long-gage fiber optic sensors. Journal of Intelligent Material Systems and Structures [J]. 2008. (In press)

[39] Wu ZS and Bin Xu. Neural networks for localized and decentralized identifications of large-scale structures. Proc. of Neural Networks and Soft Computing 2005 (Keynote)，Cracow，Poland.

[40] Xu B，Li S，Wu ZS and Yokoyama K. Performance of neural networks based damage identification strategy for noise polluted responses. International symposium on Network and Center-based Research for Smart Structures Technologies and Earthquake Engineering (SE'04)，Osaka，Japan，2004：85-90.

[41] Xu B，Wu ZS，Yokoyama K et al. A soft post-earthquake damage identification methodology using vibration time series，Smart Materials and Structures，14(4)：S116-24.

[42] Xu B，Chen G，Wu ZS. Parametric Identification for a Truss Structure Using Axial Strain，Computer-Aided Civil and Infrastructure Engineering. 2005，22(3)：210-222.

[43] Wu Z S，Xu B，Yokoyama K. Decentralized parametric damage based on neural networks. Computer-Aided Civil and Infrastructure Engineering，2002，17：175-84.

[44] Xu B. Time domain substructural post-earthquake damage detection methodology with neural networks，Lecture Notes of Computer Science，2005，3611：515-24.

[45] Xu B，Wu ZS，Chen G，Yokoyama K. Direct identification of structural parameters from dynamic responses with neural networks. Engineering Applications of Artificial Intelligence，2004，17(8)：931-43.

[46] 傅志方，华宏星．模态分析理论与应用 [M]. 上海：上海交通大学出版社，2002. 6.

[47] Xu ZD，Wu ZS. Energy Damage Detection Strategy Based on Acceleration Responses for Long-span Bridge Structures [J]. Engineering Structures，2007，29(4)：609-617.

[48] 吴智深，杨才千，李素贞。"结构健康监测中的分布式传感技术"，2005 年全国桥梁结构健康监测研讨会，9 月 21-22 日，中国南京.（特邀报告）

[49] Yang CQ, Wu ZS, Zhang YF. Structural health monitoring of an existing PC box girder bridge with distributed HCFRP sensors in a destructive test, Smart Mater. Struct., 2008, 17: 035032.

[50] 谢家全，吴赞平，华斌 等。沪宁扩建桥梁极限承载能力实桥试验研究。现代交通技术，2006，5: 77-84.

第 10 章 Chapter 10

建筑结构强震作用下的弹塑性分析*

李云贵　聂 祺

（中国建筑科学研究院，北京 10013）

摘　要：按照我国现行抗震规范的三水准设计思想，强震作用下建筑结构的弹塑性变形分析是设计的一个重要环节。而目前对建筑结构非线性分析方法研究和软件开发远不及线弹性。本文重点讨论了影响建筑结构强震作用下弹塑性分析的几个关键因素，包括结构分析模型的简化、材料的本构关系、地震动输入和非线性方程的求解等，并分析了相关问题的理论研究与工程应用现状，以及目前工程应用中存在的问题。

关键词：结构分析模型；弹塑性单元；本构关系；地震动输入；非线性动力方程

The Elasto-Plastic Analysis for Building Structures Under Unusual Earthquake

Y. G. Li　Q. Nie

(China Academy of Building Research，Beijing，100013，China)

Abstract：According to the design concept of three level in China' s current code for seismic design of buildings，the elasto-plastic deformation analysis for the building structures under the strong earthquake is an important part in structural design. But studies on the methods and software for elasto-plastic analysis of building structures is far from that on elastic analysis. Several key factors of elasto-plastic analysis under strong earthquake is discussed，including the simplification of structure analysis model，the constitutive relations of material，the input of seismic excitation and nonlinear equation solutions and so on，the state of the art of theoretical studies and its application and the existing problems in practice are analyzed，too.

Key words：Structure analysis model；the elasto-plastic element；constitutive relations；

---

* 基金项目：“十一五”国家科技支撑项目：现代建筑设计与施工关键技术研究（2006BAJ01B）

第一作者：李云贵（1962-），男，博士，教授，主要从事结构分析理论与应用软件研究，E-mail：liyungui@china.com

第二作者：聂祺（1977-），男，博士生，主要从事结构分析理论与应用软件研究，E-mail：nieqi@whut.edu.cn

seismic excitation input; elasto-plastic dynamic equation.

10.1 概述

我国幅员辽阔，约有 80%的大中城市处于地震区，突发地震多，城市人口稠密，而且处于基本建设的高潮期，建筑物集中，一旦发生地震，将导致建筑物的大量破坏或倒塌，造成严重的地震灾害。特别是 20 世纪 90 年代以来，环太平洋区域发生的几次强烈地震造成了巨大经济损失和灾难，也促使地震工程研究人员对抗震设计方法进行反思，基于性能的设计方法越来越受到重视，采用弹塑性分析方法对结构的抗震性能进行分析与评估成了人们研究的热点。

在建筑结构设计中，罕遇地震作用下的非线性变形分析是一项重要工作。《建筑抗震设计规范》[1]第 3.6.2 条规定：不规则且具有明显薄弱部位可能导致地震时严重破坏的建筑结构，应按本规范有关规定进行罕遇地震作用下的弹塑性变形分析。对各类具体工程，《建筑抗震设计规范》第 5.5.2 条给出了具体规定：结构在罕遇地震作用下薄弱层的弹塑性变形验算，应符合下列要求：

(1) 下列结构应进行弹塑性变形验算：8 度Ⅲ、Ⅳ类场地和 9 度时，高大的单层钢筋混凝土柱厂房的横向排架；7～9 度时楼层屈服强度系数小于 0.5 的钢筋混凝土框架结构；高度大于 150m 的钢结构；甲类建筑和 9 度时乙类建筑中的钢筋混凝土结构和钢结构；采用隔震和消能减震设计的结构。

(2) 下列结构宜进行弹塑性变形验算：表 5.1.2-1 所列高度范围且属于表 3.4.2-2 所列竖向不规则类型的高层建筑结构；7 度Ⅲ、Ⅳ类场地和 8 度时乙类建筑中的钢筋混凝土结构和钢结构；板柱—抗震墙结构和底部框架砖房；高度不大于 150m 的高层钢结构。

结构分析模型简化、材料的动态性能、地震动输入和结构的地震响应是决定建筑结构强震作用下弹塑性分析的关键因素。受理论研究和计算机软硬件条件的限制，过去在对建筑结构进行罕遇地震作用下的弹塑性变形分析时，一般都要做许多简化，《抗震规范》[1]第 5.5.3 条为各种简化措施规定了应用条件：结构在罕遇地震作用下薄弱层（部位）弹塑性变形计算，可采用下列方法：

(1) 不超过 12 层且层刚度无突变的钢筋混凝土框架结构、单层钢筋混凝土柱厂房可采用本节第 5.5.4 条的简化计算法；

(2) 除 1 款以外的建筑结构，可采用静力弹塑性分析方法或弹塑性时程分析等；规则结构可采用弯剪层模型或平面杆系模型，属于本规范第 3.4 节规定的不规则结构应采用空间结构模型。

对大中型钢筋混凝土结构进行强震作用下弹塑性分析计算，一般需要求解自由度达几十万甚至上百万的方程组，特别是在材料开始损伤的非线性变形阶段，混凝土的开裂和裂缝扩展对结构性能影响十分敏感，在数值分析时，单元尺寸和计算步长都不能过大，而且在每个时间步还要进行迭求解，才能够得到反映结构实际情况的分析结果，因此必须进行高性能并行计算，才能使之满足工程设计对分析精度和效率的要求。进入新世纪以来，计算机硬件资源的快速发展，特别是 PC 机性能的不断提升和网格技术的普及。使得结构的精细化设计成为结构工程领域的一种趋势，结构设计中对能够更加精确模拟建筑结构真实受力状态的分析软件的需求越来越迫切。在建筑结构地震设计中，对强震作用下的结构性能进行更加精确的分析已经成为可能。

10.2 几种非线性分析方法

10.2.1 12层以下钢筋混凝土框架结构、单层钢筋混凝土柱厂房的简化分析方法

《建筑抗震设计规范》[1]第5.5.4条规定：结构薄弱层（部位）弹塑性层间位移的简化计算，宜符合下列要求：

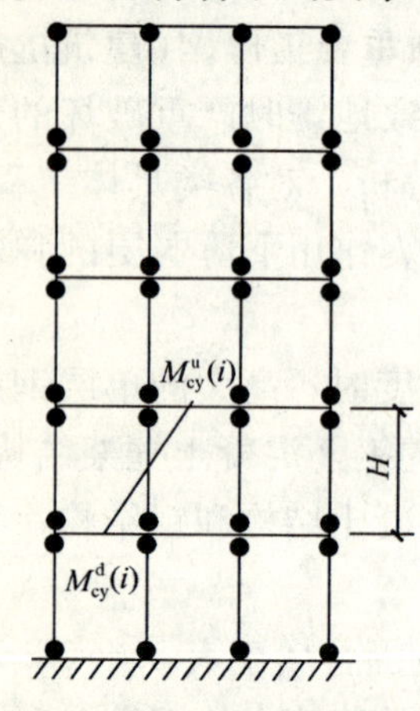

图1 拟弱柱法计算简图

（1）结构薄弱层（部位）的位置可按下列情况确定：楼层屈服强度系数沿高度分布均匀的结构，可取底层；楼层屈服强度系数沿高度分布不均匀的结构，可取该系数最小的楼层（部位）和相对较小的楼层，一般不超过2～3处；单层厂房，可取上柱。

（2）弹塑性层间位移可按下列公式计算：

$$\Delta\mu_P = \eta_P \Delta\mu_e \quad \text{或} \quad \Delta\mu_P = \mu\Delta\mu_y = \frac{\eta_P}{\xi_y}\Delta\mu_y \tag{1}$$

式中符号的含义见规范。目前，在计算楼层抗剪承载力V_y时，一般都采用“拟弱柱法”，即假定在某一楼层所有柱的柱顶和柱底都同时形成塑性铰。显然，这样的假定与抗震设计原则“强柱弱梁”不符，计算结果的精度有限。

10.2.2 静力非线性分析方法

静力非线性分析（Push-Over Analysis）方法作对建筑结构抗震性能评价方法之一，具有概念清楚、实施相对简单，能够使设计人员在某种程度上了解结构在强震作用下的性能等特点，在国外得到广泛应用。如美国联邦紧急事务管理署的《房屋抗震加固指南》（FEMA273/274/368/369/450）[3]和应用技术委员会的《混凝土建筑抗震评估和维修》（ATC-40）[4]等，都给出了采用静力非线性分析方法进行基于位移的抗震设计或评定的具体实施办法。该方法是对结构施加沿高度呈一定分布的单调递增荷载来将结构推至某一目标位移或形成倒塌机制，从而了解结构的薄弱环节和其它弹塑性状态反应。影响静力非线性分析的关键因素包括水平荷载、增量加载计算、能力谱曲线和需求谱曲线计算。

（1）水平荷载。水平荷载是用来反映地震在结构上作用的，要求水平荷载既要能够反映出地震作用下个结构层惯性力的分布特性，又应使所求结构的位移能够大体真实地反映出地震作用下结构的位移状况，但因地震作用的复杂性和结构动力特性的多样性，合理确定水平荷载已成为静力非线性分析的一个关键问题。水平荷载的分布对计算结果影响很大，常用的水平加载分布有倒三角荷载、矩形荷载和反应谱荷载三种，其中倒三角形荷载

是被广泛采用的一种荷载形式。

（2）增量加载计算。在结构上先增量施加竖向荷载，然后再增量施加沿建筑高度分布的某种水平荷载，进行非线性分析，计算结构的基底剪力 V_b 和顶点位移 U_t 曲线，即 V_b-U_t 曲线。对于每个增量步中的迭代计算，通常采用完全牛顿-拉弗逊方法（FNR）或修正牛顿-拉弗逊方法（MNR）。这是两个优秀的迭代方法，前者速度较慢但收敛性稳定好，其收敛阶数为2.0，后者速度快但收敛稳定性不如前者。在 V_b-U_t 曲线的极值点附近和下降段，采用通常的荷载增量法很难求解，采用弧长控制方法，可以较好地得到结构反应的全曲线。

（3）能力谱曲线和需求谱曲线。用等效单自由度体系代替原结构，将 V_b-U_t 曲线转换为加速度 S_a 与谱位移 S_d 曲线，即 S_a-S_d 曲线。值得注意的是，这一转换过程仅当结构层数不多，地震反应以第一振型为主时成立。根据弹性单自由度体系在地震作用下的运动方程，可得到加速度 S_a 与谱位移 S_d 的关系式，$S_a=\frac{T^2}{4\pi^2}S_d$。从而得到AD形式的需求谱曲线。需求谱曲线分为弹性和弹塑性两种，一般在典型（阻尼比为5%）弹性需求谱的基础上，通过考虑等效阻尼比或延性比得到折减的弹性需求谱或弹塑性需求谱。

（4）结构的性能点。能力谱和设计地震相应的需求谱的交点即为性能点，性能点对应的位移即为单自由度体系在该地震作用下的谱位移。利用反变换，则可得到原结构在该地震作用下的顶点位移、塑性铰分布、杆端截面曲率、总侧移，以及层间侧移等用于检测结构抗震能力的有用信息。

相对动力时程分析方法而言，静力非线性分析方法的优点是计算量小，效率较高，在获得结构弹塑性状态下的强度和变形需求、探测结构设计的薄弱部位方面是一种有效的方法。但因对结构模型和地震输入的简化较多，理论基础不严密，分析精度有限，适用范围小，仅适用于层数不多、以第一振型为主的结构。在低层规则结构分析中静力非线性分析方法能够对结构的最大反应和最大损伤做出合理的估计；但对于高阶振型影响较大的结构，静力非线性分析方法的准确性较差，在中高层结构中，随着结构高度的增加，此方法分析的结果与时程分析方法结果的误差逐渐增大，同时，对于一些非对称、不规则和扭转效应显著的结构应用方面，静力非线性分析方法的研究较少，仍需要进一步研究。

10.2.3　动力非线性分析方法

动力非线性分析方法是一种直接动力分析方法。此方法是将强震记录下来的加速度—时间曲线划分为很小的时段，然后在每个时段内逐步积分，直接求解结构的动力平衡方程：$[M_t]\cdot\{\ddot{\vec{X}}_t\}+[C_t]\cdot\{\dot{\vec{X}}_t\}+[K_t]\cdot\{\vec{X}_t\}=\{\vec{P}_t\}$，求出结构体系在各个时刻的位移、速度和加速度，进而求出结构的内力。动力非线性分析方法在20世纪60年代以来发展较快，人们越来越重视此方法的应用。相对而言，动力非线性分析方法对结构的简化假定较少，理论基础相对严密，考虑的因素较全面，适用面广，能够计算地震全过程的结构反应，给出结构的开裂和屈服次序，发现应力和塑性变形集中的部位，从而判断结构的屈服机制、薄弱环节很可能的破坏类型，因此被认为是一种可靠的、较高精度的非线性方法。

其主要缺点是计算量大，对计算机硬件资源要求高。随着计算机硬件水平的提高和非线性分析理论的不断完善，该方法将在工程设计中得到越来越广泛的应用。

10.3 弹塑性分析模型

结构分析模型的简化与计算机硬件资源密切相关。随着 PC 机从 AT、XT 到 286、386、486、586，以及现在的 Pentium、Core 等的不断发展，建筑结构弹塑性分析模型经历了从二维平面模型、层模型到三维空间模型的变化。组成建筑结构的主要受力构件包括梁、柱、支撑、剪力墙和楼板。在建筑结构的弹塑性分析中，通常采用一维单元模拟梁、柱和支撑，采用二维单元模拟剪力墙和楼板。三维块体单元因其前后处理复杂，计算工作量过于庞大，目前在工程应用中较少采用，将来随着 PC 机性能的进一步提升，三维块体单元的应用会逐渐增多。这里主要讨论一维和二维弹塑性单元。

10.3.1 一维弹塑性单元

（1）弹塑性梁元模型

常用的弹塑性梁元模型主要有两类：塑性铰模型和纤维束模型。塑性铰模型假定在构件的两端形成塑性铰，而构件的其它部位保持弹性，该模型的单元本构关系大多采用杆件的内力—变形关系，即与杆件截面相关的本构关系，如 SAP2000，ETABS 和 MIDAS 等计算程序均采用此模型。影响塑性铰模型的关键有三点：一是需要预先指定可能发生塑性铰的位置，这就存在一定的主观性；二是需要预先确定 P-M-M 屈服面相关参数，而 P-M-M屈服面与构件的截面形状密切相关；三是塑性铰区长度，已经有许多学者提出了塑性铰区的等效长度计算公式，如 Baker[5]（1956）、Corley[6]（1966）、Priestleyand[7]等，但不同公式的计算结果还是有相当大差异的[8]。

纤维束模型直接从材料的本构关系出发解决弹塑性问题。单元本构关系采用材料逐点的应力-应变关系，在单元长度方向和在截面内逐渐进入塑性，因而轴力和双向弯矩偶合作用的屈服准则和滞回规律能够得到较精确的模拟。纤维束模型与构件的截面形状无关，没有过多的假定和预先的参数计算，是对工程实际情况比较精确的模拟。目前高端的大型通用软件如 ABAQUS、ANSYS 和 NASTRAN 等和国内的 EPDA 软件均有纤维束模型。

与塑性铰模型相比，纤维模型无疑具有更高的精度。将纤维束模型用于钢结构是比较适合的，因钢材为理想弹塑性材料，不存在开裂和压碎问题，纤维束模型可以高效地模拟各种截面形式的钢构件；将纤维束模型应用于钢筋混凝土结构时，因钢筋混凝土构件开裂因素影响，各高斯积分点坐标应随截面开裂的不同而变化，计算量较大，在目前的软件实现中多采用近似处理办法，没有考虑各高斯积分点坐标随截面开裂的不同而变化的因素。此外，在应用纤维束模型时，但如何判定构件各截面是否已经进入塑性铰状态，也是一个值得商榷的问题。

上述两种单元的假定不同，与工程实际相符合程度不一致，这就反映在计算结果的精度上有较大的差异，文献[9]对塑性铰模型与纤维束模型计算的计算精度做了对比计算，结果见表 1。计算结果表明，塑性铰模型与纤维束模型的结果差异还是较大的，就模型的

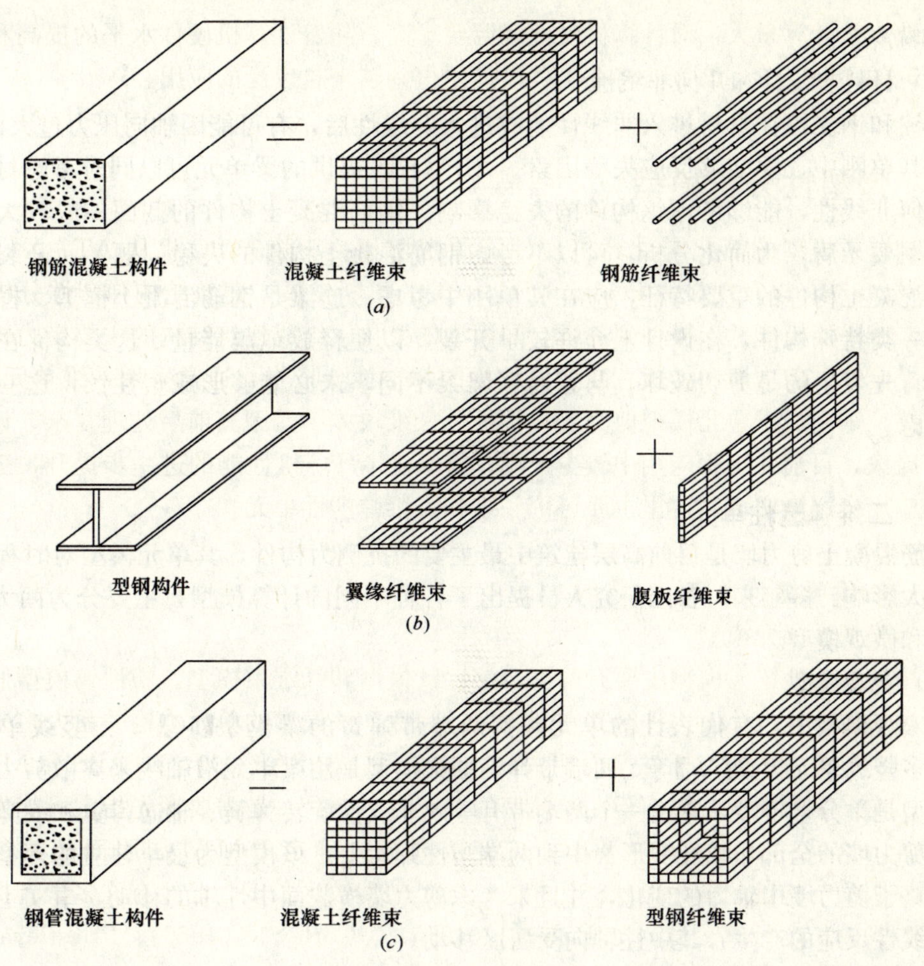

图 2　纤维束模型简图

简化程度而言，纤维束模型结果的可信度更高一些。

塑性铰模型与纤维束模型精度对比计算结果　表 1

项　目	8 度罕遇地震			9 度罕遇地震		
	塑性铰模型	纤维束模型	相对差异	塑性铰模型	纤维束模型	相对差异
层间位移（cm）	7.4	5.3	28.3%	14.63	6.35	56.6%
顶点位移（cm）	19.88	15.92	19.9%	43.75	18.5	57.7%
底部剪力（kN）	656	377	42.5%	727	410	43.6%

（2）弹塑性梁元的分析效率

按照位移模式分类，常用的弹塑性梁元有两类：铁摩辛柯梁元和欧拉梁元。在弹性分析中多采用欧拉梁元，而在非线性分析中，铁摩辛柯梁元的性能更好，但需要将单元细分。梁单元的细分对计算精度有一定影响，也决定计算效率，一般每根梁至少要划分为三个单元。算例表明，采用铁摩辛柯梁元时，选用不同的细分方案对结果影响敏感，但对收敛性影响较敏感。

（3）材料非线性和几何非线性

钢柱和钢支撑在材料进入非线性前或进入非线性后，有可能因轴向压力过大而失稳，所以在其单刚中，要能够反应失稳因素。ABAQUS提供的梁单元可以同时考虑材料非线性和几何非线性，能够考虑钢构件的失稳影响。钢筋混凝土构件的截面一般较大，失稳不是其主要矛盾，为简化分析，可以不考虑钢筋混凝土构件的失稳问题，而开裂和压碎是钢筋混凝土构件的重要特征，应在其单刚中考虑。连梁是钢筋混凝土框剪、框筒等结构中的一类特殊构件，在设计上允许较早开裂，以便释放地震能量，这类构件在地震作用下，首先发生的是剪切破坏，与普通框架梁不同，未必能够形成塑性铰，这一点也应特殊考虑。

10.3.2 二维弹塑性单元

钢筋混凝土剪力墙是目前高层建筑中最主要的抗侧力构件，其单元模型对时程分析结果有较大影响。多年来，各国研究人员提出了各种不同的计算模型，主要分为两大类：宏观模型和微观模型。

（1）宏观模型

在宏观模型中，有代表性的单元包括两端带弹簧的梁模型【10】、三竖线单元模型【11】和多竖线单元模型【12】等。两端带弹簧的梁模型是用梁单元沿轴线来离散剪力墙。梁单元模型是单分量模型，它由一个两端带有等效非线性旋转弹簧、轴向弹簧和线弹性单元组成。剪力墙的全部非线性变形集中到两端塑性弹簧上。该模型为最早建立的模型，其缺点是忽略了剪力墙中轴力的变化，并且未考虑剪力墙横截面中性轴的移动，事实上，随着墙体非线性反应的产生，其中性轴向受压区移动。

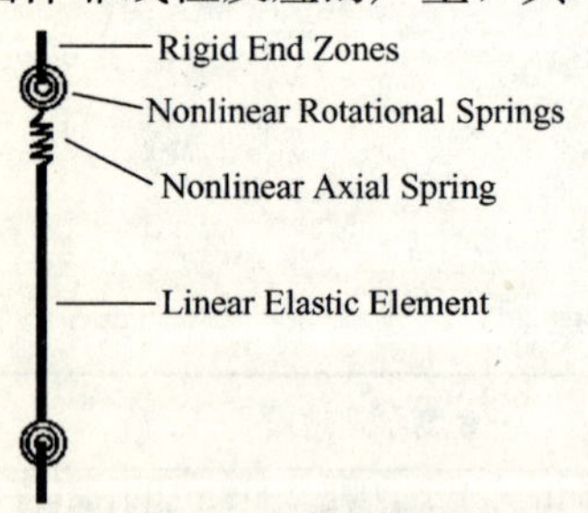

图3 两端带弹簧的梁模型

三竖线单元模型是Kabeyasawa等为解释美日合作研究的七层足尺框架结构伪动力试验结果而提出的宏观模型。该模型由三个垂直杆单元将上下两层楼板连接起来，两个外边杆元代表墙两边柱的作用，中间单元代表墙体的作用，中间杆元由垂直、水平和弯曲弹簧组成。这一模型的优点是可以模拟剪力墙横截面中性轴向压缩端的移动，而且，物理概念清晰。用此模型进行的数值模拟计算与足尺试验的结果吻合得较好。但中间单元的弯曲弹簧刚度的取值困难。

多竖线单元模型是Vulcano等于1988年提出的，目的是为解决三竖线模型的弯曲弹簧取值困难以及和两边柱单元相协调的问题。该模型用几个垂直杆代替剪力墙的弯曲弹簧，同时也代表了剪力墙的轴向刚度，水平弹簧代表剪力墙的剪切刚度。多垂直杆元模型克服了三垂直杆单元中弯曲弹簧和边柱元协调不明显的缺点，并保留了其优点。

宏观模型与在弹性分析中的薄壁杆件模型类似，其优点是自由度少、前后处理简单、计算效率高，但因对剪力墙的简化较多，分析精度有限。在PC机发展的较早阶段，采用宏观模型进行弹塑性分析的较多，但在现在的Pentium和Core机时代，随着PC机计算速度和储存能力的不断增强，宏观模型的优势已经不在，其计算精度已经不能满足工程设计的需要。

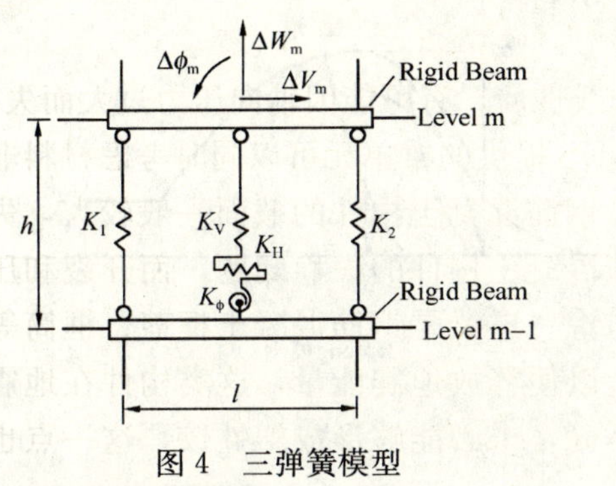

图 4　三弹簧模型

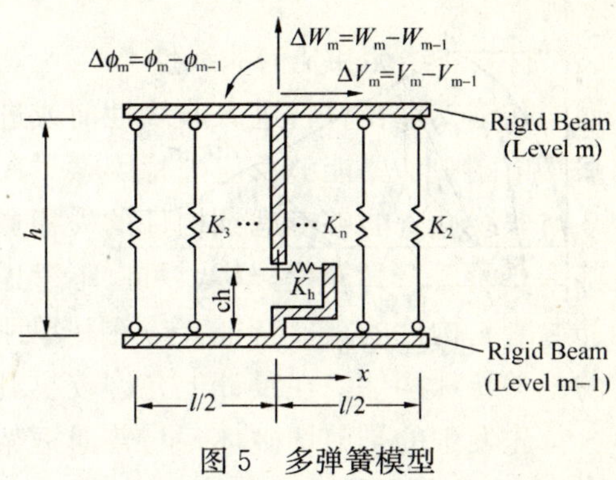

图 5　多弹簧模型

（2）微观模型

微观模型以钢筋混凝土有限元模型为代表，将剪力墙断面细分网格，根据钢筋与混凝土材料的本构关系，直接建立单元的计算模型。该模型力学概念清晰，求解精度较高；但需计算较多的自由度，需要大量的计算时间，目前用于剪力墙结构的微观模型主要有平面应力膜单元和板壳单元。

在平面应力膜单元中，每个节点有两个自由度，不考虑平面外的自由度，可采用三角形单元、矩形单元和四边形等参单元。由于三角形单元是常应变单元，矩形单元不适应曲线边界等复杂的边界形状，四边形等参单元具有较高的计算精度，且其网格的划分不受边界形状的影响，所以目前应用较多。

壳单元是平面应力问题和板弯曲问题的组合，每个节点有六个自由度，三个为平移，三个为转角，单元精度高。显然，板壳单元由于考虑了平面外的自由度，比平面应力膜单元更为精确，但计算量也增加较多。随着计算机硬件资源的快速发展，板壳单元模型的应用已经成为主导趋势。

10.4 材料本构关系

10.4.1 材料的单轴滞回本构模型

混凝土及钢筋的滞回本构模型对结构弹塑性分析的结果有很大的影响。钢筋的滞回本构模型比较简单，而混凝土的滞回本构模型要复杂的多，迄今为止，在众多学者的努力下，已经提出多个混凝土滞回规则，如 Blakeley[13]（1973），Bolong[14]（1980），朱伯龙[15]（1981），过镇海[16]（1981），Mander[17]（1988），Yankelevsky[18,19]（1989），Yassin[20]（1994）等，比较有代表性的如图 6。钢筋的滞回规则主要有 Ramberg-Osgood（1943），Kato[12]（1973），Monegotto-Pinto[4]（1973），Aktan[13]（1974）等。

混凝土滞回规则的主要区别在于：

（1）受压部分骨架曲线采用不同的基于单轴受力建立的本构曲线，或将其折线化；

（2）受拉部分骨架曲线采用不同的基于单轴受力建立的本构曲线，以不同方式考虑拉伸强化效应；

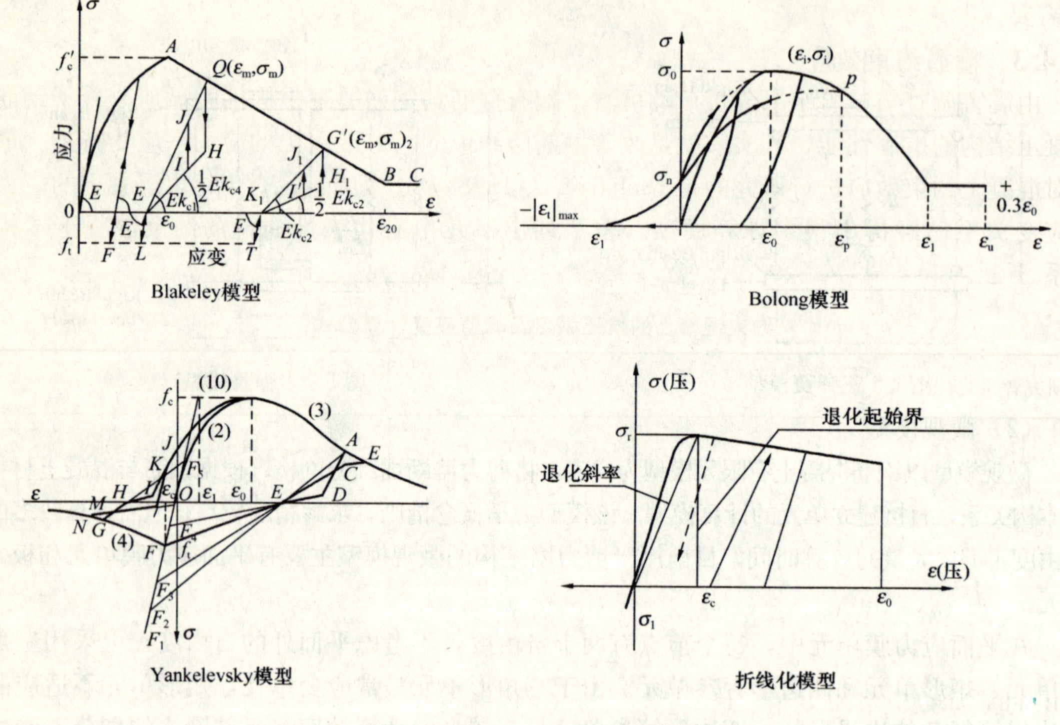

图 6 混凝土滞回本构模型

（3）卸载段和卸载后再加载段采用直线、分段折线或曲线来模拟；

（4）裂缝闭合过程中，不同程度考虑裂面接触效应。

到目前为止，材料滞回模型基本上都是对大量实验数据的拟合及简化，缺乏明确的力学基础，提出具有明确力学基础的材料滞回本构模型还需要有很多工作要做。

10.4.2 材料的二维本构关系

对于高层建筑钢筋混凝土剪力墙非线性分析所采用的二维混凝土本构关系，各国学者也提出了基于不同经典力学的模型，包括非线性弹性类（Darwin and Pecnold[23]，1976；Bathe[24]，1979（Adina 采用模型））、弹塑性类（Chen[25]，1975；Buyukozturk[26]，1977（Marc 采用模型））、弹塑性损伤类（Lublinear and Oliver[27]，1989；Lee and Fenves[28]（1998）（Abaqus 采用模型）），也有些学者专门针对剪力墙非线性分析提出二维非线性弹性和弹塑性本构模型，典型的屈服面强化模型见图 7，其中加拿大学者 Vecchio and Collins[29]（1986）和美国学者 Hsu[30]（1988）分别提出了基于转角裂缝假定的修正斜压场模型和基于定角裂缝的定角—软化桁架模型，该两类模型均能较好的计算混凝土在反复荷载作用下弯曲变形和剪切变形的耦合作用，在计算结构的非线性响应方面具有较好的准确性。

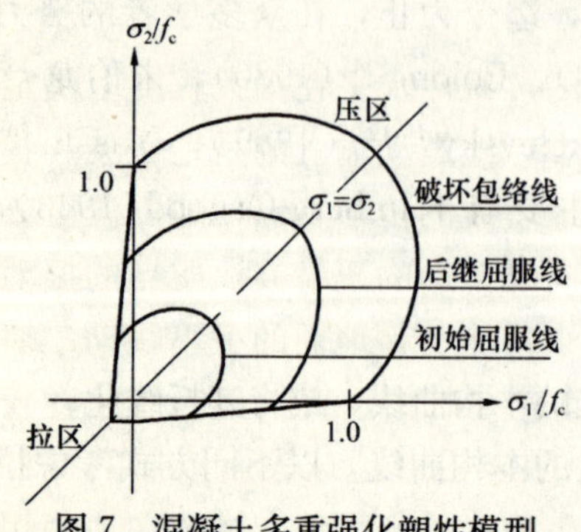

图 7 混凝土多重强化塑性模型

10.4.3 箍筋约束效应

由震害调查与试验结果分析，均可看出配有箍筋的钢筋混凝土构件明显地改善了钢筋混凝土结构的抗震性能，其实质就是改善箍筋所约束的核芯混凝土的应力—应变特征。最早对混凝土约束效应进行研究的是 Richart[23]（1928）。迄今为止，对于约束混凝土的应力—应变关系已经提出很多计算模型。表 2 列出了几个有代表性的箍筋约束混凝土本构关系。

约束混凝土的抗压强度及峰值应变计算公式 表 2

研究者及发表时间	约束抗压强度	约束峰值应变
过镇海[24]及修正[25]-[28]	$f_{cc}=(1+0.5\lambda_v)f_c$（普通） $f_{cc}=f_c(1+1.085\lambda_v\sqrt{1-s/D})$（高强）	$\varepsilon_{cc}=(1+2.5\lambda_v)\varepsilon_c$（普通） $\varepsilon_{cc}=\varepsilon_c(1+1.96\sqrt{\lambda_v(1-s/D)})$（高强）
Park	$f_{cc}=\left(1+\frac{\rho_v f_{yh}}{f'_c}\right)f_c$	同非约束峰值应变
CEB—FIP—90 模型[29]	$f_{cc}=\left(1+1.25+2.5\times\frac{1}{2}\alpha_n\alpha_s\lambda_t f_c\right)f_c$ $(\sigma_2\leqslant 0.05f_c)$ $f_{cc}=\left(1+5\times\frac{1}{2}\alpha_n\alpha_s\lambda_t f_c\right)f_c(\sigma_2\geqslant 0.05f_c)$ σ_2 为箍筋对核心混凝土的等效平均约束应力	$\varepsilon_{cc}=(f_{cc}/f_c)^2\times 2\times 10^{-3}$

10.4.4 材料性能的应变率相关性

自 1917 年 Abram 对混凝土进行压缩试验时发现混凝土抗压强度存在率敏感性后，很多学者开始对混凝土材料进行各种力学性质的动载试验研究。在地震作用下（此时应变率 $10^{-3}/s<\dot{\varepsilon}<10^{-2}/s$），混凝土的主要力学性能参数均发生变化，包括：弹性模量、极限强度等。但是，由于它的复杂性以及数值模型中实现的困难，目前，只是通过经验公式和较多的简化假定来考虑应变率的影响。具体公式可见文献［38，39］。

10.5 地震动输入

地震动输入是工程结构抗震分析的基础，地震动输入的可靠性水平在很大程度上决定了结构抗震安全评价结论的可靠性水平。在动力时程分析中，地震动输入的是地震波。我们都知道，地震是以波的形式从震源向四周传播的，根据地震波的传播特点，地震波可分为体波和面波。体波又可分为纵波和横波。纵波是由震源向外传递的压缩波，质点的振动方向与波的前进方向一致，其传播速度快、周期短、振幅小，能引起地面的上下跳动；横波是由震源向外传递的剪切波，质点的振动方向与波的前进方向相垂直，其传播速度较慢、周期较长、振幅较大，能引起地面水平晃动。面波只限于沿着地球表面传播的一种波，一般认为面波是体波经地下岩层分界面多次反射的次生波，是由纵波与横波在地表相遇后激发产生的混合波，其波速慢、波长大、振幅强，是造成建筑物强烈破坏的主要因素。强烈地震发生时，从震源发出的地震波中，纵波首先到达震中，此时震中区会感到上下颠簸；接着横波传来，震中区会感到前后左右摇晃。在远离震中的地区，纵波、横波以不同的角度与地面接触，又叠加了面波的作用，震动极为复杂。

由于地震是一个突发的随机过程，地震波具有很强的不确定性，研究结果表明，在同一类别建筑场地所测到的地震加速度记录，其峰值、波形、频谱特性和持续时间都可能大不相同，即使在同一地点，来自同一震源的先后两次地震所造成的地面运动也不会完全相同[1]。如何选择地震波是对结构进行时程分析的关键和难题，因为对某一结构来说，地震波一旦确定，其时程分析结果也就唯一确定了。《建筑抗震设计规范》（GB 50011—2001）规定：采用时程分析法时，应按建筑场地类别和设计地震分组选用不少于二组的实际强震记录和一组人工模拟的加速度时程曲线，其平均地震影响系数曲线应与振型分解反应谱法所采用的地震影响系数曲线在统计意义上相符，每条时程曲线计算所得结构底部剪力不应小于振型分解反应谱法计算结果的65%，多条时程曲线计算所得结构底部剪力不应小于振型分解反应谱法计算结果的80%。上述规定原则性很强，给予设计人员较大的灵活性，但可操作性差，给工程设计应用带来相当的困难。找到适合建筑场地条件的地震波并不容易，如何做到“统计意义上相符”，要相符到什么程度，这些都带有一定的主观性。不同人会选出不同的地震波，这是导致不同的人对同一结构时程分析结果相差很大的一个主要原因，从而也就降低了分析结果的参考价值和对设计的指导意义。一般来说，在选择地震波时，应尽可能全面反映建筑场地的地震动主要参数，包括地震动幅值、频谱特征和持续时间。

10.5.1 地震动幅值

地震动强度一般主要由地面运动加速度峰值的大小来反应，所选取的地震波最大峰值加速度应与结构所在地区的设防烈度相对应，当所选择的实际地震记录峰值加速度与结构所在地区的设防烈度相对应的地震加速度峰值不同时，可将实际地震记录的加速度按比例放大或缩小来加以修正。当仅考虑单一方向地震输入时，地震加速度峰值的变换比较简单，但若需要考虑双向或三向地震输入时，需考虑各方向地震加速度峰值的比例。《建筑抗震设计规范》[1]规定：质量和刚度分布明显不对称的结构，应计入双向地震作用下的扭转影响。采用振型分解反映谱法分析时，双向水平地震作用的扭转效应可按下列公式的较大值确定，即：$S_{eh}=\sqrt{S_x^2+(0.85S_y)^2}$或$S_{eh}=\sqrt{S_y^2+(0.85S_x)^2}$，式中$S_x$和$S_y$分别为$X$方向和$Y$方向单向水平地震作用效应。在规范条文说明中解释，当采用三维空间模型等需要双向（二个水平方向）或三向（二个水平方向和一个竖向）的地震波输入时，其加速度最大值通常按1.00（水平1）：0.85（水平2）：0.65（竖向）的比例调整，选用的实际加速度记录，可以是同一组的三个分量，也可以是不同组的记录。但美国统一建筑规范（UBC2003）[40]、美国建筑样板规范（FEMA386）、欧洲建筑抗震规范（EC8）[41]，以及上海市标准《高层建筑钢结构设计暂行规定》[42]，都采用了与《建筑抗震设计规范》不同的方法考虑双向地震作用，规定：当考虑双向地震输入时，在一个方向输入100%地震力的同时在与其垂直的方向输入30%的地震力，取其组合的较大值进行设计，即：$S_{eh}=S_x+0.3S_y$或$S_{eh}=0.3S_x+S_y$。

当采用振型分解反应谱法时，上述两种考虑方法的计算结果接近，《建筑抗震设计规范》的方法符合“在统计意义上相符”的要求，符合根据观测的强震记录确定设计地震反应谱的规律；但采用动力时程分析法时，因弹塑性分析中叠加原理不成立，不能对各分量的作用效应进行平方和开方运算，这时，《建筑抗震设计规范》的方法难以实现，后一方法虽然在理论上不符合根据观测的强震记录确定设计地震反应谱的规律，但在动力时程分析中易于实现。考虑到在采用振型分解反应谱法进行地震作用效应组合时，地震作用设计值为：$S_e=1.3S_{eh}\pm0.5S_v=1.3\ (S_{eh}\pm0.385S_v)$；又考虑到计算竖向地震作用时，竖向地震影响系数最大值取水平地震影响系数最大值的 65%，结构等效总重力荷载取其重力荷载代表值的 75%等因素，建议三向地震波加速度最大值的比例关系应取为：1.00（水平 1）：0.3（水平 2）：0.1875（竖向）。这一比例关系与现行抗震设计规范中采用振型分解反应谱法对三向地震作用的考虑是一致的，按照这一比例关系输入三向地震波，时程分析法的计算结果与振型分解反应谱法的计算结果才有可比性。

10.5.2　地震动的频谱特性

地震波的频谱特性由地震波的主要周期来表示，它受到许多因素的影响，如震源的特性、震中距离、场地条件等，选择地震波时，地震波的主要周期应尽量接近于建筑场地的卓越周期。按照上述要求选择人工波比较容易，但选择实际强震记录是一个难题，由于目前强震观测记录相对较少，在工程设计中很难找到与预定建筑场地条件一致的强震记录。在实际工程应用中，为确保所选用的地震波的平均地震影响系数曲线与振型分解反应谱法所采用的地震影响系数曲线在统计意义上相符，除控制震源特性、震中距离、场地条件等因素外，还特别关注地震波的主要周期与建筑场地卓越周期的差异，一般控制其差异不超过 20%。

10.5.3　地震动的持续时间

关于地震波的持续时间，原则上应该选用持续时间长的地震波，因为地震持续时间较长时，地震波能量大，结构反应强烈，而且当结构的变形超过弹性范围时，持续时间越长，结构在振动过程中屈服的次数越多，结构可能因塑性变形积累而破坏。《建筑抗震设计规范》[1]在条文说明中建议：输入的地震加速度时程曲线的持续时间，不论是实际的强震记录还是人工模拟波形，一般为结构基本周期的 5～10 倍。《高层建筑混凝土结构技术规程》[2]考虑到高层建筑周期较长的特点，建议在弹性动力时程分析中，地震波的持续时间不宜小于建筑结构基本自振周期的 3～4 倍，也不宜小于 12s。若对结构进行弹塑性最强震反应分析或耗能过程分析，持续时间应取得更长些。

10.6　非线性动力方程的求解

结构振动方程的求解可分为两类：一类是时程积分法，即在离散的时间域上直接求动力方程的数值解，另一类是振型迭加法。后者主要用于线性系统的分析。目前，非线性系统的动力分析主要采用时程积分法。最近几十年来，国内外学者对直接积分算法做了大量的工作，提出了很多新的算法，例如 Argyris 方法[43]（1973），Park 方法[44]（1975），

Hilber-Hughes-Taylor 方法[45]（1977），Belytschko-Mullen 方法[46]（1977），Hoff-Pahl 方法[47]（1988），高精度 ρ 方法[48]（谢闽生，1989），精细积分法[49]（钟万勰，1994），Owren-Simonsen 方法[50]（1995）等。但近 10 年来在算法上并没有本质上的进步。

时程积分法分为隐式算法和显式算法两种。所谓显式算法是指利用 t_i 时刻的平衡条件就可获得 t_{i+1}时刻的反应值。隐式算法是指必须用 t_{i+1}时刻的动平衡方程来求 t_{i+1}时刻的反应。比较有代表性的显式算法和隐式算法如下。

10.6.1 隐式算法

较有代表性的隐式算法包括 Wilson-θ 法[52]（1972）、Newmark-β 法[53]（1959）、HHT-α 法[44]（1977）等。Wilson-θ 法的优点在于是无条件稳定算法，不存在临界时间步长的限制。步长取大时，计算结果是稳定的，影响的只是精度。该方法缺点是在每一时间步积分运算中需要求解联立方程组，这将需要大量的计算工作和存贮空间。具有周期延长和振幅衰减现象，可以迅速把系统的高阶振型响应有效地从数值解中“滤去”，因而更适用于低频或低阶模态的振动和波动的计算。

Newmark-β 法的优点与 Wilson-θ 法类似，相对于 Wilson-θ 法，该方法仅导致周期延长而不导致振幅衰减，在数值解中仍然保留了系统的高频响应，在计算高频振动和波动中更加有效。该方法缺点是在每一时间步积分运算中需要求解联立方程组，这将需要大量的计算工作和存贮空间。

HHT-α 法由 Hilber、Hughes 和 Taylor 于 1977 年提出，该方法实质上是广义的 Newmark 法，当 α 取 0 时该方法退化为 Newmark 法，该方法为无条件稳定算法，通过引入少量的数值阻尼可以较为迅速的消除结构的高频噪声，而对有意义的结构低频反应则影响很小，即使是非常复杂的问题，该方法也能够很好的保证能量守恒。该方法的缺点是要求解联立方程，计算时间较长。

10.6.2 显式算法

较有代表性的显式算法包括中心差分法及 Warburton 修正[51]（1985）、Taylor 级数展开法、精细积分法[48]（1994）等。中心差分法及 Warburton 修正方法的优点在于如果阻尼矩阵是对角形式，质量矩阵也是对角形式，则整个方程是解耦的，不必形成总刚和总体质量矩阵，只需在单元一级计算即可。存储空间少，单步计算量小。该方法缺点在于是条件稳定算法，时间步长必须小于由该问题求解方程性质所决定的某个临界值，所以在时间步长上的选择是比较困难的。

Taylor 级数展开法的优点在于能够在连续的时间段内处处满足动态平衡方程。因此，与现有方法只能得到离散时刻点处的数值解不同，该方法可以得到任意时刻点处的满足精度要求的数值解，计算稳定性和精度优于中心差分法，该方法缺点在于是条件稳定算法，计算工作量大于中心差分法。

精细积分法的优点在于结合指数矩阵的精细算法能够获得高度精确的结果，几乎是计算机上的精确解，数值解精度很高。对于一般的工程结构，可以不考虑时间步长的限制。因此计算时间步长可以取得大一些，与结构刚度矩阵的性态无关，当结构的刚度矩阵奇异或近似奇异时，仍然能够求解。该方法缺点在于是条件稳定算法。由于指数矩阵不具备对称和窄带宽的特性，而且需要计算大量的矩阵相乘，需要较大的存贮空间和计算量。

10.6.3 算法比较

文献[9]对隐式算法与显式算法的计算效率和实用性做了对比计算，结果见表 3。从表 3 中的计算结果可以看出，隐式算法对于规模较小的工程效率较高，而对于规模较大的工程，显式算法的优势表现突出，而且采用显式算法更便于实现并行计算。

隐式算法与显式算法的计算效率对比计算结果　表 3

结　构	算　法	时间步长	计算耗时	占用内存	占用硬盘
十层三跨钢框架	隐　式	2e-2 (s)	197.5 (s)	820 (MB)	17 (MB)
	显　式	2e-6 (s)	5400 (s)	670 (MB)	2 (MB)
二十层十五跨钢框架	隐　式	2e-2 (s)	3.25 (h)	1024 (MB)	716 (MB)
	显　式	2e-6 (s)	1.15 (h)	720 (MB)	59 (MB)

文献[54]对某办公楼工程进行了分析，该工程地面以上 48 层，建筑高度 215m。应用 ABAQUS 有限元软件在小型机 SGI350 上对结构进行弹塑性时程分析，SGI350 小型机配置 8 个 CPU，内存 32GB，主频 800MHz。结构计算模型总计有 75849 个单元，218351 个节点，711423 个自由度。动力时程分析采用直接积分法中的显式算法，时间步长取为 8×10^{-5}s，输入地震波持时 20s，计算耗时 28h。

文献[55]在小湾拱坝计算中研究了并行系统的节点数、CUP 数配置不同时，对并行计算效率的影响规律。小湾拱坝位于云南省西部澜沧江中游，坝高 292m，是目前世界上最高的拱坝，其处于强震高发区，按 9 度设防。在小湾拱坝计算中，坝体及地基共 23862 个单元，其中，23022 个八节点六面体单元，736 个六节点五面体棱柱单元，90 个四边形边界单元，14 个三角形边界单元。25 种材料，28285 个节点，1075 个接触点。并行计算效率比较见表 4。从中可以看出，增加系统节点数和 CPU 数，可以有效减少计算耗时，但因节点间通信量增加，并行计算系统设备利用效率在降低。此外，分析经验也表明，并行计算对小规模问题不如对大规模问题有效，在选择节点机数时，要考虑到所求解问题的规模。

并行计算系统配置效率比较　表 4

系统节点数	CPU 数	计算时间 (s)	并行计算效率 (%)
1	2	17621	100
2	4	10277	86
3	6	7560	78
4	8	6337	70
5	10	5645	62

10.7 国内外非线性动力时程分析软件

近几十年来，各国研究者一直在研制结构非线性分析的计算机软件。广为人知的

DRAIN-2D 程序即是首先问世的平面二维非线性分析程序。其后的一些程序，如 DRAIN-3D、IDARC 都可视为在 DRAIN-2D 基础上发展起来的分析程序。近年来国内的研究者们也致力于开发简单实用的弹塑性动力分析软件，如建研院开发的 EPDA 等软件。下面简要介绍目前在工程应用中具有代表性的一些非线性分析软件。

10.7.1 国外非线性分析软件

1. DRAIN-2D

该程序是美国加利福尼亚大学伯克利分校最早推出的平面结构弹塑性分析程序。可以完成静力非线性分析、动力时程分析。该程序共提供八种单元，包括非线性杆单元、塑性铰梁单元、连接单元、弹性板单元、单向受力的拉压联系单元、纤维梁单元等。用户可以采用塑性铰单元或纤维梁单元来模拟钢或钢筋混凝土梁或柱，非线性连接单元可以用来模拟特殊的“骨棒”梁柱连接等情况。该程序可考虑 $P\text{-}\delta$ 效应，逐步积分求解采用的是中点加速度法。该程序曾获得过广泛的应用，但因为开发较早，前后处理功能较弱，恢复力模型比较粗糙，不能完整的反映混凝土滞回特性。

2. CANNY

该程序是由李唐宁博士开发的，由四部分构成：窗口程序、前处理部分、计算部分和后处理部分，适用于钢筋混凝土和钢结构的三维非线性分析，可处理各种复杂不规则的结构。该程序主要包括以下功能：振型分析、弹塑性静力分析、拟动力分析、动力时程反应分析等。该程序可以考虑节点和杆件之间的铰接、刚结或半刚结。程序中包含有多种单元模型可供用户选择。用户可采用两端带弹簧的等效梁单元或塑性铰梁单元模拟框架梁和柱，采用多弹簧模型模拟剪力墙，采用刚性板单元模拟混凝土楼板。

3. Y-Fiber3D

Y-Fiber3D 软件是日本 YAMATO 设计株式会社于 1996 年开发成功的，对土木结构进行非线性动力和静力分析的实用工程软件。该软件使用纤维模型模拟钢及钢筋混凝土梁柱和支撑，采用多弹簧模型模拟剪力墙。Y-Fiber3D 软件在日本获得了广泛应用，成为很多规范指定使用的标准软件，如日本土木学会、日本道路公团，东京首都高速道路公团，日本铁道（JR）等日本最主要的土木、道路桥梁政府投资机构，在他们的工程中广为采用 Y-Fiber3D 软件进行非线性计算。

4. ANSYS

ANSYS 软件是由 ANSYS 公司开发的通用有限元程序，是集结构、流体、电场、磁场、声场分析于一体的大型通用有限元分析软件。该软件提供了强大的前后处理功能，可以求解静态和瞬态非线性问题。包括材料非线性、几何非线性和接触非线性三种。该软件提供了两节点和三节点梁单元可以用来模拟钢梁，但不能直接用于模拟钢筋混凝土梁。该软件提供的各种厚壳及薄壳单元也不支持钢筋混凝土材料，该软件只提供了三维混凝土实体单元，一般需要通过二次开发，开发出钢筋混凝土梁及壳单元才能进行高层混凝土结构的静力弹塑性分析及动力弹塑性分析。

5. ABAQUS 与 MARC

ABAQUS 和 MARC 软件都是源于 Hibbitt 在 Pedro Marcal 教授指导下的博士论文，

它们都是通用有限元软件。目前 ABAQUS 提供了混凝土三维实体单元，壳单元和梁单元，但是其梁单元并不支持反复加载的混凝土本构模型，MARC 软件提供了混凝土二维，三维实体单元，未提供混凝土梁单元。作为通用软件，都能完成钢结构非线性分析，但都不能直接进行钢筋混凝土结构的动力反应分析。要想进行这类分析，必须进行二次开发，目前国内已经有学者开展这方面的工作。

10.7.2　国内非线性分析软件

1. 清华大学的弹塑性分析软件 NTAMS

清华大学结构工程研究所开发了弹塑性地震反应分析 NTAMS，该程序包括三部分，前处理模块（PPCAD）、弹塑性静力分析模块（NSAMS）、地震反应分析模块（TAMS），该程序能完成高层建筑静力推覆分析及弹塑性动力反应分析，采用塑性铰模型模拟框架梁、柱，采用可以考虑弯曲屈服和剪切屈服的多弹簧模型模拟剪力墙。在地震反应模块中利用子结构的方法形成单串或多串集中质量模型，程序中提供了四种恢复力模型供用户选用，动力微分方程求解采用 Newmark-β 法，该程序是国内较早独立开发的弹塑性地震反应分析程序。

2. 中国建筑科学研究院的弹塑性分析软件 EPDA

该软件是由中国建筑科学研究院 PKPMCAD 工程部开发的简单实用的弹塑性动力反应分析软件。采用纤维模型模拟钢及钢筋混凝土梁柱和支撑，采用弹塑性壳单元模拟剪力墙。EPDA 可以按任意给定方向计算结构的弹塑性时程响应，适用于各种材料的多、高层及超高层建筑结构，包括钢筋混凝土结构、钢结构和钢与混凝土混合结构。同时，程序中还考虑了多塔、转换层等结构特点。其前后处理功能强，应用简单。能够自动读取 SATWE、TAT、PMSAP 的几何信息、荷载信息和钢筋混凝土梁、柱、墙的实配钢筋信息。

10.8　结束语

就目前阶段而言，在强震作用下建筑结构非线性分析应用中，存在两方面问题。一是软件开发不到位，分析结果难以把握。目前应用效果较好的，基本上都是国外的通用非线性分析软件。这类软件特点是理论和技术先进，功能全面，适用领域广，在建筑结构领域的应用，仅仅是其非常小的一个方面。但因价格昂贵，操作繁琐、缺乏对建筑结构专业特点的考虑，难以在建筑结构设计领域普及应用；二是应用环境不健全。设计人员对非线性分析理论了解，远不及对线弹性分析理论的了解，对软件的分析结果准确性缺乏信心。

强震作用下建筑结构非线性分析软件的研发与应用还任重而道远，不仅需要开发出简单实用的非线性分析软件，还需要营造强震作用下建筑结构非线性分析软件应用环境，让广大的结构总工们了解非线性分析理论、能够把握非线性分析结果、能够灵活应用非线性分析结果用于指导设计。

参考文献

[1]　中华人民共和国标准.（GB 50011—2001）建筑抗震设计规范[S]，北京：中国建筑工业出版

社，2001.

[2] 中华人民共和国行业标准.（JGJ 3—2002)高层建筑混凝土结构技术规程[S]，北京：中国建筑工业出版社，2001.

[3] FEMA450，NEHRP Recommened Provisions for Seismic Regulations for New Buildings and Other Structures[S]. Federal Emergency Management Agency，2003.

[4] ATC-40，Seismic evaluation and retrofit of concrete buildings[S]. Applied Technology Council 1996.

[5] A. L. L Baker. Ultimate Load Theory Applied to the Design of Reinforced and Prestressed Concrete Frames. Concrete Publication Ltd，London，pp91，1956.

[6] Corley W. G. Rotational Capacity of Reinforced Concrete Beams[J]. Journal of Structural Division，1966，92(5)：121-146.

[7] Priestley M. J. N and Park R. Strength and Ductility of Concrete Bridge Columns under Seismic Loading[J]. ACI Structural Journal . Title No. 84-S8，pp61-67，Jan，Feb. 1987.

[8] 姬守中. 高层钢筋混凝土框—剪结构在地震作用下的弹塑性时程反应分析[D]. 上海：同济大学，2002.

[9] 孙鹏，赵宪忠. 有限元分析软件ABAQUS在结构弹塑性地震时程分析中的应用[C]. 第七届全国地震工程学术会议论文集，广州，2006：562-569.

[10] T Takayanagi，W C Schnobrich. Computed Behaviour of Reinforced Concrete Coupled Shear Walls [R]. Report No SRS 434. University of Illinois. Urban，Champaign.

[11] Kabeyasawa，Shioara T. H，Otani S. U. S. -Japan Cooperative Research on R/C Full-Scale Building Test，Part 5，Discussion of Dynamic Response System[C]. Proceeding of 8th World Conference on Earthquake Engineering：327-634.

[12] A Vulcano，V VBertero. ，V Colotti. Analytical Modeling of RC Structural Walls[C]. Proceeding of 9th World Conference on Earthquake Engineering . Tokyo. Japan. 1988：41-46.

[13] R. W. G. Blakeley，R. Park. Prestressed Concrete Sections with Cyclic Flexures[J]. Journal of Structural Engineering. ST8. 1973.

[14] Z. W. Bolong，W. Mingshun，Z. Kunlian . A Study of hysteretic Curve of Reinforced Concrete Members under Cyclic Loading[C]. 1980，Proceedings 7th World Conference on Earthquake Engineering(6). Istanbul，Turkey. 509-516.

[15] 朱伯龙，董振祥. 钢筋混凝土非线性分析[M]. 同济大学出版社，1985.

[16] 过镇海，张秀琴. 反复荷载下混凝土的应力-应变全曲线的试验研究[J]. 钢筋混凝土结构的抗震性能. 清华大学出版社，1981.

[17] J. B. Mander，M. J. N. Priestley，R. Park. Theoretical Stress-Strain Model for Confined Concrete[J]. Journal of Structural Engineering，1988. 114(8)：1804-1825.

[18] D. Z. Yankelevsky，H. W. Reinhardt. Uniaxial Cyclic Behaviors of Concrete in Tension. ASCE，1989，115(SE1)：169-176.

[19] D. Z. Yankelevsky ，H W Reinhardt. Uniaxial Cyclic Behavious of Concrete in Compression. ASCE，1987，113(SE2)：228-240.

[20] M. H. M Yassin. Nonlinear Analysis of Prestressed Concrete Structures under monotonic and cyclic loads[D]. California：University of California，1994.

[21] B. Kato. Mechanical Properties of Steel under Load Cycles Idealizing Sesmic Actions. AICAP-CEB Symposium，Rome，1979. 5.

[22] A. E. Aktan，D. A. W. Pecknold. Response of a Reinforced Concrete Section to Two Dimen-

sional Curvature Histories[J]. ACI Structural Journal，1974，71(5)：246-250.

[23] D. Darwin，D. A. Pecknold. Nonlinear Biaxial Stress Strain Law for Concrete[J]. Engineering Mechanics Division，ASCE，1977，103(2).

[24] K. J. Bathe，S. Ramaswamy. On Three Dimensional Nonlinear Analysis of Concrete Structures [J]. Nuclear Engineering and Design，1979，52(3).

[25] W. F Chen. Plasticity in Reinforced Concrete[M]. McGraw-Hill. 1982.

[26] O. Buyukozturk. Nonlinear Analysis of Reinforced Concrete Structures[J]. Computers and Structures 1977，7：149-156.

[27] J. Lubliner，J. Oliver，S. Oller and E. Oñate. A Plastic-Damage Model for Concrete[J]. International Journal of Solids and Structures. 1989，25(1)：299-329.

[28] J. Lee and G. L. Fenves，Plastic-Damaged Model for Cycling Loading of Concrete Structures，Journal of Engineering Mechanics.，1998，124(8).

[29] F. J. Vecchio. The Modified Compression-filed Theory for Reinforcement Concrete Element Subjected to Shear[J] ACI Structural Journal. 1986，83(2)：219-231.

[30] T. T. C. Hsu. Softened Truss Model Theory for Shear and Torsion[J]. ACI Structural Journal 1988，85(6)：624-635.

[31] F. E Richart，A Brandtzage，R. L. Brown. A Study of the Failure of Concrete UnderCombined Compressive Stresses. University of Illinois Eng. Exp. Station，Bulletin No 185，1928，16(12)：1-105.

[32] 过镇海，张秀琴，翁义军. 箍筋约束混凝土的强度和变形[M]. 城乡建设部抗震办公室编. 唐山地震十周年中国抗震防灾论文集. 北京. 1986：143-150.

[33] 支云芳，牛绍仁，张义琢. 箍筋约束高强硅短柱受力性能的试验研究[J]. 重庆建筑大学学报，1996，18(2)：53-60.

[34] 胡海涛，叶知满. 复合箍筋约束高强混凝土应力应变性能[J]. 工业建筑，1997. 27(10)：23-28.

[35] 关萍，王清湘，赵国藩. 高强约束混凝土应力—应变本构关系的试验研究[J]. 工业建筑，1997. 27(11)：26-29.

[36] 钱稼茹，程丽荣，周栋梁. 普通箍筋约束混凝上柱的中心受压性能[J]. 清华大学学报(自然科学版) 2002，42(10)：1369-1373.

[37] CEB-FIP MODEL CODE 1990(DESIGN CODE). Comite Euro-International du Beton. Bulletin Dinformation No. 231-214，Lausanne，May 1993.

[38] 尚仁杰. 混凝土动态本构行为研究[D]. 大连：大连理工大学，1994.

[39] 董毓利，谢和平，赵鹏. 不同应变率下混凝土受压全过程的实验研究及其本构模型[J]. 水利学报，1997.

[40] UBC 2003. Structural Engineering Design Provision[S]，International Conference of Building Official，2003.

[41] Eurocode 8：Design of structures for earthquake resistance[S] European Committee for Standardization 2001.

[42] 上海市标准(DBJ 08-32-92). 高层建筑钢结构设计暂行规定[S]，上海：1993.

[43] J. H. Argyris，T Dunne Non-Linear Oscillations Using the Finite Element Technique[J]. Computer Methods in Applied Mechanics and Engineering，1973，2：203-250.

[44] K C Park. An Improved Stiffly Stable Method for Direct Integration of Nonlinear Structural Dynam-

ics Equations[J]. Journal of Applied Mechanics，1975，42：464-470.

[45] H M Hilber，T J R Hughes，R L Taylor. Improved Numerical Dissipation for Time Integration Algorithms Structural Dynamics [J]. Earthquake Engineering and Structural Dynamics，1977，6：99-117.

[46] T. Belytschko，H. J. Yen，R. Mullen. Mixed Methods for Time Integration[J]. Computer Methods in Applied Mechanics and Engineering，1979，17(18)：259-275.

[47] C. Hoff，P. J. Pahl. Development of An Implicit Method with Numerical Dissipation From A Generalized Single-Step Algorithm for Structural Dynamics[J]. Computer Methods in Applied Mechanics and Engineering，1988，67：367-385.

[48] 谢闽生. 结构动力分析的高精度ρ法[J]. 力学学报，1989，21(4)：459-468.

[49] 钟万勰. 结构动力方程的精细时程积分法[J]. 大连理工大学学报，1994，34(2)：131-136.

[50] B. Owren，H. H. Simonsen. Alternative Integration Methods for Problems in Structural Dynamics[J]. Computer Methords in Applied Mechanics and Engineeriong，1995，122：1-10.

[51] G. B. Warburton. Some Recent Advances in Structural Vibrations . In Vibration of Engineering Structures(Edited by I. A. Brebbia and S. A. Orszag)，Lecture Notes in Engineering，No 10，Springer，Berlin 1985 ，No，10：2125-223.

[52] E. L. Wilson. A Compute Program for the Dynamic Stress Analysis of Underground Structures [R]. SESM Report No. 68-1，Department of Civil Engineering，University of California，1968，Berkeley.

[53] N. M. Newmark. A Method of Computation for Structural Dynamics[J]. Journal of the Engineering Mechanic Division，1959，85(3)：249-260.

[54] 汪大绥等. 应用高性能计算平台进行结构弹塑性时程分析. 首届工程设计高性能计算(HPC)技术应用论坛论文集，上海，2007.

[55] 陈厚群等. 大坝抗震中的高性能计算[C]. 第七届全国地震工程学术会议论文集，广州，2006：14-23.

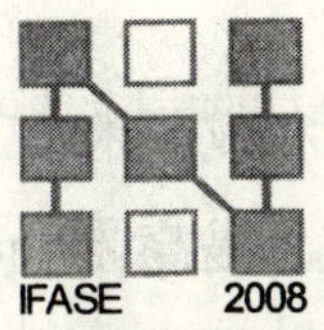

International Forum on Advances In Structural Engineering（2008）

第二届“结构工程新进展国际论坛”简介

论坛主题——结构防灾、监测与控制

会议地点：中国·大连

会议时间：2008 年 10 日至 11 日

支持单位：中华人民共和国住房和城乡建设部

中华人民共和国住房和城乡建设部　建筑节能与科技司

主办单位： 中国建筑工业出版社

 同济大学《建筑钢结构进展》编辑部

 香港理工大学《结构工程进展》编委会

承办单位： 大连理工大学土木水利学院

关于论坛

“结构工程新进展国际论坛”旨在促进我国结构工程界对学术成果和工程经验的总结及交流，汇集国内外结构工程各方面的最新科研信息，提高专业学术水平，推动我国建筑行业科技发展。论坛划分为四个领域，以每年一个主题的形式轮流出现。分别为：

- **新型结构材料与体系（包括组合结构、膜结构等）**
- **钢结构（包括空间结构、大跨度结构）**
- **混凝土结构**
- **结构防灾、监测与控制（包括抗震、抗风、抗火等）**

论坛主题循环往复，充分而及时地体现研究成果的更新与交替。论坛有两大特点。其一，论坛会议采取特邀报告人专题演讲的形式，其人选在81位包括院士、长江学者、国家杰出青年基金获得者及业内专家学者所组成的论坛学术委员会推荐的基础上产生，从而保证演讲内容的学术性、代表性及领先水平。其二，“论坛文集”的学术定位：文集由特邀报告人对演讲论文进行深度展开或延伸，充分表达其学术成果，按主题每届一册，于论坛召开之前由中国建筑工业出版社正式出版。论坛文集不仅是论坛的重要标志，更重要的是，它将成为结构工程研究领域的风向标，为教学科研人员、在校研究生提供一个既有前瞻性又有可行性的、能引导研究方向的重要文献资料。同时，该书的连续出版，也将成为结构工程科研历程的全部记录，对整个研究工作的沿革起着承上启下的作用。

组织委员会

滕锦光（香港理工大学协理副校长，《结构工程进展》主编）

李宏男（大连理工大学土木水利学院院长，长江学者特聘教授）

委　员： 胡永旭（中国建筑工业出版社　副总编辑兼社长助理）

伊廷华（大连理工大学　博士）

陆　烨　何亚楣　刘玉姝　（同济大学）

赵梦梅　刘婷婷　（中国建筑工业出版社）

学术委员会（按姓氏笔画排序）

丁大钧	仇保兴	牛荻涛	王永昌	王光远	王国周
叶可明	叶列平	石永久	龙驭球	刘锡良	吕西林
吕志涛	孙利民	江　亿	江欢成	吴　波	吴智深
张佑启	张慕胜	李　杰	李　惠	李宏男	李国强
李宗津	李忠献	李秋胜	李爱群	杨永斌	汪大绥
沈世钊	沈祖炎	肖　岩	陈宜明	陈建飞	陈绍蕃
陈惠发	陈锦顺	陈肇元	周绪红	周福霖	周锡元
岳清瑞	欧进萍	范立础	金伟良	柯长华	胡庆昌
赵国藩	赵晓林	郝　洪	钟善桐	项海帆	容柏生
徐　建	徐　磊	徐幼麟	徐有邻	徐培福	柴　昶
聂建国	袁　驷	钱七虎	顾　明	顾　强	高赞明
崔鸿超	梁以德	傅学怡	曾庆元	程懋堃	葛汉彬
董石麟	谢礼立	韩林海	蓝　天	赖　明	鲍广鑑
蔡克铨	蔡春声	蔡益燕	滕锦光		

执行委员会

主　席： 李宏男

副主席： 伊廷华　王国新

委　员： 霍林生　李　钢　任　亮　李东升　肖诗云　孙　丽　王　杨

第二届论坛特邀报告人简介（按论坛报告顺序）

沈世钊　中国工程院院士，哈尔滨工业大学教授，博士生导师。长期从事结构工程领域的研究和教学，20世纪80年代以来致力于大跨空间结构新兴学科的开拓，在“悬索结构体系及其解析理论”、“网壳结构非线性稳定”、“大跨柔性屋盖风致动力响应及抗风设计”、“网壳结构动力稳定性及强震失效机理”等前沿研究中取得重要成果；并结合一些重大工程创造性地设计了多项具有典型意义的新型空间结构，为我国大跨度空间结构的发展作出了重要贡献。其成果获国家科技进步二等奖、建设部科技进步一等奖、黑龙江省科技进步一等奖等10余项奖励。迄今发表论文200余篇，出版《网壳结构稳定性》、《悬索结构设计》、《钢结构构件稳定理论》等著作，培养硕士、博士90余人。

周福霖　中国工程院院士，广州大学工程抗震研究中心主任，教授，博士生导师，联合国工业发展组织隔震技术顾问，国际隔震减震控制学会副理事长，中国建筑学会抗震减灾研究会常务理事，结构减震控制专业委员会主任委员，全国高校建筑工程专业指导委员会委员。曾获得全国五一劳动奖章、建设部劳动模范、广东省五一奖章获得者、广东省有突出贡献专家、省优秀教师、市模范教师等10多个光荣称号。主要从事土木工程结构、结构抗震、结构隔震减震控制等领域的研究。曾主持多项国家科学基金项目、美国国家科学基金项目、中-美科技合作项目、联合国科学技术研究开发项目。其研究成果“多层房屋橡胶垫基础隔震技术”被誉为“世界隔震技术发展的第三个里程碑”。出版著作4部，发表论文近100篇。其成果先后曾获得国家科技进步奖二等奖、建设部科技进步奖一、二等奖、广东省科技进步奖一等奖等10多个奖项。先后培养硕士、博士、博士后等几十名。

欧进萍　中国工程院院士，结构监测、控制与防灾减灾专家，大连理工大学校长，国家自然科学基金学科评审组组长，国家自然科学基金重大研究计划—重大工程的动力灾变指导专家组组长，国际结构控制与监测学会理事及中国分会主席，国际智能结构健康监测学会执行理事，以及多个学术刊物编委和副主任编委。在结构地震损伤与动力可靠性、结构振动控制、结构健康监测和海洋平台结构安全保障技术等方面取得系统性研究成果，部分已被我国有关规范和规程采纳，并应用于实际工程。发表学术论文200余篇、出版著作4部；获得国家科技进步二等奖2项（排名第一）、省部级科技进步一等奖4项（3项排名第一、1项排名第二）、二等奖3项；获霍英东教育基金会青年教师研究基金和研究一等奖各一次（独立）；获国家发明专利10项；获“做出突出贡献的中国博士学位获得者”、“国家级中青年有突出贡献专家”和“全国优秀博士后”等称号。

李　杰　“长江学者奖励计划”首批特聘教授，同济大学建筑结构研究所所长，上海防灾救灾研究所所长，国家自然科学基金委杰出青年基金学部评审委员会委员、中国建筑学会结构计算理论与工程应用专业委员会主任、混凝土结构理论与工程应用专业委员会副主任、中国振动工程学会随机振动专业委员会副理事长、中国城市安全与工程防灾学术委员会副主任等20余个学术团体及学术期刊的学术职务。1996年获得国务院颁发的政府特殊津贴，1998年入选“百千万人才工程”第一，第二层次人选，同年，被授予国家级“有突出贡献的中青年专家”称号，2001年被授予“全国优秀教师”称号。研究内容包括生命线工程抗震、混凝土结构基本理论与工程应用、随机动力系统分析与控制、结构健康监测与智能结构等。先后主持国家杰出青年科学基金、国家自然科学基金、国家“九五”攀登计划和“十五”科技攻关等重点项目30余项。作为第一完成人获得国家科技进步三等奖1项，省部级科技进步一等奖3项；作为第一或主要完成人获得部省级科技进步二等奖5项，三等奖7项。出版学术著作6部，发表学术论文近300篇，其中期刊论文200余篇，SCI、EI收录90余篇。

W. D. Iwan　博士，加州理工学院应用力学系教授，地震工程研究实验室主任，美国工程院院士，国际结构控制与监测学会主席，美国自然灾害研究会成员，美国地球科学研究所成员，美国能源部核反应堆高级评论组成员，美国核调整委员会核管道系统地震标准评论组成员，日本灾害预防评估组成员，美国大学联盟地震工程研究会创始人、主任。Iwan博士1957年毕业于加州理工学院，获得学士学位；1958年获得硕士学位；1961年获得博士学位。主要研究方向包括：地震安全；结构动力学；结构的地震响应；振动与动力学；非结构系统；离岸工程的优化抗震设计；实时监测；预警系统。获得奖励多项，包括加州地震安全基金会授予的著名Alquist奖，拥有多项专利，编写、合著了诸多地震工程和地震学的著作。

李宏男　大连理工大学土木水利学院院长，“长江学者奖励计划”特聘教授，国家杰出青年科学基金获得者，国家级有突出贡献的中青年科技专家。兼任国际结构控制与监测学会中国分会副主席，中国土木工程学会和中国振动工程学会理事、中国振动工程学会结构控制分会副理事长等10余个学术团体的常务理事、理事和委员。主要从事工程结构的教学与研究工作，从事多维地震动理论、结构多维抗震设计理论及其应用、结构多维随机振动理论、房屋结构质量的动态检测及其工程应用、大跨越输电塔体系的抗震计算理论、结构减震控制与健康监测等领域的研究工作。主持纵向科研项目30余项，包括国家“十五”科技攻关项目、国家“十一五”科技支撑计划项目、国际合作项目、国家自然科学基金重点项目等。是国家“十一五”科技支撑计划重点项目专家组组长。撰写了210余篇期刊学术论文，SCI和EI收录200余篇，著作7部，被国内外学者引用1000余次。获国家专利10项。获国家科技进步二等奖1项，省部级一等奖5项、二等奖3项，均为项目负责人。

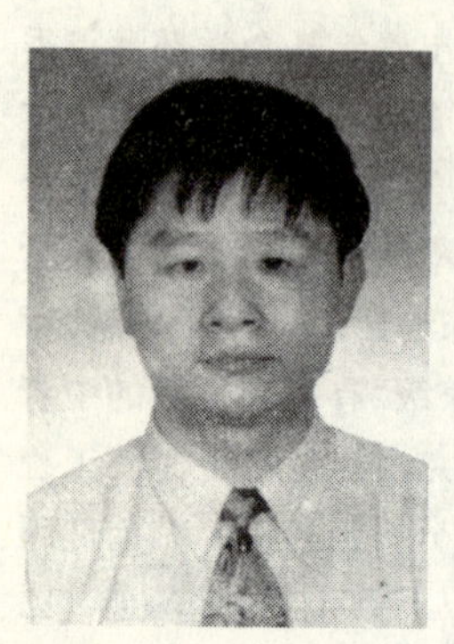

李国强　同济大学副校长，教授，博士生导师。中国钢结构协会副会长、中国工程建设标准化协会副理事长、中国钢结构协会防火与防腐分会理事长、中国建筑学会抗震防灾分会常务理事，中国建筑学会抗震防灾分会结构抗火专业委员会主任，中日建筑结构技术交流委员会副会长，担任 Journal of Constructional Steel Research 等 5 本国际学术期刊编委。李国强教授主要从事多高层建筑钢结构分析与设计理论和钢结构抗火计算与设计理论研究，主持完成国家自然科学基金、国家杰出青年科学基金、国家 863 计划、国家攀登计划等重要科研项目 30 余项，出版研究专著 10 部，发表国内外期刊论文近 400 篇。主编国家标准《建筑钢结构防火技术规范》(编制中)和参编国家标准《建筑抗震设计规范》(GB 50011)等国家和地方建设规范 10 部。研究成果获国家科技进步二等奖 1 项(第 9 完成人)、省部科技进步一等奖 3 项(第一完成人 2 项，第二完成人 1 项)、二等奖 5 项(第一完成人 4 项、第六完成人 1 项)。

徐幼麟　香港理工大学土木与结构工程系系主任，结构工程讲座教授，城市城灾研究中心主任。中国结构控制学会副主席，美国土木工程学会会员，澳大利亚风工程学会会员，中国风工程学会执行委员，香港工程师学会资深会员。徐教授长期从事结构风工程，结构健康监测，结构振动控制和智能结构领域的研究与教学工作。在国际学术期刊上发表 150 篇文章，其中 137 篇被 SCI 收录。先后承担了 60 多个研究项目，包括香港青马悬索桥、香港昂船洲斜拉桥和中央新电视台的研究工作。曾获 2001/2002 香港理工大学杰出青年教授基金、2003 年中国教育部的自然科学提名一等奖及 2006 香港裘槎基金会“优秀科研者奖”等多个奖项。

吴智深　日本茨城大学教授，智能结构及保障系统研究机构主任，“长江学者奖励计划”讲座教授，东南大学城市工程科学国际研究中心、东南大学—横店集团玄武岩纤维研发中心主任以及南京大学地球环境计算工程研究所所长，国际建筑 FRP 学会副会长，国际结构健康监测与智能结构学会常务理事，日本土木学会混凝土结构健康监测委员会主委等有关学术团体的要职 30 余项，8 种国际期刊及其他多种核心期刊的副主编、编委或特邀主编，中国科协第 6 和 7 届全国代表大会列席代表、全国科学技术大会特邀代表。研究领域包括：纤维复合材料、智能传感技术、结构与桥梁工程、非线性计算力学、城市安全与节能等。主持国家级重大项目、科研及人才基金项目 40 多项，包括日本文部科学省重大科研项目，国家杰出青年科学基金(B 类)，中科院杰出青年基金项目等。发表期刊论文 200 多篇，国际会议论文 200 多篇，获得日本和国内专利 30 余项。

李云贵 中国建筑科学研究院建筑工程软件研究所副所长，中国建筑学会建筑结构分会理事、计算机应用专业委员会主任，中国土木工程学会结构可靠度委员会副主任，建设部信息化专家。主要从事高层结构分析与设计、建筑结构计算机辅助设计、信息技术在建设领域应用、工程结构可靠度等理论研究及软件开发工作。负责开发的“高层建筑结构空间有限元分析与设计软件 SATWE”获国家科技进步二等奖，全国 80%以上的设计单位都在应用该软件，并已推广到新加坡、越南、香港等东南亚国家和地区。主持了多项重大项目，如国家“十一五”科技支撑计划项目等。获得国家科技进步二等奖三项、建设部科技进步一、二等奖各三项，国家教委科技进步二、三等奖各一项，发表国内外学术论文 90 余篇，1997 年入选国家“百千万人才工程”(第一、二层次)，1998 年获政府特殊津贴。